Advanced
Mathematics
and Mechanics
Applications Using

Second Edition

Advanced Mathematics and Mechanics Applications Using

Second Edition

Howard B. Wilson
University of Alabama

Louis H. Turcotte
U.S. Army Engineer Waterways Experiment Station

CRC Press
Boca Raton   Boston   London   New York   Washington, D.C.

Library of Congress Cataloging-in-Publication Data

Wilson, H. B. (Howard B.)
    Advanced mathematics and mechanics applications using MATLAB / by
Howard B. Wilson, Louis H. Turcotte. – – 2nd ed.
        p.   cm.
    Includes bibliographical references and index.
    ISBN 0-8493-1686-3
    1. MATLAB. 2. Engineering mathematics– –Data processing.
3. Mechanics, Applied– –Data processing. I. Turcotte, Louis H.
II. Title.
TA345.W55 1997
620′.00151—dc21   97-48444
                                                                                                              CIP

Preface

This book uses MATLAB® to analyze various applications in mathematics and mechanics. The authors hope to encourage engineers and scientists to consider this modern programming environment as an excellent alternative to languages such as FORTRAN. MATLAB[1] embodies an interactive environment for technical computing which includes a high level programming language and remarkably comprehensive graphics commands facilitating two- and three-dimensional data presentation. The wealth of intrinsic functions to handle matrix algebra, Fourier series, and complex-valued functions makes straightforward calculator operations of many tasks previously requiring complicated subroutine libraries with lengthy and cumbersome argument lists.

Various problem solutions, drawn from the teaching and research interests of the authors, emphasize linear and nonlinear differential equation methods. Linear partial differential equations and linear matrix differential equations are analyzed using eigenfunctions and series solutions. Several types of physical problems are considered. Among these are heat conduction, fluid flow, harmonic response of strings, beams and trusses, geometrical properties of areas and volumes, flexure and buckling of indeterminate beams, elastostatic stress analysis, and multi-dimensional optimization.

The excellent capability of MATLAB to numerically integrate matrix differential equations is applied to a number of cases illustrating the power of the methodology as well as essential aspects of numerical approximation. Attention is restricted to the Runge-Kutta method which is adequate to handle most situations. Space limitation led us to omit some interesting MATLAB features concerning predictor-corrector

---

[1]MATLAB is a registered trademark of The MathWorks, Inc. For additional information contact:

The MathWorks, Inc.
24 Prime Park Way
Natick, MA 01760-1500
(508) 653-1415, Fax: (508) 653-2997
Email: info@mathworks.com

methods, stiff systems, and event locations.

This book is not an introductory numerical analysis text. It is most useful as a reference book or a supplementary text in computationally oriented courses emphasizing physical applications. The authors have previously solved many of the examples in FORTRAN. Our MATLAB solutions consume over one hundred pages (over ten thousand lines). Although few books published recently present this much code, comparable FORTRAN versions would probably be four to five times as long. In fact, the conciseness achievable in MATLAB was a primary motivation for writing the book.

The programs have extensive comments and are intended for study as separate entities without an additional reference. Consequently, some deliberate redundancy exists between program comments and text discussions. We have also used a program listing style we feel will be helpful to most readers. The source listings have line numbers adjacent to the MATLAB code. MATLAB does not employ line numbers, nor does it permit use of infamous *goto* statements. (MATLAB does give line numbers in reference to error diagnostic messages.) Line numbers are included because they provide convenient reference points during discussions of particular program segments. We have also concatenated multiple MATLAB statements on the same line whenever possible without interrupting the logical flow.

All the programs presented are designed to operate under the 5.x version of MATLAB. The intended operating system environment is either Microsoft Windows 95 or UNIX X-Windows, with both the text and graphics windows simultaneously being visible. Many of the programs will run without changes under MATLAB 4.x. However, utilization of the windowed operating environment and the many powerful new graphics functions of MATLAB 5.x sometimes leads to programs requiring extensive changes to operate in MATLAB 4.x. Furthermore, the programs are available from the CRC Press web site http://www.crcpress.com.

This new edition differs extensively from the previous one. Improved graphics features of MATLAB 5.x are used in the majority of programs. Some problem solutions are general enough to be stand-alone analysis tools. Typical examples include programs on geometrical properties, beam flexure, and truss dynamics. Four new chapters on eigenvalue applications, beam flexure, analytic functions, and nonlinear optimization have been added. Among the most interesting features of MATLAB is its simplicity of handling the eigenvalue analysis complemented by graphics facilitating natural frequency calculations and animation of dynamical response. Several of our examples utilize those capabilities.

We emphasize that our primary target audience involves people interested in physical applications. A thorough grounding in ideas of Euclidean geometry, Newtonian mechanics, and some mathematics beyond calculus is essential to understand many of the topics.

Finally, the authors appreciate receiving comments, suggestions, and corrections which students, teachers, and researchers have shared with us concerning the first edition. We particularly enjoy interacting with people who apply advanced mathematics to real world problems. If this book encourages interest in MATLAB among engineers who only spend part of their time using computers, our primary goals will have been achieved.

Howard B. Wilson
hwilson@ua1vm.ua.edu

Louis H. Turcotte
turcotte@wes.army.mil

For my father who always loved learning.

Howard B. Wilson

For my loving wife, Evelyn, our departed cat, Patches, and my parents.

Louis H. Turcotte

Contents

Chapter 1

Introduction

1.1 MATLAB: A Tool for Engineering Analysis

This book illustrates various MATLAB applications in mechanics and applied mathematics. Our objective is to employ numerical methods in examples emphasizing the appeal of MATLAB as a programming tool. The programs are intended for study as a primary component of the text. The numerical methods include interpolation, numerical integration, finite differences, linear algebra, Fourier analysis, roots of nonlinear equations, linear differential equations, nonlinear differential equations, linear partial differential equations, analytic functions, and optimization methods. Many intrinsic functions included in MATLAB are used along with extensions developed by the authors. The physical applications vary widely from solution of linear and nonlinear differential equations in system dynamics to geometric property calculations for areas and volumes.

During the last twenty-five years FORTRAN has been the favorite language for solving mathematical and engineering problems on digital computers. FORTRAN is valuable for analyzing complex problems. Nevertheless, it is limited by an artificial and sometimes awkward mismatch between concise formulations typified by modern matrix methods and the FORTRAN code needed to implement such calculations. The modern programming language MATLAB greatly reduces this computational barrier by providing a highly interactive medium, including many advanced mathematical tools as intrinsic functions. Advanced software features such as dynamic memory allocation and interactive error tracing reduce the time needed to develop problem solutions. The powerful but simple graphics commands in MATLAB also facilitate preparation of high quality graphs and surface plots appropriate for technical papers and books. Because the graphics features can be used interactively, most of the frustration encountered in debugging compiled FORTRAN code employing binary libraries is overcome. Experience of the authors indicates that a MATLAB problem solution which has some mathematical structure and employs graphics may require as little as one fifth as much code as that needed to accomplish the task in FORTRAN. Consequently, more time can be devoted to the primary purpose of computing, namely, improved understanding of physical system behavior.

Most of the mathematical background needed to understand the topics presented in this book is typically covered in an undergraduate engineering curriculum. This should include a thorough grounding in calculus, differential equations, and knowledge of a procedure oriented programming language such as FORTRAN. An additional course on advanced engineering mathematics covering linear algebra, matrix differential equations, and eigenfunction solutions of partial differential equations will also be valuable. The MATLAB programs in this book were written primarily to serve as instructional examples. The greatest benefit to the reader will probably be derived through careful study of the programs. Furthermore, we believe that several of the MATLAB functions are useful for practical applications. Typical examples include spline routines to interpolate, differentiate, and integrate; area and inertial moments for general plane shapes; and volume and inertial properties of arbitrary polyhedra.

The expanded use of advanced computing tools like MATLAB is occurring rapidly. Continued growth will be fueled by a large community of users familiar with sophisticated analytical methods. Furthermore, advances in software development techniques and remarkable decreases in hardware costs will accelerate these developments. The authors are hopeful that this book will motivate analysts already comfortable with traditional languages like FORTRAN to learn MATLAB. The rewards of such efforts will be considerable.

1.2   Use of MATLAB Commands and Related Reference Materials

MATLAB has a rich command vocabulary embracing numerous mathematical topics. The current section presents instructions on: a) how to learn MATLAB commands, b) how to examine and understand MATLAB's lucidly written and easily accessible "demo" programs, and c) how to expand the command language by writing new functions and programs. A comprehensive online help system is included and provides lengthy documentation of all the operators and commands. Additional capabilities are provided by auxiliary toolboxes. The reader is encouraged to study the command summary to get a feeling for the language structure and to have an awareness of powerful operations such as **null**, **orth**, **eig**, and **fft**.

The manual for *The Student Edition of MATLAB* [97] should be read thoroughly and kept handy for reference. Other references [46, 95, 101] also provide valuable supplementary information. This book expands the standard MATLAB documentation to include additional examples which we believe are complementary to more basic instructional materials.

Learning to use **help, type, demo**, and **diary** is an important first step to mastering MATLAB. **help** lists a summary of available commands with a short description of each command. **help** *function-name* (such as **help** *plot*) lists available documentation on a command or function generically called "function-name". MATLAB responds by printing internal comments contained at the start of the relevant function (comments are printed until the first blank or executable statement occurs). This feature allows users to create online help for functions they write by simply inserting appropriate comments at the top of the function. The instruction **type** *function-name* lists the entire source code for any function for which source code is available (the code for some intrinsic functions cannot be listed because it is stored in binary form for computational efficiency). Consider the following list of typical examples.

help help   discusses use of help command.

help demos   lists names of various demo programs.

type linspace   lists the source code for the function which generates a vector of equidistant data values.

type plot   outputs a message indicating that **plot** is a built-in function.

intro   executes the source code in a script file called **intro** which illustrates various MATLAB functions.

type intro   lists the source code for the **intro** demo program. By studying this example, readers can quickly learn to use many MATLAB commands.

graf2d   demonstrates X-Y graphing.

graf3d   demonstrates X-Y-Z graphing.

help diary   provides instructions on how results appearing on the command screen can be saved into a file for later printing, editing, or merging with other text.

diary *fil_name*   instructs MATLAB to record, into a file called *fil_name*, all text appearing on the command screen until the user types **diary off**. The **diary** command is especially useful for getting copies of library programs such as **zerodemo**.

It can also be used to save a sequence of commands performed interactively. Then the saved text can be edited to create a MATLAB function for repeated use.

demo initiates access to a lengthy set of demo programs. Users should run all of the demos to fully appreciate the functionality of MATLAB. It is also helpful to obtain source listings of the demo programs. Programs used in the demos which should receive detailed study are: **intro**, **zerodemo**, **fitdemo**, **quaddemo**, **odedemo**, **ode45**, **fftdemo**, and **truss**. These functions and script files utilize most of the fundamental commands in MATLAB.

A program was written to illustrate various coding aspects in MATLAB. Users experienced with another procedure oriented language can learn to code in MATLAB by studying this instructional example along with other code segments included in the demo programs. Our example analyzes a problem of finding a root of a function when the function itself is defined by an integral. The analysis tools needed are a root finder and a numerical integrator. Although the intrinsic MATLAB root finder **fzero** and the integrator **quad8** are both excellent, we have included functions using interval halving and Simpson's rule to illustrate several programming features. The basic relationships occurring in the solution process are that the main program calls a root finder which repeatedly evaluates a function, the values of which are obtained by numerical integration. Our function is $F(x) = 1 - \int_1^x t^{-1}dt$. Since $F(x) = 1 - ln(x)$, it evidently equals zero when x equals e, the base of natural logarithms.[1] This value is compared with results obtained by the root finder. In addition to the numerical analysis process illustrated by this example, the program includes a handy function for reading several variables on a single line, and demonstrates the use of formatted output. The steps in the process are as follows:

1. Read search limits and error tolerances governing the root search.

2. Call a root finder (we use a simple bisection algorithm) to find a root of $F(x)$. The root finder must repeatedly evaluate $F(x)$ during the search.

3. To evaluate $F(x)$, a numerical integration routine (based on Simpson's rule) is called.

---

[1]MATLAB uses a value of e given by $\exp(1) = 2.71828182845904$.

4. The numerical integration routine repeatedly evaluates the integrand $1/x$.

This process involves program modules nested five deep. The modules of the program are:

rootest	Read data, call root finder, print results.
bisect	Searches for a root between prescribed limits by interval halving.
fnc	Evaluate the function $1 - \int_1^x \frac{dt}{t}$.
simpson	Simpson rule integrator used to compute **fnc**.
oneovx	The integrand $1/x$ called by **simpson**.
read	A utility function which reads several data items on the same line.

A listing of the program appears below. We have utilized several different modes of data input and output for instructional purposes.[2] The following notes direct attention to salient points in the program.

Table 1.1. Description of Code in Example

Routine	Line	Operation
rootest	1-21	Typing **help rootest** would print the first 21 comment lines.
	25,26	Three global variables are declared and one initialized. Such variables should have unique names (appending an underscore to the variable name is a commonly employed method) to avoid unintentional conflict with local variables in other functions. Once a variable is declared global, it is known in other modules where these variables are also defined global.
	32-35	**fprintf** is used to write headings. Note use of ... to continue lines and \n to issue a line feed.
		continued on next page

[2] General utility functions, like **simpson**, which are not intrinsic to MATLAB are only listed the first time they are used in an application program. Whenever a function needed in the code for an article is not shown, the reader should consult Appendix A.

Routine	Line	Operation
		continued from previous page
	37-43	Character strings are created for printing later.
	46	Start a loop which will continue until a **break** or **return** is reached. See line 54.
	51	Function **read** is used to input three variables.
	61-64	**fprintf** is used to print several lines requesting data input.
	67-70	**num2str** and **int2str** convert numbers to strings, which are used to compose a longer string containing data parameters.
	73-78	Logical sequence employing **if, else, end**.
	83	Use **cputime** to check computation time.
	85	Evaluate elapsed computation time.
	94-104	Print formatted results.
bisect	26	Specify global variables.
simpson	1	*funct* is a character string passed as the name of a function being integrated.
	22	**feval** is used to compute a function value.
	23,24	Vector indexing and **sum** are used to accumulate terms.
fnc	15	Use **round** to compute the number of integration subintervals.
oneovx	18	Increment a global variable to track the number of function evaluations.

1.2.1 Example Program to Compute the Value of e

Output from Example

```
rootest

** CALCULATING THE BASE OF NATURAL LOGARITHMS **
**               BY ROOT SEARCH                 **

Input lower search limit, upper search limit
and an error tolerance for the root
(typical values are 2.5, 3.0, 1.e-5)
Use 0,0,0 to terminate execution
? > 2.5,3.0,1.e-5

Input the allowable number of iterations
(Use a negative value if intermediate root
estimates are to be printed)
? > -100

Search limits are from 2.5 to 3
Root tolerance = 1e-005 with max iterations = 100

     i        e-approx       funct. val.
     1      2.7500000000    -1.1601e-002
     2      2.6250000000     3.4919e-002
     3      2.6875000000     1.1389e-002
     4      2.7187500000    -1.7222e-004
     5      2.7031250000     5.5915e-003
     6      2.7109375000     2.7055e-003
     7      2.7148437500     1.2656e-003
     8      2.7167968750     5.4643e-004
     9      2.7177734375     1.8704e-004
    10      2.7182617188     7.3980e-006
    11      2.7185058594    -8.2413e-005
    12      2.7183837891    -3.7509e-005
    13      2.7183227539    -1.5056e-005
    14      2.7182922363    -3.8288e-006
    15      2.7182769775     1.7846e-006
    16      2.7182846069    -1.0221e-006

Iterations            =  16
e-approximation       =  2.718280792
Percent error         = -3.81205e-005
CPU time              =  0.17 seconds
Integrand evaluations =  12977
```

```
Input lower search limit, upper search limit
and an error tolerance for the root
(typical values are 2.5, 3.0, 1.e-5)
Use 0,0,0 to terminate execution
? > 2,4,1e-10

Input the allowable number of iterations
(Use a negative value if intermediate root
estimates are to be printed)
? > 100

Search limits are from 2 to 4
Root tolerance = 1e-010 with max iterations = 100

Iterations              =   34
e-approximation         =   2.718281829
Percent error           =   1.77576e-009
CPU time                =   0.07 seconds
Integrand evaluations =   25403

Input lower search limit, upper search limit
and an error tolerance for the root
(typical values are 2.5, 3.0, 1.e-5)
Use 0,0,0 to terminate execution
? > 0,0,0

All Done
```

Script File rootest

```
 1: % Example: rootest
 2: % ~~~~~~~~~~~~~~~~~
 3: % This instructional program illustrates nested
 4: % function calls.  The base of natural
 5: % logarithms (e) is approximated by finding a
 6: % value of x which makes the function
 7: %
 8: %     fnc(x)=1-integral( (dt)/t ; t=1, => ,x )
 9: %
10: % equal zero.  The sequence of function calls
11: % is as follows: 1) The main program calls
12: % bisect to find roots by interval halving.
13: % 2) bisect calls fnc(x) which equals one
14: % minus the integral of 1/x over limits from
15: % 1 to x. 3) fnc(x) calls simpson to perform
16: % the numerical integration. 4) simpson
17: % calls the integrand oneovx = 1/x
18: %
19: % User m functions required:
20: %    bisect, fnc, simpson, oneovx, read
21: %-------------------------------------------------
22:
23: % Place selected parameters in global storage
24: % for accessibility by other routines
25: global kount_ ifprnt_ trouble_
26: trouble_=0;
27:
28: % Typical values for program parameters
29: % a=2.5; b=3.0; xtol=1.e-5; nmax/100;
30:
31: sgn_chng=0;
32: fprintf('\n** CALCULATING THE BASE OF ')
33: fprintf('NATURAL LOGARITHMS **')
34: fprintf('\n**                BY ROOT SEARCH')
35: fprintf('                  **\n\n')
36:
37: str1=['Input lower search limit, upper ', ...
38:       'search limit \nand an error '];
39: str2=['tolerance for the root \n(typical ', ...
40:       'values are 2.5, 3.0, 1.e-5)', ...
41:       '\nUse 0,0,0 to terminate execution\n'];
42: str3=['\nNo sign change between given ', ...
```

```
43:             'limits. Choose data again\n'];
44:
45: % Repeatedly read intervals for root search
46: while 1
47:    kount_=0;
48:    ifprnt_=0;
49:    while sgn_chng==0
50:       fprintf(str1); fprintf(str2);
51:       [a,b,xtol]=read('? > ');
52:       if norm([a,b,xtol]) == 0
53:          fprintf('\nAll Done\n');
54:          return;
55:       end
56:       sgn_chng=((fnc(a)*fnc(b)) <= 0 );
57:       if sgn_chng==0
58:          fprintf(str3);
59:       end
60:    end
61:    fprintf('\nInput the allowable number of ')
62:    fprintf('iterations \n(Use a negative value')
63:    fprintf(' if intermediate root \nestimates')
64:    fprintf(' are to be printed)\n');
65:    nmax=input('? > ');
66:    disp(' '),
67:    disp(['Search limits are from ', ...
68:          num2str(a),' to ',num2str(b)])
69:    s1=num2str(xtol); s2=int2str(abs(nmax));
70:    s3=int2str(abs(nmax));
71:    fprintf(['Root tolerance = ',s1])
72:    fprintf([' with max iterations = ',s2,'\n'])
73:    if nmax < 0
74:       ifprnt_=1;
75:       nmax=abs(nmax);
76:    else
77:       ifprnt_=0;
78:    end
79:    if ifprnt_==1
80:       fprintf('\n     i          e-approx')
81:       fprintf('        funct. val.')
82:    end
83:    cptim=cputime;
84:    [rt,ntotl]=bisect('fnc',a,b,xtol,nmax);
85:    cptim=cputime-cptim;
86:
87:    if trouble_==1
```

```
88:      fprintf(['\nRerun program using a ', ...
89:                'larger number of iterations\n'])
90:    else
91:  % Evaluate e to precision given by the computer
92:      acurat=exp(1);
93:      pcter=100*(rt-acurat)/acurat;
94:      fprintf('\n\nIterations           =  ')
95:      fprintf('%g',ntotl)
96:      fprintf('\ne-approximation      =  ')
97:      fprintf('%11.9f',rt)
98:      fprintf('\nPercent error        =  ')
99:      fprintf('%g',pcter)
100:      fprintf('\nCPU time             =  ')
101:      fprintf('%g seconds',cptim)
102:      fprintf('\nIntegrand evaluations =  ')
103:      fprintf('%g',kount_)
104:      fprintf('\n\n')
105:    end
106:    sgn_chng=0;
107:  end
```

Function bisect

```
1:  function [rt,ntotl]=bisect(fn,a,b,xtol,nmax)
2:  %
3:  % [rt,ntotl]=bisect(fn,a,b,xtol,nmax)
4:  % ~~~~~~~~~~~~~~~~~~~~~~~~~~~~~~~~~~~
5:  % Determine a root of function fn by interval
6:  % halving
7:  %
8:  % fn      - character string giving the name of
9:  %           the function for which a root is
10:  %           sought
11:  % a,b     - limits for the root search
12:  % xtol    - uncertainty tolerance for the root
13:  %           value
14:  % nmax    - maximum allowable number of function
15:  %           values to find the root to desired
16:  %           accuracy
17:  % rt      - computed approximation to the root
18:  %           accurate within a tolerance of xtol
19:  % ntotl   - number of function evaluations made
20:  %           to compute the root
```

```
21: %
22: % User m functions called:  argument fn
23: %-----------------------------------------------
24:
25: % Global variables
26: global ifprnt_ trouble_
27:
28: % Check the number of function values needed to
29: % compute the root
30: ntotl=round(log(abs(a-b)/xtol)/log(2));
31: if ntotl > nmax
32:    str=['The root tolerance of ', ...
33:         num2str(xtol),' takes '];
34:    str=[str,int2str(ntotl),' iterations.'];
35:    disp(str);
36:    str=['\nThis value exceeds the ' ...
37:         'allowed maximum of '];
38:    str=[str,int2str(nmax),' iterations\n'];
39:    fprintf(str);
40:    trouble_=1; return
41: end
42:
43: % Perform iteration to compute the root
44: fa=feval(fn,a);
45: for i=1:ntotl
46:    rt=(a+b)/2;
47:    frt=feval(fn,rt);
48:    if ifprnt_==1
49:       fprintf('\n %4i    %13.10f    %13.4e', ...
50:                i, rt, frt)
51:    end
52:    if frt*fa > 0, a=rt; fa=frt; else, b=rt; end
53: end
54: rt=(a+b)/2;
```

Function simpson

```
1: function ansr=simpson(funct,a,b,neven)
2: %
3: % ansr=simpson(funct,a,b,neven)
4: % ~~~~~~~~~~~~~~~~~~~~~~~~~~~~~~
5: %
6: % This function integrates "funct" from
```

```
 7: % "a" to "b" by Simpson's rule using
 8: % "neven+1" function values.  Parameter
 9: % "neven" should be an even integer.
10: %
11: % Example use:  ansr=simpson('sin',0,pi/2,4)
12: %
13: % funct   -  character string name of
14: %              function integrated
15: % a,b     -  integration limits
16: % neven   -  an even integer defining the
17: %              number of integration intervals
18: % ansr    -  Simpson rule estimate of the
19: %              integral
20: %
21: % User m functions called: argument funct
22: %-----------------------------------------------
23:
24: ne=max(2,2*round(.1+neven/2)); d=(b-a)/ne;
25: x=a+d*(0:ne); y=feval(funct,x);
26: ansr=(d/3)*(y(1)+y(ne+1)+4*sum(y(2:2:ne))+...
27:      2*sum(y(3:2:ne-1)));
```

Function fnc

```
 1: function f=fnc(x)
 2: %
 3: % f=fnc(x)
 4: % ~~~~~~~~
 5: % This function integrates to get the function
 6: % for which roots are to be computed.
 7: %
 8: % x    - argument vector for the function
 9: % f    - approximation for
10: %         f=1-integral( (dt)/t ; t=1, => ,x )
11: %
12: % User m functions called:  simpson, oneovx
13: %-----------------------------------------------
14:
15: ne=round(400*abs(x-1));
16: f=1-simpson('oneovx',1,x,ne);
```

Function oneovx

```
1: function v=oneovx(x)
2: %
3: % v=oneovx(x)
4: % ~~~~~~~~~~~
5: % This function is the integrand passed to
6: % simpson. The number of function values is
7: % passed as a global variable.
8: %
9: % x        - argument vector x
10: % v        - vector with components of 1./x
11: %
12: % User m functions called:  none.
13: %-----------------------------------------------
14:
15: global kount_
16:
17: v=1 ./x;
18: kount_=kount_+length(x);
```

Function read

```
1: function [a1,a2,a3,a4,a5,a6,a7,a8,a9,a10, ...
2:           a11,a12,a13,a14,a15,a16,a17,a18, ...
3:           a19,a20]=read(labl)
4: %
5: % [a1,a2,a3,a4,a5,a6,a7,a8,a9,a10,a11,a12, ...
6: %   a13,a14,a15,a16,a17,a18,a19,a20]=read(labl)
7: %~~~~~~~~~~~~~~~~~~~~~~~~~~~~~~~~~~~~~~~~~~~~~~
8: %
9: % This function reads up to 20 variables on one
10: % line. The items should be separated by commas
11: % or blanks. Using more than 20 output
12: % variables will result in an error.
13: %
14: % labl               - Label preceding the
15: %                      data entry.  It is set
16: %                      to '? ' if no value of
17: %                      labl is given.
18: % a1,a2,...,a_nargout - The output variables
19: %                      which are created
```

```
20: %                          (cannot exceed 20)
21: %
22: % A typical function call is:
23: % [A,B,C,D]=read('Enter values of A,B,C,D: ')
24: %
25: % User m functions required: none
26: %-----------------------------------------------
27:
28: if nargin==0, labl='? '; end
29: n=nargout;
30: str=input(labl,'s'); str=['[',str,']'];
31: v=eval(str);
32: L=length(v);
33: if L>=n, v=v(1:n);
34:   else, v=[v,zeros(1,n-L)]; end
35: for j=1:nargout
36:   eval(['a',int2str(j),'=v(j);']);
37: end
```

Chapter 2

Elementary Aspects of MATLAB Graphics

2.1 Introduction

MATLAB's capabilities for plotting curves and surfaces are versatile and easy to understand. In fact, the effort required to learn MATLAB would be rewarding even if it were only used to construct plots, save graphic images, and output publication quality graphs on a laser printer. Numerous help features and well-written demo programs are included with MATLAB. By executing the demo programs and studying the relevant code, users can quickly understand the techniques necessary to implement graphics within their programs. This chapter discusses a few of the graphics commands. These commands are useful in many applications and do not require extensive time to master. This next section will provide a quick overview of the basics of using MATLAB's graphics. The subsequent sections in this chapter present four additional examples (summarized in the table below) involving interesting applications which use these graphics primitives.

Example	Purpose
Polynomial Interpolation	2-D graphics and polynomial interpolation functions
Conformal Mapping	2-D graphics and some aspects of complex numbers
String Vibration	2-D and 3-D graphics for a function of form $y(x,t)$
Properties of Curves and Surfaces	2-D and 3-D graphics related to a spiral curve and surfaces

2.2   Overview of Graphics

The following commands should be executed since they will accelerate the understanding of graphics functions, and others, included within MATLAB.

help help	discusses use of **help** command.
help	lists categories of help.
help general	lists various utility commands.
help more	describes how to control output paging.
help diary	describes how to save console output to a file.
help plotxy	describes 2D plot functions.
help plotxyz	describes 3D plot functions.
help graphics	describes more general graphics features.
help demos	lists names of various demo programs.
intro	executes the **intro** program showing MATLAB commands including fundamental graphics capabilities.
help funfun	describes several numerical analysis programs contained in MATLAB.
type humps	lists a function employed in several of the MATLAB demos.
fplotdemo	executes program **fplotdemo** which plots the function named **humps**.
help peaks	describes a function **peaks** used to illustrate surface plots.
peaks	executes the function **peaks** to produce an interesting surface plot.
spline2d	executes a demo program to draw a curve through data input interactively.

The example programs can be studied interactively by issuing the command **more on** and then using the **type** command to list programs of interest. Furthermore, a code listing can be saved in a diary file for routing to any convenient print device. For example, making a source listing of the demo program **spline2d** is achieved by the following command sequence:

```
diary spline2d.sav;
type spline2d;
diary off
```

This creates, in the current working directory, an ASCII file named *spline2d.sav*. The user may then manipulate this file appropriately (e.g., edit, print, etc.).

More advanced features of MATLAB graphics, including handle graphics, control of shading and light sources, creation of movies, etc., exceed the scope of the present text. Instead we concentrate on using the basic commands listed below and on producing simple animations. The more advanced graphics features can be mastered by studying the MATLAB manuals and relevant demo programs.

The principal graphing commands discussed here are

Command	Purpose
plot	draw two-dimensional graphs
xlabel, **ylabel**, **zlabel**	define axis labels
title	define graph title
axis	set various axis parameters (min, max, etc.)
text	place text at selected locations
grid	draw grid lines
mesh	draw surface plot with mesh
surface	draw surface plot
hold	fix the graph limits between successive plots
view	change surface viewing position
drawnow	empty graphics buffer immediately
clf	clear graphics window
contour	draw contour plot
ginput	read coordinates interactively

All of these commands, along with numerous others, are extensively documented by the help facilities in MATLAB. The user can get an introduction to these capabilities by typing "**help plot**" and by running the demo programs. The accompanying code for the demo program should be examined since it provides worthwhile insight into how MATLAB graphics is used.

2.3   Polynomial Interpolation Example

Many familiar mathematical functions such as $\arctan(x)$, $\exp(x)$, $\sin(x)$, etc. can be represented well near $x = 0$ by Taylor series expansions. If a series expansion converges rapidly, taking a few terms in the series may yield good polynomial approximations. Assuming such a procedure is plausible, one approach to polynomial approximation results by taking some data points, say (x_i, y_i), $1 \leq i \leq n$ and determining the polynomial of degree $(n - 1)$ which passes through those points. It appears reasonable that using evenly spaced data is appropriate and that increasing the number of polynomial terms should improve the accuracy of the approximating function. However, it has actually been found that passing a polynomial through points on a function $y(x)$, where x values are evenly spaced, typically gives approximations which are not smooth between the data points and tend to oscillate at the ends of the interpolating interval [20]. Attempting to reduce the oscillation by raising the polynomial order makes matters worse. Surprisingly, a special set of unevenly spaced points bunching data near the interval ends according to

$$x_j = \frac{a+b}{2} + \left(\frac{a-b}{2}\right) \cos\left[\frac{\pi}{n}\left(j - \frac{1}{2}\right)\right] \qquad 1 \leq j \leq n$$

for the interval $a \leq x \leq b$ turns out to be preferable. This formula gives the so-called Chebyshev points which are optimally chosen in the sense described by Conte and de Boor [20].

The interpolation program **polyplot** employs MATLAB functions **polyfit** and **polyval** to compute approximations of the function $y = 1/(1 + x^4)$. Function **polyfit** evaluates coefficients in a polynomial of degree n which fits m data points by least square approximation. For a unique solution we need $m \geq n + 1$. The polynomial fits the data perfectly when $m = n + 1$. The function **polyval** can then be used to evaluate the polynomials for arbitrary arguments. The data points are computed using function **linspace** for equidistant data and **chbpts** for Chebyshev points. The graphic functions used in this example are **plot**, **title**, **xlabel**, **ylabel**, **get**, and **text**. It is worthwhile to study selected lines in the program. These lines and their operation are summarized below.

The program produces the graph in Figure 2.1. Notice that the interpolation function through equidistant data points oscillates severely near ± 4, whereas the Chebyshev points produce more reasonable results. By changing the program data the reader can verify that using more interpolation points improves the Chebyshev approximation but makes the results from equidistant data much worse.

Line	Operation
20	$\mathbf{xe} = \mathbf{xe}(:)$ insures $\mathbf{xe}$ is a column
24	$\mathbf{ye} = 1./(1 + \mathbf{xe}. \wedge \mathbf{p})$ raises each component of $\mathbf{xe}$ to power $\mathbf{p}$, adds one and reciprocates the results componentwise.
50-52	creates a string which is continued between lines by using ...
53	**title** prints string **titl**. Shorter strings can be directly included as an argument in the function.
59-61	creates a legend for the plot.
63	**genprint** saves the graphics image to a file for later printing.

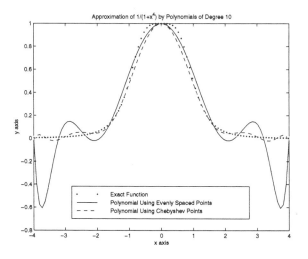

Figure 2.1. Approximation by Polynomials of Degree 10

MATLAB Example

Script File polyplot

```
 1: % Example: polyplot
 2: % ~~~~~~~~~~~~~~~~~~~
 3: % This program illustrates the use of various
 4: % graphics commands to show how the location
 5: % of interpolation points affects the accuracy
 6: % and smoothness of polynomial approximations
 7: % to the function 1/(1+x^p).
 8: %
 9: % User m functions required:
10: %    chbpts
11: %-------------------------------------------------
12:
13: % Set data parameters. Functions linspace and
14: % chbpts generate data with even and Chebyshev
15: % spacing
16: clear; n=11; a=-4; b=4; p=4;
17: xe=linspace(a,b,n); xc=chbpts(a,b,n);
18:
19: % Make sure all vectors are columns
20: xe=xe(:); xc=xc(:);
21: xx=linspace(a,b,91); xx=xx(:);
22:
23: % Compute function values
24: ye=1 ./(1+xe.^p);
25: yc=1 ./(1+xc.^p);
26: yy=1 ./(1+xx.^p);
27:
28: % Use function polyfit to compute polynomial
29: % coefficients. Since the number of data
30: % points is one greater than the polynomial
31: % order, the polynomials will pass exactly
32: % through the data points.
33: cofe=polyfit(xe,ye,n-1);
34: cofc=polyfit(xc,yc,n-1);
35:
36: % Use function polyval to evaluate the
37: % polynomials
38: yye=polyval(cofe,xx); yyc=polyval(cofc,xx);
39:
40: % Plot the exact function and interpolating
```

```
41: % polynomials. Note that ... is used to
42: % continue a line
43: plot(xx,yy,'.',xx,yye,'-',xx,yyc,'--',...
44: xe,ye,'wo',xc,yc,'wo');
45:
46: % Form a title and place it on the graph. Note
47: % that the functions num2str and int2str
48: % convert real and integer numbers into
49: % strings.
50: titl=['Approximation of 1/(1+x^', ...
51:     num2str(p),') by ',...
52:     'Polynomials of Degree ',int2str(n-1)];
53: title(titl)
54:
55: % Place labels on the x and y axes
56: xlabel('x axis'); ylabel('y axis');
57:
58: % Use a legend to identify different curves
59: legend('Exact Function',...
60: 'Polynomial Using Evenly Spaced Points',...
61: 'Polynomial Using Chebyshev Points',4);
62:
63: %pause; genprint('polyplot');
64: disp(' '); disp('All Done');
```

Function chbpts

```
1: function x=chbpts(xmin,xmax,n)
2: %
3: % x=chbpts(xmin,xmax,n)
4: % ~~~~~~~~~~~~~~~~~~~~~~
5: % Determine n points with Chebyshev spacing
6: % between xmin and xmax.
7: %
8: % User m functions called:  none
9: %------------------------------------------------
10:
11: x=(xmin+xmax)/2+((xmin-xmax)/2)* ...
12:   cos(pi/n*((0:n-1)'+.5));
```

Function genprint

```
 1: function genprint(fname,append)
 2: %
 3: % genprint(fname,append)
 4: % ~~~~~~~~~~~~~~~~~~~~~~~~
 5: % This function saves a plot to a file.  If
 6: % the file exists, it is erased first unless
 7: % the append option is specified.
 8: %
 9: % fname  - name of file to save plot to
10: %          without a filename extension
11: % append - optional, if included plot is
12: %          appended to file fname
13: %
14: % SYSTEM DEPENDENT ROUTINE
15: %
16: % User m functions called:  none
17: %-------------------------------------------------
18:
19: %...Define these appropriately
20: ext=['.eps']; % filename extension to use
21: opt=['eps'];  % option for print command
22:
23: %...Append extension to filename
24: file_name=[fname,ext];
25:
26: %...Determine computer type
27: system_type=computer;
28:
29: %...Use correct command for different systems
30: if strcmp(system_type(1:2),'PC')
31:   erase_cmd=['delete ', file_name];
32: elseif strcmp(system_type(1:3),'SGI')
33:   erase_cmd=['!rm ', file_name];
34: else
35:   disp(' ');
36:   disp('Unknown system type in genprint');
37:   break;
38: end
39:
40: % Save to encapsulated postscript file
41: if nargin == 1
42:   if exist(file_name)==2
```

```
43:      eval(erase_cmd);
44:    end
45:    eval(['print -d',opt,' ',fname]);
46: else
47:    eval(['print -d',opt,' -append ',fname]);
48: end
```

2.4   Conformal Mapping Example

This example involves analytic functions and conformal mapping. The complex function $w(z)$ which maps $|z| \leq 1$ onto the interior of a square of side length 2 can be written in power series form as

$$w(z) = \sum_{k=0}^{\infty} b_k z^{4k+1}$$

where

$$b_k = c \left[\frac{(-1)^k (\frac{1}{2})_k}{k!(4k+1)} \right] \qquad \sum_{k=0}^{\infty} b_k = 1$$

and c is a scaling coefficient chosen to make $z = 1$ map to $w = 1$ [73]. Truncating the series after some finite number of terms, say m, produces an approximate square with rounded corners. Increasing m reduces the corner rounding but convergence is rather slow so that using even a thousand terms still gives perceptible inaccuracy. The purpose of the present exercise is to show how a polar coordinate region characterized by

$$z = re^{i\theta} \qquad r_1 \leq r \leq r_2 \qquad \theta_1 \leq \theta \leq \theta_2$$

transforms and to exhibit an undistorted plot of the region produced in the w-plane. The exercise also emphasizes the utility of MATLAB for handling complex arithmetic and complex functions. The program has a short driver **squarrun** and a function **squarmap** which computes points in the w region and coefficients in the series expansion. Salient features of the program are summarized in the table below.

Results produced when $0.5 \leq r \leq 1$ and $0 \leq \theta \leq 2\pi$ by a twenty-term series appear in Figure 2.2. The reader may find it interesting to run the program using several hundred terms and take $0 \leq \theta \leq \pi/2$. The corner rounding remains noticeable even when $m = 1000$ is used. Later in this book we will visit the mapping problem again to show that a better approximation is obtainable using rational functions.

Routine	Line	Operation
squarrun	20-41	functions **input**, **disp**, **fprintf**, and **read** are used to input data interactively. Several different methods of printing were used for purposes of illustration rather than necessity.
	45	function **squarmap** generates results.
	49	function **genprint** is a system dependent routine which is used to create plot files for later printing.
squarmap	31-33	functions **linspace** and **ones** are used to generate points in the z-plane.
	43-45	series coefficients are computed using **cumprod** and the mapping is evaluated using **polyval** with a matrix argument.
	48-51	scale limits are calculated to allow an undistorted plot of the geometry. Use is made of MATLAB functions **real** and **imag**.
	57-73	loops are executed to plot the circumferential lines first and the radial lines second.
cubrange		function which determines limits for a square or cube shaped region.

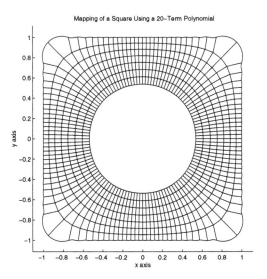

Figure 2.2. Mapping of a Square Using a 20-Term Polynomial

MATLAB Example

Script File squarrun

```
 1: % Example: squarrun
 2: % ~~~~~~~~~~~~~~~~~~~
 3: %
 4: % Driver program to plot the mapping of a
 5: % circular disk onto the interior of a square
 6: % by the Schwarz-Christoffel transformation.
 7: %
 8: % User m functions required:
 9: %    squarmap, read, cubrange, genprint
10: %-----------------------------------------------
11:
12: % Illustrate use of the function input to
13: % interactively read data on the same line
14:
15: fprintf('\nCONFORMAL MAPPING OF A SQUARE ')
16: fprintf('BY USE OF A\n')
17: fprintf('TRUNCATED SCHWARZ-CHRISTOFFEL ')
18: fprintf('SERIES\n\n')
19:
20: fprintf('Input the number of series ')
21: fprintf('terms used')
22: m=input('(try 20)? ');
23:
24: % Illustrate use of the function disp
25: disp('')
26: str=['\nInput the inner radius, outer ' ...
27:       'radius and number of increments ' ...
28:       '\n(try .5,1,8)\n'];
29: fprintf(str);
30:
31: % Use function read to input several variables
32: [r1,r2,nr]=read;
33:
34: % Use function fprintf to print more
35: % complicated heading
36: str=['\nInput the starting value of ' ...
37:       'theta, the final value of theta \n' ...
38:       'and the number of theta increments ' ...
39:       '(the angles are in degrees) ' ...
40:       '\n(try 0,360,120)\n'];
```

```
41: fprintf(str); [t1,t2,nt]=read;
42:
43: % Call function squarmap to make the plot
44: hold off; clf;
45: [w,b]=squarmap(m,r1,r2,nr,t1,t2,nt+1);
46:
47: % Remove an existing plot file if it exists.
48: % Then save the plot.
49: %genprint('squarplt');
50: disp(' '); disp('All Done');
```

Function squarmap

```
1: function [w,b]=squarmap(m,r1,r2,nr,t1,t2,nt)
2: %
3: % [w,b]=squarmap(m,r1,r2,nr,t1,t2,nt)
4: % ~~~~~~~~~~~~~~~~~~~~~~~~~~~~~~~~~~~~~
5: % This function evaluates the conformal mapping
6: % produced by the Schwarz-Christoffel
7: % transformation w(z) mapping abs(z)<=1 inside
8: % a square having a side length of two.  The
9: % transformation is approximated in series form
10: % which converges very slowly near the corners.
11: %
12: % m          - number of series terms used
13: % r1,r2,nr - abs(z) varies from r1 to r2 in
14: %              nr steps
15: % t1,t2,nt - arg(z) varies from t1 to t2 in
16: %              nt steps (t1 and t2 are measured
17: %              in degrees)
18: % w          - points approximating the square
19: % b          - coefficients in the truncated
20: %              series expansion which has the
21: %              form
22: %
23: %          w(z)=sum({j=1:m},b(j)*z*(4*j-3))
24: %
25: % User m functions called:  cubrange
26: %-----------------------------------------------
27:
28: % Generate polar coordinate grid points for the
29: % map.  Function linspace generates vectors
30: % with equally spaced components.
```

```
31: r=linspace(r1,r2,nr)';
32: t=pi/180*linspace(t1,t2,nt);
33: z=(r*ones(1,nt)).*(ones(nr,1)*exp(i*t));
34:
35: % Use high point resolution for the
36: % outer contour
37: touter=pi/180*linspace(t1,t2,10*nt);
38: zouter=r2*exp(i*touter);
39:
40: % Compute the series coefficients and
41: % evaluate the series
42: k=1:m-1;
43: b=cumprod([1,-(k-.75).*(k-.5)./(k.*(k+.25))]);
44: b=b/sum(b); w=z.*polyval(b(m:-1:1),z.^4);
45: wouter=zouter.*polyval(b(m:-1:1),zouter.^4);
46:
47: % Determine square window limits for plotting
48: uu=real([w(:);wouter(:)]);
49: vv=imag([w(:);wouter(:)]);
50: rng=cubrange([uu,vv],1.1);
51: axis('square'); axis(rng); hold on
52:
53: % Plot orthogonal grid lines which represent
54: % the mapping of circles and radial lines
55:
56: % First draw the circle maps
57: for j=1:nr-1
58:   wj=w(j,:);
59:   plot(real(wj),imag(wj));
60: end
61: plot(real(wouter),imag(wouter))
62:
63: % Then draw the radial line maps
64: for j=1:nt
65:   wj=w(:,j);
66:   plot(real(wj),imag(wj));
67: end
68:
69: % Add a title and axis labels
70: title(['Mapping of a Square Using a ', ...
71:         num2str(m),'-Term Polynomial'])
72: xlabel('x axis'); ylabel('y axis')
73: figure(gcf); hold off;
```

Function cubrange

```
 1: function range=cubrange(xyz,ovrsiz)
 2: %
 3: % range=cubrange(xyz,ovrsiz)
 4: % ~~~~~~~~~~~~~~~~~~~~~~~~~~~~
 5: % This function determines limits for a square
 6: % or cube shaped region for plotting data values
 7: % in the columns of array xyz to an undistorted
 8: % scale
 9: %
10: % xyz    - a matrix of the form [x,y] or [x,y,z]
11: %          where x,y,z are vectors of coordinate
12: %          points
13: % ovrsiz - a scale factor for increasing the
14: %          window size. This parameter is set to
15: %          one if only one input is given.
16: %
17: % range  - a vector used by function axis to set
18: %          window limits to plot x,y,z points
19: %          undistorted. This vector has the form
20: %          [xmin,xmax,ymin,ymax] when xyz has
21: %          only two columns or the form
22: %          [xmin,xmax,ymin,ymax,zmin,zmax]
23: %          when xyz has three columns.
24: %
25: % User m functions called:  none
26: %------------------------------------------------
27:
28: if nargin==1, ovrsiz=1; end
29: pmin=min(xyz); pmax=max(xyz); pm=(pmin+pmax)/2;
30: pd=max(ovrsiz/2*(pmax-pmin));
31: if length(pmin)==2
32:   range=pm([1,1,2,2])+pd*[-1,1,-1,1];
33: else
34:   range=pm([1 1 2 2 3 3])+pd*[-1,1,-1,1,-1,1];
35: end
```

2.5   String Vibration Example

The next example investigates a familiar type of problem involving a function $y(x, t)$ known for $a \leq x \leq b$ and $t > 0$. For the case studied, y represents transverse deflection in a taut string. However, it could just as well describe current flow in a transmission line or pressure in a porous medium. This example has a concise series solution and illustrates well the dynamical ideas of interest. A more general problem using Fourier series to handle arbitrary initial deflection is presented in Chapter 9.

In the string vibration problem, we use dimensionless variables to simplify interpreting the results. The string is at rest initially. The left end is fixed and the right end is suddenly given a sinusoidally varying motion causing waves which propagate back and forth along the string. The related boundary value problem in dimensionless variables is

$$\frac{\partial^2 y}{\partial x^2} = \frac{\partial^2 y}{\partial t^2} \qquad 0 \leq x \leq 1 \qquad t > 0$$

$$y(0, t) = 0 \qquad y(1, t) = \sin(\omega t)$$

$$y(x, 0) = 0 \qquad \frac{\partial y}{\partial t}(x, 0) = 0$$

As long as ω does not equal an integer multiple of π, this problem has a simple series solution given by

$$y(x, t) = \frac{\sin(\omega x) \sin(\omega t)}{\sin(\omega)} - $$

$$2\omega \left\{ \sum_{n=1}^{\infty} \left[\frac{(-1)^n}{\omega^2 - n^2 \pi^2} \right] \sin(n\pi t) \sin(n\pi x) \right\}$$

We need graphical results to help us visualize the dynamics of the system. Note that at any fixed position x_0, $y(x_0, t)$ represents the time history of that point on the string. Similarly, at any particular time t_0, then $y(x, t_0)$ for $0 \leq x \leq 1$ represents the deflection pattern of the string. Suppose $\omega \approx k\pi$ for integer k, then a nearly zero denominator occurs in the solution, thereby producing large displacement amplitudes. This results in transverse deflections which build up from zero initially to large values as translating waves repeatedly reflect from the end boundaries of the string. Terms in the solution such as

$$\sin(n\pi x) \sin(n\pi t) = \frac{1}{2} \left\{ \cos\left[n\pi(x - t)\right] - \cos\left[n\pi(x + t)\right] \right\}$$

represent components which simultaneously translate to the left and to the right with unit speed. Consequently, no displacement will occur near the end $x = 0$ until $t \geq 1$. By the time $t = 10$, for example, several

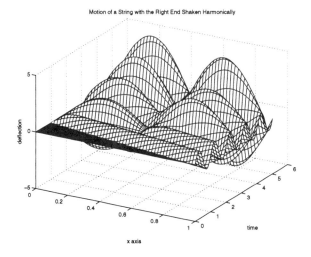

Figure 2.3. String Motion: One End is Shaken Harmonically

reflections will have occurred from each end. This produces a complex deflection pattern. A function, **shkstrng**, was written to evaluate the series. Input parameters allow the user to select the number of series terms, the forcing frequency ($\omega \neq k\pi$) and the x and t limits. The MATLAB function **mesh** was employed to plot the surface $y(x, t)$ shown in Figure 2.3. The wave propagation mentioned above is clearly evident. A nearly resonant frequency $\omega = 1.95\pi$ was used. In Figure 2.4, we plot the deflection when the wave has moved halfway along the string. In Figure 2.5, we plot the deflection history at $x = 0.25$. That motion builds up from zero to fairly large amplitude by $t = 10$. A function called **motion** is also given which animates the string displacement. The animation is quite helpful for visualizing how waves travel along the string and reflect off the boundaries. Furthermore, progressive build up of the forced oscillations is evident. We leave these functions for the reader to study.

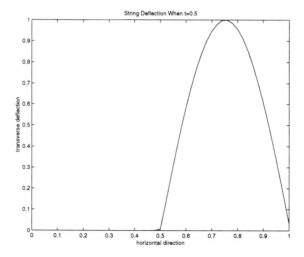

Figure 2.4. String Deflection at $t = 0.5$

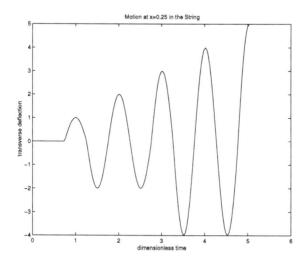

Figure 2.5. Motion at Quarterpoint of the String

MATLAB Example

Script File stringmo

```
 1: % Example: stringmo
 2: % ~~~~~~~~~~~~~~~~~~~
 3: % This is a driver program to illustrate motion
 4: % of a string having one end subjected to
 5: % harmonic oscillation.
 6: %
 7: % User m functions required:
 8: %    shkstrng, ploteasy, pauz
 9: %-------------------------------------------------
10:
11: fprintf('\nFORCED MOTION OF A VIBRATING ');
12: fprintf('STRING\n');
13: wf=1.98*pi; tmax=(2*pi)*.8;
14: [y,t,x]=shkstrng(wf,80,0,tmax,75,0,1,51);
15: surf(x,t,y); ylabel('time')
16: view([30,30]); xlabel('x axis');
17: zlabel('deflection'); % colormap([1 1 1]);
18: title(['Motion of a String with the Right', ...
19:        ' End Shaken Harmonically']);
20: fprintf('\nPress [Enter] for the\n');
21: fprintf('deflection when t=0.5\n');
22: figure(gcf); % genprint('strngsrf');
23: pauz;
24: [yp5,tp5,xp5]=shkstrng(wf,80,.5,.5,1,0,1,51);
25: ploteasy(xp5,yp5,'horizontal direction', ...
26:          'transverse deflection', ...
27:          'String Deflection When t=0.5');
28: fprintf('Press [Enter] for the deflection\n');
29: fprintf('history at x=0.25\n');
30: %genprint('dflatep5');
31: pauz;
32:
33: [yxc,txc,xc]= ...
34:    shkstrng(wf,80,0,tmax,151,.25,.25,1);
35: ploteasy(txc,yxc,'dimensionless time', ...
36:          'transverse deflection', ...
37:          'Motion at x=0.25 in the String');
38: %genprint('motnqrtp');
39: fprintf('Press [Enter] to animate')
40: fprintf('\nthe string motion\n'); pauz;
```

```
41: motion(x,y);
42: disp(' '); disp('All Done');
```

Function shkstrng

```
1: function [y,t,x]= ...
2:            shkstrng(w,nsum,t1,t2,nt,x1,x2,nx)
3: %
4: % [y,t,x]=shkstrng(w,nsum,t1,t2,nt,x1,x2,nx)
5: % ~~~~~~~~~~~~~~~~~~~~~~~~~~~~~~~~~~~~~~~~~~~~
6: % Simulation of the motion of a string having
7: % one end fixed and the other end shaken
8: % harmonically.
9: %
10: % w     - forcing frequency
11: % t1,t2 - minimum and maximum times
12: % nt    - number of time values
13: % x1,x2 - minimum and maximum x values
14: %         lying between zero and one
15: % nx    - number of x values
16: %
17: % t,x   - vectors of time and position values
18: % y     - matrix of transverse deflection
19: %         values having nt rows and nx
20: %         columns
21: %
22: % User m functions called:  none
23: %-------------------------------------------------
24:
25: t=linspace(t1,t2,nt)'; x=linspace(x1,x2,nx);
26: np=pi*(1:nsum); y=sin(w*t)/sin(w)*sin(w*x);
27: a=2*w*ones(nt,1)*(cos(np)./(np.^2-w^2));
28: y=y+a.*sin(t*np)*sin(np'*x);
```

Function ploteasy

```
1: function ploteasy(x,y,xlabl,ylabl,titl)
2: %
3: % ploteasy(x,y,xlabl,ylabl,titl)
4: % ~~~~~~~~~~~~~~~~~~~~~~~~~~~~~~~~
5: % Easy plot function with a simple
6: % argument list
```

```
 7: %
 8: % x,y   - data to be plotted
 9: % xlabl - horizontal axis label for the graph
10: % ylabl - vertical axis label for the graph
11: % titl  - title for the graph
12: %
13: % User m functions called:  none
14: %-------------------------------------------------
15:
16: plot(x,y);
17: if nargin==2, figure(gcf); return, end
18: if nargin>2, xlabel(xlabl); end
19: if nargin>3, ylabel(ylabl); end
20: if nargin>4, title(titl); end
21: figure(gcf);
```

Function motion

```
 1: function motion(x,y,inct,trac)
 2: %
 3: % motion(x,y,inct,trac)
 4: % ~~~~~~~~~~~~~~~~~~~~~
 5: % This function animates the motion history
 6: % of the string.
 7: %
 8: % x    - horizontal position coordinates
 9: %        corresponding to various columns
10: %        of matrix y
11: % y    - matrix with row j specifying the
12: %        string position at the j'th time
13: %        value
14: % inct - the number of row increments used
15: %        to select positions for plotting.
16: %        Using inct=2 would plot every other
17: %        row of y. inct=1 is the default value.
18: % trac - if this parameter is present,
19: %        successive plot images are left on
20: %        the screen. Otherwise, each
21: %        configuration is shown and removed
22: %        before the next image is shown. The
23: %        default choice is to remove
24: %        successive images.
25: %
```

```
26: % User m functions called:  none
27: %---------------------------------------------
28:
29: if nargin ==2, inct=1; trac=0; end
30: if nargin ==3, trac=0; end
31: if inct > 1
32:   [nt,nx]=size(y); y=y(1:inct:nx,:);
33: end
34:
35: xmin=min(x); xmax=max(x);
36: ymin=min(y(:)); ymax=max(y(:)); clf;
37: axis([xmin,xmax,2*ymin,2*ymax]);
38: [nt,nx]=size(y); axis off; hold on
39: nt
40: for j=1:nt-1
41:   plot(x,y(j,:)); drawnow; figure(gcf);
42:   if trac ==0, cla; end
43: end
44: plot(x,y(nt,:)); figure(gcf); hold off;
```

Function pauz

```
 1: function pauz(strng)
 2: %
 3: % pauz(strng)
 4: % ~~~~~~~~~~~
 5: % On some systems MATLAB will not let you
 6: % invoke the print menu option when using a
 7: % pause statement. This routine gets around
 8: % that problem by invoking another MATLAB
 9: % function (input) which does not have that
10: % characteristic.
11: %
12: % User m functions called:  none
13: %---------------------------------------------
14:
15: if nargin==0, strng=' '; end
16: dumy=input(strng,'s');
```

2.6   Properties of Curves and Surfaces

In this section some properties of space curves and surfaces are studied.
Examples illustrating the graphics capabilities of MATLAB to describe
three-dimensional geometries are given. Readers should also study the
demo examples and intrinsic documentation on functions such as **plot3**,
surf, and **mesh** to appreciate the wealth of plotting options available.

2.6.1   Curve Properties

A space curve is a one-dimensional region representable in parametric
form as

$$R(t) = \hat{\imath}\, x(t) + \hat{\jmath}\, y(t) + \hat{k}\, z(t) \qquad a < t < b$$

where $\hat{\imath}$, $\hat{\jmath}$, $\hat{k}$ are cartesian base vectors, and t is a scalar parameter
such as arc length s or time. At each point on the curve, differential
properties naturally lead to a triad of orthonormal base vectors $\hat{T}$, $\hat{N}$,
and $\hat{B}$ called the tangent, the principal normal, and the binormal. The
normal vector points toward the center of curvature and the binormal is
defined by $\hat{T} \times \hat{N}$ to complete the triad. Coordinate planes associated
with the triad are the normal plane containing $\hat{N}$ and $\hat{B}$, the tangent
plane containing $\hat{T}$ and $\hat{B}$, and the osculating plane containing $\hat{T}$ and
$\hat{N}$. Two other scalar properties of interest are the curvature κ (the
reciprocal of the curvature radius) and the torsion τ, which quantifies
the rate at which the triad twists about the direction of $\hat{T}$ as a generic
point moves along the curve. When a curve is parameterized in terms of
arc length s, the five quantities just mentioned are related by the Frenet
formulas [89] which are

$$\frac{d\hat{T}}{ds} = \kappa \hat{N} \qquad \frac{d\hat{B}}{ds} = -\tau \hat{N} \qquad \frac{d\hat{N}}{ds} = -\kappa \hat{T} + \tau \hat{B}$$

Since most curves are not easily parameterized in terms of arc length,
more convenient formulas are needed for computing $\hat{T}$, $\hat{N}$, $\hat{B}$, κ, and
τ. All the desired quantities can be found in terms of $R'(t)$, $R''(t)$, and
$R'''(t)$. Among the five properties, only torsion, τ, depends on $R'''(t)$.
The pertinent formulas are

$$\hat{T} = \frac{R'(t)}{|R'(t)|} \qquad \hat{B} = \frac{R'(t) \times R''(t)}{|R'(t) \times R''(t)|}$$

$$\hat{N} = \hat{B} \times \hat{T} \qquad \kappa = \frac{|R'(t) \times R''(t)|}{|R'(t)|^3}$$

$$\tau = \frac{\hat{B} \cdot R'''(t)}{|R'(t) \times R''(t)|}$$

When the independent variable t means time we get

$$V = \text{velocity} = \frac{dR}{dt} = \frac{ds}{dt}\frac{dR}{ds} = v\hat{T}$$

where v is the magnitude of velocity called speed. Differentiating again leads to

$$\frac{dV}{dt} = \text{acceleration} = \frac{dv}{dt}\hat{T} + \kappa v^2 \hat{N}$$

so the acceleration involves a tangential component with magnitude equal to the time rate of change of speed, and a normal component of magnitude κv^2 directed toward the center of curvature. The torsion is only encountered when the time derivative of acceleration is considered. This is seldom of interest in Newtonian mechanics.

A function **crvprp3d** was written to evaluate $\hat{T}$, $\hat{N}$, $\hat{B}$, κ, and τ in terms of $R'(t)$, $R''(t)$, and $R'''(t)$. Another function **aspiral** applies **crvprp3d** to the curve described by

$$R(t) = [(r_o + kt)\cos(t); \ (r_o + kt)\sin(t); \ ht]$$

where t is the polar coordinate angle for cylindrical coordinates. Figure 2.6 depicts results generated from the default data set where

$$r_o = 2\pi \qquad k = 1 \qquad h = 2 \qquad 2\pi \le t \le 8\pi$$

with 101 data points being used. A cross section normal to the surface would produce a right angle describing the directions of the normal and binormal at a typical point. The spiral itself passes along the apex of the right angle. This surface illustrates how the intrinsic triad of base vectors changes position and direction as a point moves along the curve.

An additional function **crvprpsp** was written to test how well cubic spline interpolation approximates curve properties for the spiral. MATLAB provides function **spline** to connect data points by a piecewise cubic interpolation curve having continuous first and second derivatives [26]. This function utilizes other intrinsic functions[1] such as **unmkpp**, **mkpp**, and **ppval**. Although basic MATLAB does not include functions for spline differentiation, this can be remedied by the short function **splined** which computes first and second derivatives of the interpolation curve defined by function **spline**. In our example using spline interpolation, approximation of τ was not obtained because a cubic spline only has its first two derivatives continuous. Approximations for $R'''(t)$ could have been generated by interpolating the computed values of $R'(t)$ and differentiating the results twice. That idea was not explored. To assess the accuracy of the spline interpolation, values for $\mathbf{norm}(\hat{B} - \hat{B}_{\text{approx}})$

---

[1] These functions are included with MATLAB and are a subset of the more comprehensive *Spline Toolbox* also available from The MathWorks.

and $|(k - k_{approx})/k|$ were obtained at 101 sample points along the curve. Results depicted in Figure 2.7 show errors in the third decimal place except near the ends of the interpolation interval where a "not a knot" boundary condition is employed [26].

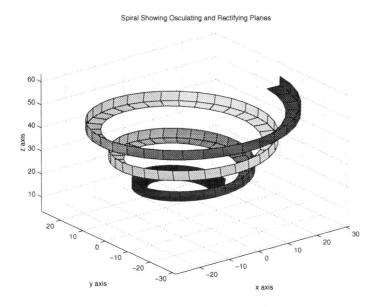

Figure 2.6. Spiral Showing Osculating and Rectifying Planes

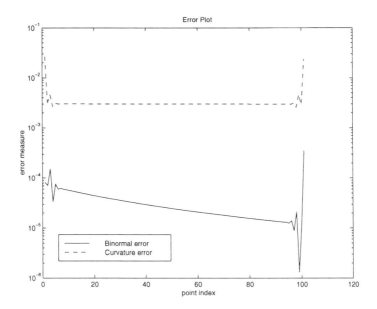

Figure 2.7. Error Plot

2.6.2   Program Output and Code

Script File splinerr

```
 1: % Example: splinerr
 2: % ~~~~~~~~~~~~~~~~~~
 3: %
 4: % This program calculates the binormal and
 5: % curvature error for the spiral configuration.
 6: %
 7: % User m functions called:
 8: %      aspiral, crvprpsp, genprint
 9: %-----------------------------------------------
10:
11: clear; hold off; clf;
12: [R,T,N,B,KAP]=aspiral; m=size(R,2);
13: [r,t,n,b,k]=crvprpsp(R,m);
14: disp(' '); disp('Press [Enter] to continue');
15: pause; disp(' ');
16:
17: errv=sqrt(sum((B-b).^2));
18: errk=abs((KAP-k)./KAP);
19: hold off; clf;
20: semilogy(1:m,errv,'-',1:m,errk,'--');
21: xlabel('point index'); ylabel('error measure');
22: title('Error Plot');
23: legend('Binormal error','Curvature error');
24: figure(gcf);
25: disp('Use mouse to locate legend block');
26: disp('Press [Enter] when finished'); disp(' ');
27: pause; %genprint('binerr');
```

Function aspiral

```
 1: function [R,T,N,B,kap,tau,arclen]= ...
 2:                          aspiral(r0,k,h,t)
 3: %
 4: % [R,T,N,B,kap,tau,arclen]=aspiral(r0,k,h,t)
 5: % ~~~~~~~~~~~~~~~~~~~~~~~~~~~~~~~~~~~~~~~~~~~~
 6: %
 7: % This function computes geometrical properties
 8: % of a spiral curve having the parametric
 9: % equation
10: %
11: %   R = [(r0+k*t)*cos(t);(r0+k*t)*sin(t);h*t)]
12: %
13: % A figure showing the curve along with the
14: % osculating plane and the rectifying plane
15: % at each point is also drawn.
16: %
17: % r0,k,h - parameters which define the spiral
18: % t      - a vector of parameter values at
19: %          which the curve is evaluated from
20: %          the parametric form.
21: %
22: % R      - matrix with columns containing
23: %          position vectors for points on the
24: %          curve
25: % T,N,B  - matrices with columns containing the
26: %          tangent,normal,and binormal vectors
27: % kap    - vector of curvature values
28: % tau    - vector of torsion values
29: % arclen - value of arc length approximated as
30: %          the sum of chord values between
31: %          successive points
32: %
33: % User m functions called:
34: %          crvprp3d, cubrange, genprint
35: %-------------------------------------------------
36:
37: if nargin==0
38:   k=1; h=2; r0=2*pi; t=linspace(2*pi,8*pi,101);
39: end
40:
41: % Evaluate R, R'(t), R''(t) and R'''(t) for
42: % the spiral
```

```
43: t=t(:)'; s=sin(t); c=cos(t); kc=k*c; ks=k*s;
44: rk=r0+k*t; rks=rk.*s; rkc=rk.*c; n=length(t);
45: R=[rkc;rks;h*t]; R1=[kc-rks;ks+rkc;h*ones(1,n)];
46: R2=[-2*ks-rkc;2*kc-rks;zeros(1,n)];
47: R3=[-3*kc+rks;-3*ks-rkc;zeros(1,n)];
48:
49: % Obtain geometrical properties
50: [T,N,B,kap,tau]=crvprp3d(R1,R2,R3);
51: arclen=sum(sqrt(sum((R(:,2:n)-R(:,1:n-1)).^2)));
52:
53: % Generate points on the osculating plane and
54: % the rectifying plane along the curve.
55: w=arclen/100; Rn=R+w*N;   Rb=R+w*B;
56: X=[Rn(1,:);R(1,:);Rb(1,:)];
57: Y=[Rn(2,:);R(2,:);Rb(2,:)];
58: Z=[Rn(3,:);R(3,:);Rb(3,:)];
59:
60: % Draw the surface
61: v=cubrange([X(:),Y(:),Z(:)]); hold off; clf; close;
62: surf(X,Y,Z); axis(v); xlabel('x axis');
63: ylabel('y axis'); zlabel('z axis');
64: title(['Spiral Showing Osculating and ', ...
65:        'Rectifying Planes']); grid on; drawnow;
66: figure(gcf); %genprint('spiral');
```

Function crvprp3d

```
1: function [T,N,B,kap,tau]=crvprp3d(R1,R2,R3)
2: %
3: % [T,N,B,kap,tau]=crvprp3d(R1,R2,R3)
4: % ~~~~~~~~~~~~~~~~~~~~~~~~~~~~~~~~~~~~~
5: %
6: % This function computes the primary
7: % differential properties of a three-dimensional
8: % curve parameterized in the form R(t) where t
9: % can be arc length or any other convenient
10: % parameter such as time.
11: %
12: % R1  - the matrix with columns containing R'(t)
13: % R2  - the matrix with columns containing R''(t)
14: % R3  - the matrix with columns containing
15: %       R'''(t).  This matrix is only needed
16: %       when torsion is to be computed.
```

```
17: %
18: % T    - matrix with columns containin the
19: %          unit tangent
20: % N    - matrix with columns containing the
21: %          principal normal vector
22: % B    - matrix with columns containing the
23: %          binormal
24: % kap  - vector of curvature values
25: % tau  - vector of torsion values. This equals
26: %          [] when R3 is not given
27: %
28: % User m functions called:  none
29: %-----------------------------------------------
30:
31: nr1=sqrt(dot(R1,R1)); T=R1./nr1(ones(3,1),:);
32: R12=cross(R1,R2); nr12=sqrt(dot(R12,R12));
33: B=R12./nr12(ones(3,1),:); N=cross(B,T);
34: kap=nr12./nr1.^3;
35:
36: % Compute the torsion only when R'''(t) is given
37: if nargin==3
38:    tau=dot(B,R3)./nr12;
39: else
40:    tau=[];
41: end
```

Function crvprpsp

```
1: function [R,T,N,B,kappa]=crvprpsp(Rd,n)
2: %
3: % [R,T,N,B,kappa]=crvprpsp(Rd,n)
4: % ~~~~~~~~~~~~~~~~~~~~~~~~~~~~~~~
5: %
6: % This function computes spline interpolated
7: % values for coordinates, base vectors and
8: % curvature obtained by passing a spline curve
9: % through data values given in Rd.
10: %
11: % Rd    - a matrix containing x,y and z values
12: %          in rows 1, 2 and 3.
13: % n     - the number of points at which
14: %          properties are to be evaluated along
15: %          the curve
```

```
16: %
17: % R      - a 3 by n matrix with columns
18: %          containing coordinates of interpolated
19: %          points on the curve
20: % T,N,B - matrices of dimension 3 by n with
21: %          columns containing components of the
22: %          unit tangent, unit normal, and unit
23: %          binormal vectors
24: % kappa - a vector of curvature values
25: %
26: % User m functions called:
27: %          splined, crvprp3d
28: %-----------------------------------------------
29:
30: % Create a spline curve through the data points,
31: % and evaluate the derivatives of R.
32: nd=size(Rd,2); td=0:nd-1; t=linspace(0,nd-1,n);
33: ud=Rd(1,:)+i*Rd(2,:); u=spline(td,ud,t);
34: u1=splined(td,ud,t); u2=splined(td,ud,t,1);
35: z=spline(td,Rd(3,:),t);
36: z1=splined(td,Rd(3,:),t);
37: z2=splined(td,Rd(3,:),t,1);
38: R=[real(u);imag(u);z];
39: R1=[real(u1);imag(u1);z1];
40: R2=[real(u2);imag(u2);z2];
41:
42: % Get curve properties from crvprp3d
43: [T,N,B,kappa]=crvprp3d(R1,R2);
```

Function splined

```
1: function val=splined(xd,yd,x,if2)
2: %
3: % val=splined(xd,yd,x,if2)
4: % ~~~~~~~~~~~~~~~~~~~~~~~~~
5: %
6: % This function evaluates the first or second
7: % derivative of the piecewise cubic
8: % interpolation curve defined by the intrinsic
9: % function spline provided in MATLAB.
10: %
11: % xd,yd - data vectors determining the spline
12: %          curve produced by function spline
```

```
13: % x        - vector of values where the first or
14: %            the second derivative are desired
15: % if2      - a parameter which is input only if
16: %            y''(x) is required. otherwise, y'(x)
17: %            is returned.
18: %
19: % val      - the first or second derivative values
20: %            for the spline
21: %
22: % User m functions called: none
23: %------------------------------------------------
24:
25: [b,c]=unmkpp(spline(xd,yd)); n=length(b)-1;
26: if nargin < 4
27:   c=[zeros(n,1),3*c(:,1),2*c(:,2),c(:,3)];
28: else
29:   c=[zeros(n,2),6*c(:,1),2*c(:,2)];
30: end
31: val=ppval(mkpp(b,c),x);
```

2.6.3   Surface Properties

Surfaces are two-dimensional regions described parametrically as

$$R(u, v) = \hat{\imath}x(u, v) + \hat{\jmath}u(u, v) + \hat{k}z(u, v)$$

where u and v are scalar parameters. This parametric form is helpful for generating a grid of points on the surface as well as for computing surface tangents and the surface normal. Holding v fixed while u varies generates a curve in the surface called a u coordinate line. A tangent vector to the u-line is given by

$$g_u = \frac{\partial R}{\partial u} = \hat{\imath}\frac{\partial x}{\partial u} + \hat{\jmath}\frac{\partial y}{\partial u} + \hat{k}\frac{\partial z}{\partial u}$$

Similarly, holding u fixed and varying v produces a v-line with tangent vector

$$g_v = \frac{\partial R}{\partial v} = \hat{\imath}\frac{\partial x}{\partial v} + \hat{\jmath}\frac{\partial y}{\partial v} + \hat{k}\frac{\partial z}{\partial v}$$

Consider the following cross product.

$$g_u \times g_v \, du \, dv = \hat{n} \, dS$$

In this equation $\hat{n}$ is the unit surface normal and dS is the area of a parallelogram shaped surface element having sides defined by $g_u \, du$ and $g_v \, dv$.

The intrinsic functions surf(X,Y,Z) and mesh(X,Y,Z) depict surfaces by showing a grid network and related surface patches characterized when parameters u and v are varied over constant limits. Thus, values

$$(u_\imath, v_\jmath) \qquad 1 \leq \imath \leq n \qquad 1 \leq \jmath \leq m$$

lead to matrices

$$X = [x(u_\imath, v_\jmath)] \qquad Y = [y(u_\imath, v_\jmath)] \qquad Z = [z(u_\imath, v_\jmath)]$$

from which surface plots are obtained. Function **surf** colors the surface patches whereas **mesh** colors the grid lines.

As a simple example, consider the ellipsoidal surface described parametrically as

$$x = a\cos\theta\cos\phi \qquad y = b\cos\theta\sin\phi \qquad z = c\sin\theta$$

$$-\frac{\pi}{2} \leq \theta \leq \frac{\pi}{2} \qquad -\pi \leq \phi \leq \pi$$

The surface equation evidently satisfies the familiar equation

$$\left(\frac{x}{a}\right)^2 + \left(\frac{y}{b}\right)^2 + \left(\frac{z}{c}\right)^2 = 1$$

for an ellipsoid. The function elipsoid(a,b,c) called with $a = 2$, $b = 1.5$, $c = 1$ produces the surface plot in Figure 2.8.

Many types of surfaces can be parameterized in a manner similar to the ellipsoid. We will examine two more problems involving a torus and a conical frustum. Consider a circle of radius b lying in the xz-plane with its center at [a,0,0]. Rotating the torus about the z-axis produces a torus having the surface equation

$$x = [a + b \cos \theta] \cos \phi \qquad y = [a + b \cos \theta] \sin \phi \qquad z = b \sin \phi$$

$$-\pi \le \theta \le \pi \qquad -\pi \le \phi \le \pi$$

This type of equation is used below in an example involving several bodies. Let us also produce a surface covering the ends and side of a conical frustum (a cone with the top cut off). The frustum has base radius r_b, top radius r_t, and height h, with the symmetry axis along the z-axis. The surface can be parameterized using an azimuthal angle θ and an arc length parameter relating to the axial direction. The lateral side length is

$$r_s = \sqrt{h^2 + (r_b - r_t)^2}$$

Let us take $0 \le s \le (r_b + r_s + r_t)$ and describe the surface $R(s, \theta)$ by coordinate functions

$$x = r(s) \cos \theta \qquad y = r(s) \sin \theta \qquad z = z(s)$$

where $0 \le \theta \le 2\pi$ and

$$r(s) = s \qquad 0 \le s \le r_b$$

$$r(s) = r_b + \frac{(r_t - r_b)(s - r_b)}{r_s} \qquad z = \frac{h(s - r_b)}{r_s} \qquad r_b \le s \le (r_b + r_s)$$

$$r(s) = r_b + r_s + r_t - r \qquad z = h \qquad (r_b + r_s) \le s \le (r_b + r_s + r_t)$$

The function **frus** produces a grid of points on the surface in terms of r_b, r_t, h, the number of increments on the base, the number of increments on the side, and the number of increments on the top. Figure 2.9 shows the plot generated by **frus**. The function **frustum**, employed later in the chapter on optimization methods to determine the closest point on a frustum surface to an arbitrary point in space, is fundamentally the same as **frus**.

An example called **srfex** employs the ideas just discussed and illustrates how MATLAB represents several interesting surfaces. Points on the surface of an annulus symmetric about the z-axis are created, and two more annuli are created by interchanging axes. A pyramid with a square base is also created and the combination of four surfaces is plotted by finding a data range to include all points and then plotting each surface in succession using the hold instruction (See Figure 2.10). Although the rendering of surface intersections is not perfect, a useful description of a fairly involved geometry results.

This section is concluded with a discussion of how a set of coordinate points can be moved to a new position by translation and rotation of axes. Suppose a vector

$$r = \hat{i}x + \hat{j}y + \hat{k}z$$

undergoes a coordinate change which moves the initial coordinate origin to (X_o, Y_o, Z_o) and moves the base vectors $\hat{i}$, $\hat{j}$, $\hat{k}$ into $\hat{e}_1$, $\hat{e}_2$, $\hat{e}_3$. Then the endpoint of r passes to

$$R = \hat{i}X + \hat{j}Y + \hat{k}Z = R_o + \hat{e}_1 x + \hat{e}_2 y + \hat{e}_3 z$$

where

$$R_o = \hat{i}X_o + \hat{j}Y_o + \hat{k}Z_o$$

Let us specify the directions of the new base vectors by employing the columns of a matrix V where we take

$$\hat{e}_3 = \frac{V(:,1)}{\mathbf{norm}[V(:,11)]}$$

If $V(:,2)$ exists we take $V(:,1) \times V(:,2)$ and unitize this vector to produce $\hat{e}_2$. The triad is completed by taking $\hat{e}_1 = \hat{e}_2 \times \hat{e}_3$. In the event that $V(:,2)$ is not provided, we use [1;0;0] and proceed as before. The functions **rgdbodmo** and **rotatran** can be used to transform points in the manner described above.

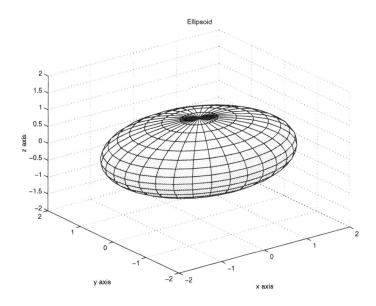

Figure 2.8. Ellipsoid

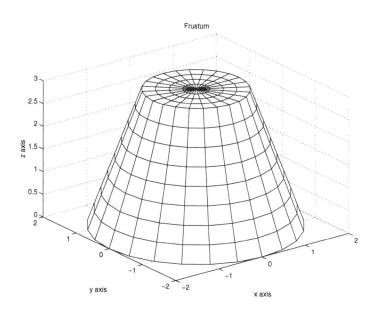

Figure 2.9. Frustum

Intersecting Surfaces

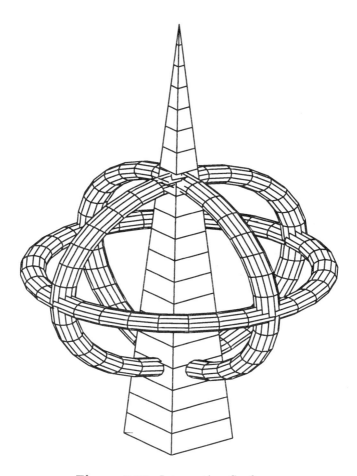

Figure 2.10. Intersecting Surfaces

2.6.4 Program Output and Code

Function elipsoid

```
1: function [x,y,z]=elipsoid(a,b,c)
2: %
3: % [x,y,z]=elipsoid(a,b,c)
4: % ~~~~~~~~~~~~~~~~~~~~~~~~
5: %
6: % This function plots an ellipsoid having semi-
7: % diameters a,b,c.
8: %
9: % User m functions called: genprint
10: %------------------------------------------------
11:
12: th=linspace(-pi/2,pi/2,17)';
13: ph=linspace(-pi,pi,33);
14: x=a*cos(th)*cos(ph); y=b*cos(th)*sin(ph);
15: z=c*sin(th)*ones(size(ph)); surf(x,y,z);
16: axis([-1,1,-1,1,-1,1]*max([a,b,c]));
17: title('Ellipsoid'); xlabel('x axis');
18: ylabel('y axis'); zlabel('z axis'); grid on;
19: figure(gcf); %genprint('ellipse');
```

Function frus

```
1: function [X,Y,Z]=frus(rb,rt,h,n,noplot)
2: %
3: % [X,Y,Z]=frus(rb,rt,h,n,noplot)
4: % ~~~~~~~~~~~~~~~~~~~~~~~~~~~~~~~
5: %
6: % This function computes points on the surface
7: % of a conical frustum which has its axis along
8: % the z axis.
9: %
10: % rb,rt,h - the base radius,top radius and
11: %           height
12: % n       - a vector containing the number
13: %           of tabulation increments on the
14: %           base, side, top, and circumference
15: % noplot  - parameter input when no plot is
16: %           desired
17: %
```

```
18: % X,Y,Z   - points on the surface
19: %
20: % User m functions called: lintrp, genprint
21: %-----------------------------------------------
22:
23: if nargin==0
24:    rb=2; rt=1; h=3; n=[6,8,4];
25: end
26:
27: nb=n(1); ns=n(2); nt=n(3);
28: if length(n)==3, n=[n(:);24]; end
29: nc=n(4)+1; na=nb+ns+nt;
30: x=linspace(0,na,na+1)';ax=[0,nb,nb+ns,na]';
31: r=lintrp(ax,[0,rb,rt,0],x);
32: Z=lintrp(ax,[0,0,h,h],x)*ones(1,nc);
33: xy=r*exp(i*linspace(0,2*pi,nc));
34: X=real(xy); Y=imag(xy);
35: if nargin<5
36:    surf(X,Y,Z); title('Frustum');
37:    xlabel('x axis'); ylabel('y axis');
38:    zlabel('z axis'); grid on;
39:    figure(gcf); %genprint('frustum');
40: end
```

Function srfex

```
1: function [x1,y1,x2,y2,x3,y3,xf,yf,zf,v]= ...
2:                             srfex(da,na,df,nf)
3: %
4: % [x1,y1,x2,y2,x3,y3,xf,yf,zf,v]= ...
5: %                             srfex(da,na,df,nf)
6: % ~~~~~~~~~~~~~~~~~~~~~~~~~~~~~~~~~~~~~~~~~~~~~~~~~
7: %
8: % This graphics example draws three annuli
9: % intersecting a spike.
10: %
11: % User m functions called: genprint
12: %-----------------------------------------------
13:
14: if nargin==0
15:    da=[4,.35]; na=[32,13];
16:    df=[1.5,0,18]; nf=[2,1,20,4];
17: end
```

```
18:
19: % Create a torus having polygonal cross section.
20: % Data for the torus is stored in da and na
21: r0=da(1); r1=da(2); nfaces=na(1); nlat=na(2);
22:
23: t=linspace(0,2*pi,nlat)';
24: xz=[r0+r1*cos(t),r1*sin(t)];
25: z1=xz(:,2); z1=z1(:,ones(1,nfaces+1));
26: th=linspace(0,2*pi,nfaces+1);
27: x1=xz(:,1)*cos(th); y1=xz(:,1)*sin(th);
28: y2=x1; z2=y1; x2=z1; y3=x2; z3=y2; x3=z2;
29:
30: % Create a frustum of a pyramid. Data for the
31: % frustum is stored in df and nf
32: rb=df(1); rt=df(2); h=df(3); nb=nf(1);
33: nt=nf(2); ns=nf(3); nc=nf(4);
34:
35: % Generate radius values for rotation about
36: % the z axis
37: R=[linspace(0,rb*(nb-1)/nb,nb), ...
38:     linspace(rb,rt,ns+1),...
39:     linspace(rt*(nt-1)/nt,0,nt)]';
40: zf=[zeros(1,nb),linspace(0,h,ns+1), ...
41:     h*ones(1,nt)]';
42: zf=zf(:,ones(1,nc+1))-0.4*h;
43:
44: % Make a surface of revolution by rotation
45: % about the z axis
46: th=linspace(pi/nc,pi/nc+2*pi,nc+1);
47: xf=R*cos(th); yf=R*sin(th);
48:
49: % Compute a data range to make an
50: % undistorted plot
51: x=[x1(:);x2(:);x3(:);xf(:)];
52: y=[y1(:);y2(:);y3(:);yf(:)];
53: z=[z1(:);z2(:);z3(:);zf(:)];
54: xmn=min(x); xmx=max(x); ymn=min(y); ymx=max(y);
55: zmn=min(z); zmx=max(z); dz=zmx-zmn;
56: zmd=(zmn+zmx)/2; dx=xmx-xmn; dy=ymx-ymn;z=z(:);
57: dz=zmx-zmn; xmd=(xmn+xmx)/2; ymd=(ymn+ymx)/2;
58: zmd=(zmn+zmx)/2; d=max([dx,dy,dz])/2;
59: v=[xmd-d; xmd+d; ymd-d; ymd+d; zmd-d; zmd+d];
60:
61: % Plot four figures on top of each other
62: hold off; clf; close;
```

```
63: surf(x1,y1,z1); hold on;
64: surf(x2,y2,z2); surf(x3,y3,z3); surf(xf,yf,zf);
65: xlabel('x axis'); ylabel('y axis');
66: zlabel('z axis');
67: title('Intersecting Surfaces');
68: axis(.7*v); axis('off');
69: colormap hsv; figure(gcf); hold off;
70: %genprint('intsurf');
```

Function rgdbodmo

```
1: function [X,Y,Z]=rgdbodmo(x,y,z,v,R0)
2: %
3: % [X,Y,Z]=rgdbodmo(x,y,z,v,R0)
4: % ~~~~~~~~~~~~~~~~~~~~~~~~~~~~~~
5: %
6: % This function transforms coordinates x,y,z to
7: % new coordinates X,Y,Z by rotating and
8: % translating the reference frame
9: %
10: % x,y,z - initial coordinate matrices referred
11: %          to base vectors [1;0;0], [0;1;0] and
12: %          [0;0;1]. Columns of v are used to
13: %          create new basis vectors i,j,k such
14: %          that a typical point [a;b;c] is
15: %          transformed into [A;B;C] according
16: %          to the equation
17: %             [A;B;C]=R0(:)+[i,j,k]*[a;b;c]
18: % v        - a matrix having three rows and either
19: %          one or two columns used to construct
20: %          the new basis [i,j,k] according to
21: %          methods employed function rotatran
22: % R0       - a vector which translates the rotated
23: %          coordinates when R0 is input.
24: %          Otherwise no translation is imposed.
25: %
26: % X,Y,Z - matrices containing the transformed
27: %          coordinates
28: %
29: % User m functions called: rotatran
30: %-------------------------------------------------
31:
32: [n,m]=size(x); XYZ=[x(:),y(:),z(:)]*rotatran(v)';
```

```
33: X=XYZ(:,1); Y=XYZ(:,2); Z=XYZ(:,3);
34: if nargin==5
35:    X=X+R0(1); Y=Y+R0(2); Z=Z+R0(3);
36: end
37: if m==1, return, end
38: X=reshape(X,n,m); Y=reshape(Y,n,m);
39: Z=reshape(Z,n,m);
```

Function rotatran

```
 1: function mat=rotatran(v)
 2: %
 3: % mat=rotatran(v)
 4: % ~~~~~~~~~~~~~~~~
 5: % This function creates a rotation matrix based
 6: % on the columns of v.
 7: %
 8: % v    - a matrix having three rows and either
 9: %        one or two columns which are used to
10: %        create an orthonormal triad [i,j,k]
11: %        returned in the columns of mat. The
12: %        third base vector k is defined as
13: %        v(:,1)/norm(v(:,1)). If v has two
14: %        columns then, v(:,1) and v(:,2) define
15: %        the xz plane with the direction of j
16: %        defined by cross(v(:,1),v(:2)). If only
17: %        v(:,1) is input, then v(:,2) is set
18: %        to [1;0;0].
19: %
20: % mat - the matrix having columns containing
21: %        the basis vectors [i,j,k]
22: %
23: % User m functions called: none
24: %----------------------------------------------------
25:
26: k=v(:,1)/norm(v(:,1));
27: if size(v,2)==2, p=v(:,2); else, p=[1;0;0]; end
28: j=cross(k,p); nj=norm(j);
29: if nj~=0
30:    j=j/nj; mat=[cross(j,k),j,k];
31: else
32:    mat=[[0;1;0],cross(k,[0;1;0]),k];
33: end
```

Chapter 3

Summary of Concepts From Linear Algebra

3.1 Introduction

This chapter briefly reviews important concepts of linear algebra. We assume the reader already has some experience working with matrices, and linear algebra applied to solving simultaneous equations and eigenvalue problems. MATLAB has excellent capabilities to perform matrix operations using the fastest and most accurate algorithms currently available. The books by Strang [94] and Golub and Van Loan [46] give comprehensive treatments of matrix theory and of algorithm developments accounting for effects of finite precision arithmetic. One beautiful aspect of matrix theory is that fairly difficult proofs often lead to remarkably simple results valuable to users not necessarily familiar with all of the theoretical developments. For instance, the property that every real symmetric matrix of order n has real eigenvalues and a set of n orthonormal eigenvectors can be understood and used by someone unfamiliar with the proof. The current chapter summarizes a number of fundamental matrix properties and some of the related MATLAB functions. These intrinsic functions are largely based on algorithms from the LINPACK and EISPACK software libraries [33, 41, 87]. Professor Cleve Moler actively contributed to development of these libraries and later he initiated development of an interactive computing environment which he named MATLAB. Readers should simultaneously study the current chapter and the MATLAB demo program on linear algebra.

3.2 Vectors, Norms, Linear Independence, and Rank

Consider an n by m matrix

$$A = [a_{ij}] \qquad 1 \leq i \leq n \qquad 1 \leq j \leq m$$

having real or complex elements. The shape of a matrix is computed by $\mathbf{size}(A)$ which returns a vector containing n and m. The matrix obtained by conjugating the matrix elements and interchanging columns and rows is called the transpose. Transposition is accomplished with a $'$ operator, so that

$$A\_transpose = A'$$

Transposition without conjugation of the elements can be performed as $A.'$ or as $\mathbf{conj}(A')$. Of course, whenever A is real, A' is simply the traditional transpose.

The structure of a matrix A is characterized by the matrix rank and sets of basis vectors spanning four fundamental subspaces. The rank r is the maximum number of linearly independent rows or columns in the matrix. We discuss these spaces in the context of real matrices. The basic subspaces are:

1. The column space containing all vectors representable as a linear combination of the columns of A. The column space is also referred to as the range or the span.

2. The null space consisting of all vectors perpendicular to every row of A.

3. The row space consisting of all vectors which are linear combinations of the rows of A.

4. The left null space consisting of all vectors perpendicular to every column of A.

MATLAB has intrinsic functions to compute rank and subspace bases

- matrix_rank = $\mathbf{rank}(A)$

- column_space = $\mathbf{orth}(A)$

- null_space = $\mathbf{null}(A)$

- row_space = $\mathbf{orth}(A')'$

- left_null_space = $\mathbf{null}(A')'$

The basis vectors produced by $\mathbf{null}$ and $\mathbf{orth}$ are orthonormal. They are generated using the SVD algorithm [46].

3.3   Systems of Linear Equations, Consistency, and Least Square Approximation

Let us discuss the problem of solving systems of simultaneous equations. Representing a vector B as a linear combination of the columns of A requires determination of a vector X to satisfy

$$AX = B \quad \Longleftrightarrow \quad \sum_{j=1}^{m} A(:,j)\, x(j) = B$$

where the j'th column of A is scaled by the j'th component of X to form the linear combination. The desired representation is possible if and only if B lies in the column space of A. This implies the consistency requirement that A and $[A, B]$ must have the same rank. Even when a system is consistent, the solution will not be unique unless all columns of A are independent. When matrix A, with n rows and m columns, has rank r less than m, the general solution of $AX = B$ is expressible as any particular solution plus an arbitrary linear combination of $m - r$ vectors forming a basis for the null space. MATLAB gives the solution vector as $X = A \backslash B$. When r is less than m, MATLAB produces a least square solution having as many components as possible set equal to zero.

In instances where the system is inconsistent, regardless of how X is chosen, the error vector defined by

$$E = AX - B$$

can never be zero. An approximate solution can be obtained by making E normal to the columns of A. We get

$$A'AX = A'B$$

which is known as the system of normal equations. They are also referred to as least square error equations. It is not difficult to show that the same equations result by requiring E to have minimum length. The normal equations are always consistent and are uniquely solvable when **rank**$(A) = m$. A comprehensive discussion of least square approximation and methods for solving overdetermined systems is presented by Lawson and Hanson [61]. It is instructive to examine the results obtained from the normal equations when A is square and nonsingular. The least square solution would give

$$X = (A'A)^{-1}A'B = A^{-1}(A')^{-1}A'B = A^{-1}B$$

Therefore, the least square solution simply reduces to the exact solution of $AX = B$ for a consistent system. MATLAB handles both consistent and inconsistent systems as $X = A \backslash B$. However, it is only sensible to use the least square solution of an inconsistent system when AX produces an acceptable approximation to B. This implies

$$\mathbf{norm}(AX - B) < tol * \mathbf{norm}(B)$$

where tol is suitably small.

A simple but important application of overdetermined systems arises in curve fitting. An equation of the form

$$y(x) = \sum_{j=1}^{m} f_j(x)c_j$$

involving known functions $f_j(x)$, such as x^{j-1} for polynomials, must approximately match data values (X_i, Y_i), $1 \leq i \leq n$, with $n > m$. We simply write an overdetermined system

$$\sum_{j=1}^{n} f_j(X_i)c_j \approx Y_i \qquad 1 \leq i \leq n$$

and obtain the least square solution. The approximation is acceptable if the error components

$$e_i = \sum_{j=1}^{m} f_j(X_i)c_j - Y_i$$

are small enough and the function $y(x)$ is also acceptably smooth between the data points.

Let us illustrate how well MATLAB handles simultaneous equations by constructing the steady-state solution of the matrix differential equation

$$M\ddot{x} + C\dot{x} + Kx = F_1 \cos(\omega t) + F_2 \sin(\omega t)$$

where M, C, and K are constant matrices and F_1 and F_2 are constant vectors. The steady-state solution has the form

$$x = X_1 \cos(\omega t) + X_2 \sin(\omega t)$$

where X_1 and X_2 are chosen so that the differential equation is satisfied. Evidently

$$\dot{x} = -\omega X_1 \sin(\omega t) + \omega X_2 \cos(\omega t)$$

and

$$\ddot{x} = -\omega^2 x$$

Substituting the assumed form into the differential equation and comparing sine and cosine terms on both sides yields

$$(K - \omega^2 M)X_1 + \omega C X_2 = F_1$$

$$-\omega C X_1 + (K - \omega^2 M)X_2 = F_2$$

The equivalent partitioned matrix is

$$\left[\begin{array}{c|c} (K - \omega^2 M) & \omega C \\ \hline -\omega C & (K - \omega^2 M) \end{array} \right] \left[\begin{array}{c} X_1 \\ X_2 \end{array} \right] = \left[\begin{array}{c} F_1 \\ F_2 \end{array} \right]$$

A simple MATLAB function to produce $X1$ and $X2$ when M, C, K, $F1$, and $F2$ are known is

```
function [x1,x2,xmax]=forcresp(m,c,k,f1,f2,w)
kwm=k-(w*w)*m; wc=w*c;
x=[kwm,wc;-wc,kwm]\[f1;f2]; n=length(f1);
x1=x(1:n); x2=x(n+1:2*n);
xmax=sqrt(x1.*x1+x2.*x2);
```

The vector, **xmax**, defined in the last line of the function above, has components specifying the maximum amplitude of each component of the steady-state solution. The main computation in this function occurs in the third line, where matrix concatenation is employed to form a system of $2n$ equations with x being the concatenation of X_1 and X_2. The last line uses vector indexing to extract X_1 and X_2 from x. The notational simplicity of MATLAB is elegantly illustrated by these features: a) any required temporary storage is assigned and released dynamically, b) no looping operations are needed, c) matrix concatenation and inversion are accomplished with intrinsic functions using matrices and vectors as sub-elements of other matrices, and d) extraction of sub-vectors is accomplished by use of vector indices. The important idea communicated by this function is that mathematical notation and MATLAB language constructs are essentially identical in this important application.

3.4 Applications of Least Square Approximation

The idea of solving an inconsistent system of equations in the least square sense, so that some required condition is approximately satisfied, has numerous applications. Typically, we are dealing with a large number of equations (several hundred is common) involving a much smaller number of parameters used to closely fit some constraint. Linear boundary value problems often require the solution of a differential equation applicable in the interior of a region while the function values are known on the boundary. This type of problem can sometimes be handled by using a series of functions which satisfy the differential equation exactly. Weighting the component solutions to approximately match the remaining boundary condition may lead to useful results. Below, we examine three instances where least square approximation is helpful.

3.4.1 A Membrane Deflection Problem

Let us illustrate how least square approximation can be used to compute the transverse deflection of a membrane subjected to uniform pressure. The transverse deflection u for a membrane which has zero deflection on

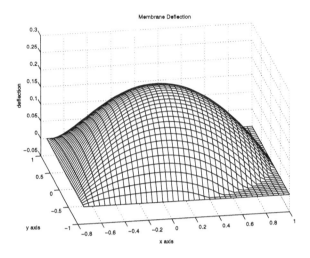

Figure 3.1. Surface Plot of Membrane

a boundary L satisfies the differential equation

$$\frac{\partial^2 u}{\partial x^2} + \frac{\partial^2 u}{\partial y^2} = -\gamma \qquad \text{(x,y) inside L}$$

where γ is a physical constant. Properties of harmonic functions [18] imply that the differential equation is satisfied by a series of the form

$$u = \gamma \left[\frac{-|z|^2}{4} + \sum_{j=0}^{n} c_j \, \mathbf{real}(z^{j-1}) \right]$$

where $z = x + \imath y$ and constants c_j are chosen to make the boundary deflection as small as possible, in the least square sense. As a specific example, we analyze a membrane consisting of a rectangular part on the left joined with a semicircular part on the right. The surface plot in Figure 3.1 and the contour plot in Figure 3.2 were produced by the function **membran** listed below. This function generates boundary data, solves for the series coefficients, and constructs plots depicting the deflection pattern. The results obtained using a twenty-term series satisfy the boundary conditions quite well.

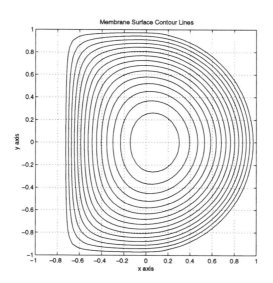

Figure 3.2. Membrane Surface Contour Lines

MATLAB Example

Function membran

```
 1: function [dfl,cof]=membran(h,np,ns,nx,ny)
 2: %
 3: % [dfl,cof]=membran(h,np,ns,nx,ny)
 4: % ~~~~~~~~~~~~~~~~~~~~~~~~~~~~~~~~~~
 5: % This function computes the transverse
 6: % deflection of a uniformly tensioned membrane
 7: % which is subjected to uniform pressure. The
 8: % membrane shape is a rectangle of width h and
 9: % height two joined with a semicircle of
10: % diameter two.
11: %
12: % Example use:  membran(0.75,100,50,40,40);
13: %
14: % h         - the width of the rectangular part
15: % np        - the number of least square points
16: %             used to match the boundary
17: %             conditions in the least square
18: %             sense is about 3.5*np
19: % ns        - the number of terms used in the
20: %             approximating series evaluate
21: %             deflections. The series has the
22: %             form
23: %
24: %             dfl = abs(z)^2/4 +
25: %                   sum({j=1:ns},cof(j)*
26: %                   real(z^(j-1)))
27: %
28: % nx,ny     - the number of x points and y points
29: %             used to compute deflection values
30: %             on a rectangular grid
31: % dfl       - computed array of deflection values
32: % cof       - coefficients in the series
33: %             approximation
34: %
35: % User m functions called:  pauz, genprint,
36: %                           versn
37: %-------------------------------------------------
38:
39: % Generate boundary points for least square
40: % approximation
```

```
41: z=[exp(i*linspace(0,pi/2,round(1.5*np))),...
42:    linspace(i,-h+i,np),...
43:    linspace(-h+i,-h,round(np/2))];
44: z=z(:);
45:
46: % Form the least square equations and solve
47: % for series coefficients
48: a=ones(length(z),ns);
49: for j=2:ns, a(:,j)=a(:,j-1).*z; end
50: cof=real(a)\(z.*conj(z))/4;
51:
52: % Generate a rectangular grid for evaluation
53: % of deflections
54: xv=linspace(-h,1,nx); yv=linspace(-1,1,ny);
55: [x,y]=meshgrid(xv,yv); z=x+i*y;
56:
57: % Evaluate the deflection series on the grid
58: dfl=-z.*conj(z)/4+ ...
59:    real(polyval(cof(ns:-1:1),z));
60:
61: % Set values outside the physical region of
62: % interest to zero
63: dfl=real(dfl).*(1-((abs(z)>=1)&(real(z)>=0)));
64:
65: % Make surface and contour plots
66: hold off; close;
67: mesh(x,y,dfl); view(-10,30);
68: xlabel('x axis'); ylabel('y axis');
69: zlabel('deflection');
70: title('Membrane Deflection'); disp(' ');
71: disp('Press [Enter] to show a contour plot');
72: if versn==5
73:    colormap([0 0 0]);
74: else
75:    colormap([1 1 1]);
76: end
77: figure(gcf); pauz; %genprint('membdefl');
78: contour(x,y,dfl,15);
79: axis([-1,1,-1,1]); axis('square');
80: xlabel('x axis'); ylabel('y axis');
81: title('Membrane Surface Contour Lines');
82: figure(gcf); %genprint('membcntr');
```

Function versn

```
 1: function v=versn
 2: %
 3: % v=versn
 4: % ~~~~~~~
 5: % This function returns the most significant
 6: % integer value for the current version of
 7: % MATLAB.
 8: %
 9: % User m functions called:  none
10: %-----------------------------------------------
11:
12: v=version; v=eval(v(1));
```

3.4.2 Mixed Boundary Value Problem for a Function Harmonic Inside a Circular Disk

Problems where a partial differential equation is to be solved inside a region with certain conditions imposed on the boundary occur in many situations. Often the differential equation is solvable exactly in a series form containing arbitrary linear combinations of known functions. An approximation procedure imposing the boundary conditions to compute the series coefficients produces a satisfactory solution if the desired boundary conditions are found to be well satisfied. Consider a mixed boundary value problem in potential theory [72] pertaining to a circular disk of unit radius. We seek $u(r, \theta)$ where function values are specified on one part of the boundary and normal derivative values are specified on the remaining part. The mathematical formulation is

$$\frac{\partial^2 u}{\partial r^2} + \frac{1}{r}\frac{\partial u}{\partial r} + \frac{1}{r^2}\frac{\partial^2 u}{\partial \theta^2} = 0 \qquad 0 \le r < 1 \qquad 0 \le \theta \le 2\pi$$

$$u(1, \theta) = f(\theta) \qquad -\alpha < \theta < \alpha$$

$$\frac{\partial u}{\partial r}(1, \theta) = g(\theta) \qquad \alpha < \theta < 2\pi - \alpha$$

The differential equation has a series solution of the form

$$u(r, \theta) = c_0 + \sum_{n=1}^{\infty} r^n [c_n \cos(n\theta) + d_n \sin(n\theta)]$$

where the boundary conditions require

$$c_0 + \sum_{n=1}^{\infty} [c_n \cos(n\theta) + d_n \sin(n\theta)] = f(\theta) \qquad -\alpha < \theta < \alpha,$$

and

$$\sum_{n=1}^{\infty} n[c_n \cos(n\theta) + d_n \sin(n\theta)] = g(\theta) \qquad \alpha < \theta < 2\pi - \alpha$$

The series coefficients can be obtained by least square approximation. Let us explore the utility of this approach by considering a particular problem for a field which is symmetric about the x-axis. We want to solve

$$\nabla^2 u = 0 \qquad r < 1$$

$$u(1, \theta) = \cos(\theta) \qquad |\theta| < \pi/2$$

$$\frac{\partial u}{\partial r}(1, \theta) = 0 \qquad \pi/2 < |\theta| \le \pi$$

This problem characterizes steady-state heat conduction in a cylinder with the left half insulated and the right half held at a known temperature. The appropriate series solution is

$$u = \sum_{n=0}^{\infty} c_n r^n \cos(n\theta)$$

subject to

$$\sum_{n=0}^{\infty} c_n \cos(n\theta) = \cos(\theta) \qquad |\theta| < \pi/2$$

$$\sum_{n=0}^{\infty} n c_n \cos(n\theta) = 0 \qquad \pi/2 < |\theta| \leq \pi$$

We solve the problem by truncating the series after a hundred or so terms and forming an overdetermined system derived by imposition of both boundary conditions. The success of this procedure depends on the series converging rapidly enough so that a system of least square equations having reasonable order and satisfactory numerical condition results. It can be shown by complex variable methods (see Muskhelishvili [72]) that the exact solution of our problem is given by

$$u = \mathbf{real} \left[z + z^{-1} + (1 - z^{-1})\sqrt{z^2 + 1} \right] / 2 \qquad |z| \leq 1$$

where the square root is defined for a branch cut along the right half of the unit circle with the chosen branch being that which equals $+1$ at $z = 0$. Readers familiar with analytic function theory can verify that the boundary values of u yield

$$u(1, \theta) = \cos(\theta) \qquad |\theta| \leq \pi/2$$

$$u(1, \theta) = \cos(\theta) + \sin(|\theta|/2)\sqrt{2|\cos(\theta)|} \qquad \pi/2 \leq |\theta| \leq \pi$$

A least square solution is presented in function **mbvp**. Results from a series of 100 terms are shown in Figure 3.3. The series solution is accurate within about one percent error except for points near $\theta = \pi/2$. Although the results are not shown here, using 300 terms gives a solution error which is almost imperceptible on a graph. Hence the least square series solution provides a reasonable method to handle the mixed boundary value problem.

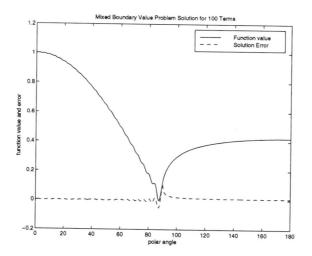

Figure 3.3. Mixed Boundary Value Problem Solution

MATLAB Example

Script file mbvprun

```
 1: % Example: mbvprun
 2: % ~~~~~~~~~~~~~~~~~~~
 3: % Mixed boundary value problem for a function
 4: % harmonic inside a circle.
 5: %
 6: % User m functions required:
 7: %    mbvp, genprint
 8: %------------------------------------------------
 9:
10: disp('Calculating');
11:
12: % Set data for series term and boundary
13: % condition points
14: nser=100; nf=200; ng=200; neval=500;
15:
16: % Compute the series coefficients
17: [cof,y]=mbvp('cos',pi/2,nser,nf,ng,neval);
18:
19: % Evaluate the exact solution for comparision
20: thp=linspace(0,pi,neval)';
21: y=cos(thp*(0:nser-1))*cof;
22: ye=cos(thp)+sin(thp/2).* ...
23:    sqrt(2*abs(cos(thp))).*(thp>=pi/2);
24:
25: % Plot results showing the accuracy of the
26: % least square solution
27: thp=thp*180/pi; plot(thp,y,'k-',thp,y-ye,'k--');
28: xlabel('polar angle');
29: ylabel('function value and error')
30: title(['Mixed Boundary Value Problem ', ...
31:       'Solution for ',int2str(nser),' Terms']);
32: legend('Function value','Solution Error');
33: figure(gcf); %genprint('mbvp');
```

Function mbvp

```
 1: function [cof,y]= ...
 2:            mbvp(func,alp,nser,nf,ng,neval)
 3: %
```

```
 4: % [cof,y]=mbvp(func,alp,nser,nf,ng,neval)
 5: % ~~~~~~~~~~~~~~~~~~~~~~~~~~~~~~~~~~~~~~~~~~
 6: % This function solves approximately a mixed
 7: % boundary value problem for a function which
 8: % is harmonic inside the unit disk, symmetric
 9: % about the x axis, and has boundary conditions
10: % involving function values on one part of the
11: % boundary and zero gradient elsewhere.
12: %
13: % func       - function specifying the function
14: %                value between zero and alp
15: %                radians
16: % alp        - angle between zero and pi which
17: %                specifies the point where
18: %                boundary conditions change from
19: %                function value to zero gradient
20: % nser       - number of series terms used
21: % nf         - number of function values
22: %                specified from zero to alp
23: % ng         - number of points from alp to pi
24: %                where zero normal derivative is
25: %                specified
26: % neval      - number of boundary points where
27: %                the solution is evaluated
28: % cof        - coefficients in the series
29: %                solution
30: % y          - function values for the solution
31: %
32: % User m functions called:  none.
33: %-------------------------------------------------
34:
35: % Create evenly spaced points to impose
36: % boundary conditions
37: th1=linspace(0,alp,nf);
38: th2=linspace(alp,pi,ng+1); th2(1)=[];
39:
40: % Form an overdetermined system based on the
41: % boundary conditions
42: yv=feval(func,th1);
43: cmat=cos([th1(:);th2(:)]*(0:nser-1));
44: [nr,nc]=size(cmat);
45: cmat(nf+1:nr,:)=...
46:    (ones(ng,1)*(0:nser-1)).*cmat(nf+1:nr,:);
47: cof=cmat\[yv(:);zeros(ng,1)];
48:
```

```
49: % Evaluate the solution on the boundary
50: thp=linspace(0,pi,neval)';
51: y=cos(thp*(0:nser-1))*cof;
```

3.4.3 Using Rational Functions to Conformally Map a Circular Disk Onto a Square

Another problem illustrating the value of least square approximation arises in connection with an example discussed earlier in Section 2.4 where a slowly convergent power series was used to map the interior of a circle onto the interior of a square [73]. It is sometimes possible for slowly convergent power series of the form

$$w = f(z) = \sum_{j=0}^{N} c_j z^j \qquad |z| \le 1$$

to be replaceable by a rational function

$$w = \frac{\displaystyle\sum_{j=0}^{n} a_j z^j}{1 + \displaystyle\sum_{j=1}^{m} b_j z^j}$$

Of course, the polynomial is simply a special rational function form with $m = 0$ and $n = N$. This rational function implies

$$\sum_{j=0}^{n} a_j z^j - w \sum_{j=1}^{m} b_j z^j = w$$

Coefficients a_j and b_j can be computed by forming least square equations based on boundary data. In some cases, the resulting equations are rank deficient and it is safer to solve a system of the form $UY = V$ as $Y = \mathbf{pinv}(U) * V$ rather than using $Y = U \backslash V$. The former solution uses the pseudo inverse function **pinv** which automatically sets to zero any solution components that are undetermined.

Two functions **ratcof** and **raterp** were written to compute rational function coefficients and to evaluate the rational function for general matrix arguments. These functions are useful to examine the conformal mapping of the circular disk $|z| \le 1$ onto the square defined by $|\mathbf{real}(w)| \le 1$, $|\mathbf{imag}(w)| \le 1$. A polynomial approximation of the mapping function has the form

$$w/z = \sum_{j=0}^{N} c_j (z^4)^j$$

where N must be quite large in order to avoid excessive corner rounding. If we evaluate w versus z on the boundary for large N (500 or more), and then develop a rational function fit with $n = m = 10$, a

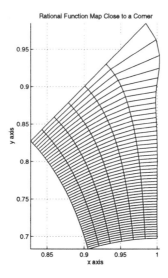

Figure 3.4. Rational Function Map Close to a Corner

reasonably good representation of the square results without requiring a large number of series terms. The following program illustrates the use of functions[1] **ratcof** and **raterp**. It also includes a function **sqmp** to generate coefficients in the Schwarz-Christoffel series. Figure 3.4 shows the geometry mapping produced in the vicinity of a corner.

---

[1] In Chapter 12 a rational function is given to handle mapping onto either the interior or the exterior of a square.

MATLAB Example

Script file makratsq

```
 1: % Example:  makratsq
 2: % ~~~~~~~~~~~~~~~~~~
 3: % Create a rational function map of a unit disk
 4: % onto a square.
 5: %
 6: % User m functions required:
 7: %     sqmp, ratcof, raterp, genprint
 8: %-----------------------------------------------
 9:
10: disp(' ');
11: disp('RATIONAL FUNCTION MAPPING OF A CIRCULAR');
12: disp('         DISK ONTO A SQUARE'); disp(' ');
13: disp('Calculating'); disp(' ');
14:
15: % Generate boundary points given by the
16: % Schwarz-Christoffel transformation
17: nsc=501; np=401; ntop=10; nbot=10;
18: z=exp(i*linspace(0,pi/4,np));
19: w=sqmp(nsc,1,1,1,0,45,np);
20: w=mean(real(w))+i*imag(w);
21: z=[z,conj(z)]; w=[w,conj(w)];
22:
23: % Compute the series coefficients for a
24: % rational function fit to the boundary data
25: [top,bot]=ratcof(z.^4,w./z,ntop,nbot);
26: top=real(top); bot=real(bot);
27:
28: % The above calculations produce the following
29: % coefficients
30: % [top,bot]=
31: %            1.0787     1.4948
32: %            1.5045     0.1406
33: %            0.0353    -0.1594
34: %           -0.1458     0.1751
35: %            0.1910    -0.1513
36: %           -0.1797     0.0253
37: %            0.0489     0.2516
38: %            0.2595     0.1069
39: %            0.0945     0.0102
40: %            0.0068     0.0001
```

```
41:
42: % Generate a polar coordinate grid to describe
43: % the mapping near the corner of the square.
44: % Then evaluate the mapping function.
45: r1=.95; r2=1; nr=7; t1=.9*pi/4; t2=pi/4; nt=51;
46: [r,th]=meshgrid(linspace(r1,r2,nr), ...
47:          linspace(t1,t2,nt));
48: z=r.*exp(i*th); w=z.*raterp(top,bot,z.^4);
49:
50: % Plot the mapped geometry
51: mesh(real(w),imag(w),zeros(size(w)));
52: axis equal; view(2);
53: title('Rational Function Map Close to a Corner');
54: xlabel('x axis'); ylabel('y axis');
55: figure(gcf); %genprint('ratsqmap');
```

Function sqmp

```
 1: function [w,b]=sqmp(m,r1,r2,nr,t1,t2,nt)
 2: %
 3: % [w,b]=sqmp(m,r1,r2,nr,t1,t2,nt)
 4: % ~~~~~~~~~~~~~~~~~~~~~~~~~~~~~~~~~
 5: % This function evaluates the conformal
 6: % mapping produced by the Schwarz-Christoffel
 7: % transformation w(z) mapping abs(z)<=1 inside
 8: % a square having a side length of two.  The
 9: % transformation is approximated in series form
10: % which converges very slowly near the corners.
11: % This function is the same as squarmap of
12: % chapter 2 with no plotting.
13: %
14: % m         - number of series terms used
15: % r1,r2,nr  - abs(z) varies from r1 to r2 in
16: %             nr steps
17: % t1,t2,nt  - arg(z) varies from t1 to t2 in
18: %             nt steps (t1 and t2 are
19: %             measured in degrees)
20: % w         - points approximating the square
21: % b         - coefficients in the truncated
22: %             series expansion which has
23: %             the form
24: %
25: %             w(z)=sum({j=1:m},b(j)*z*(4*j-3))
```

```
26: %
27: % User m functions called:  none.
28: %-----------------------------------------------
29:
30: % Generate polar coordinate grid points for the
31: % map. Function linspace generates vectors with
32: % equally spaced components.
33: r=linspace(r1,r2,nr)';
34: t=pi/180*linspace(t1,t2,nt);
35: z=(r*ones(1,nt)).*(ones(nr,1)*exp(i*t));
36:
37: % Compute the series coefficients and evaluate
38: % the series
39: k=1:m-1;
40: b=cumprod([1,-(k-.75).*(k-.5)./(k.*(k+.25))]);
41: b=b/sum(b); w=z.*polyval(b(m:-1:1),z.^4);
```

Function ratcof

```
 1: function [a,b]=ratcof(xdata,ydata,ntop,nbot)
 2: %
 3: % [a,b]=ratcof(xdata,ydata,ntop,nbot)
 4: % ~~~~~~~~~~~~~~~~~~~~~~~~~~~~~~~~~~~~~
 5: %
 6: % Determine a and b to approximate ydata as
 7: % a rational function of the variable xdata.
 8: % The function has the form:
 9: %
10: %     y(x) = sum(1=>ntop) ( a(j)*x^(j-1) ) /
11: %            ( 1 + sum(1=>nbot) ( b(j)*x^(j)) )
12: %
13: % xdata,ydata - input data vectors (real or
14: %                complex)
15: % ntop,nbot   - number of series terms used in
16: %                the numerator and the
17: %                denominator.
18: %
19: % User m functions called: none
20: %-----------------------------------------------
21:
22: ydata=ydata(:); xdata=xdata(:);
23: m=length(ydata);
24: if nargin==3, nbot=ntop; end;
```

```
25: x=ones(m,ntop+nbot); x(:,ntop+1)=-ydata.*xdata;
26: for i=2:ntop, x(:,i)=xdata.*x(:,i-1); end
27: for i=2:nbot
28:   x(:,i+ntop)=xdata.*x(:,i+ntop-1);
29: end
30: ab=x\ydata;
31: a=ab(1:ntop); b=ab(ntop+1:ntop+nbot);
```

Function raterp

```
1:  function y=raterp(a,b,x)
2:  %
3:  % y=raterp(a,b,x)
4:  % ~~~~~~~~~~~~~~~~
5:  % This function interpolates using coefficients
6:  % from function ratcof.
7:  %
8:  % a,b - polynomial coefficients from function
9:  %       ratcof
10: % x   - argument at which function is evaluated
11: % y   - computed rational function values
12: %
13: % User m functions called:  none.
14: %-----------------------------------------------
15:
16: a=flipud(a(:)); b=flipud(b(:));
17: y=polyval(a,x)./(1+x.*polyval(b,x));
```

3.5 Eigenvalue Problems

3.5.1 Statement of the Problem

Another important linear algebra problem involves the computation of nonzero vectors X and numbers λ such that

$$AX = \lambda X$$

where A is a square matrix of order n. The number λ, which can be real or complex, is called an eigenvalue corresponding to an eigenvector X. The eigenvalue equation implies

$$[I\lambda - A]X = 0$$

so that λ values must be selected to make $I\lambda - A$ singular. The polynomial

$$f(\lambda) = \det(I\lambda - A) = \lambda^n + c_1\lambda^{n-1} + \ldots + c_n\lambda^0$$

is called the characteristic equation and its roots are the eigenvalues. It can be factored into

$$f(\lambda) = (\lambda - \lambda_1)(\lambda - \lambda_2)\cdots(\lambda - \lambda_n)$$

The eigenvalues are generally complex numbers and some of the roots may be repeated. In the usual situation, distinct roots $\lambda_1, \cdots, \lambda_n$ yield n linearly independent eigenvectors obtained by solving

$$(A - \lambda_j I)X_j = 0 \qquad 1 \leq j \leq n$$

The case involving repeated eigenvalues is more complicated. Suppose a particular eigenvalue such as λ_1 has multiplicity k, then the general solution of

$$(A - \lambda_1 I)X = 0$$

will yield as few as one, or as many as k, linearly independent vectors. If fewer than k independent eigenvectors are found for any root of multiplicity k, then matrix A is called defective. Occurrence of a defective matrix is not typical. It usually implies special behavior of the associated physical system. In the general situation the complete set of eigenvectors can be written as

$$A[X_1, \cdots, X_n] = [X_1\lambda_1, \cdots, X_n\lambda_n]$$

$$= [X_1, \cdots, X_n]\,\mathbf{diag}(\lambda_1, \cdots, \lambda_n)$$

or

$$AU = U\Lambda$$

where U contains the eigenvectors as columns and Λ is a diagonal matrix with eigenvalues on the diagonal. When the eigenvalues are linearly independent, the matrix U, which is known as the modal matrix, is nonsingular. This allows A to be expressed as

$$A = U\Lambda U^{-1}$$

which is convenient for various computational purposes. Unfortunately, this decomposition is not always possible. It does exist whenever the eigenvalues are distinct and for special types of matrices such as symmetric matrices which are discussed below.

A matrix A is symmetric if $A = A'$ where A' is obtained by interchanging columns and rows, and conjugating all elements. It can be shown that symmetric matrices have eigenvalues which are all real and a linearly independent set of eigenvectors. Furthermore, the eigenvectors X_j and X_k for any two unequal eigenvalues turn out to satisfy an orthogonality condition

$$X_j' X_k = 0 \qquad j \neq k$$

Eigenvectors belonging to the same repeated eigenvalue are not automatically orthogonal. Nevertheless, they can be replaced by an equivalent orthogonal set by applying a process called Gram-Schmidt orthogonalization [46]. In cases of interest here, the symmetric matrix A can be assumed to have real elements. Therefore the eigenvalues are real with eigenvectors satisfying $X_i' X_j = \delta_{ij}$, where δ_{ij} is the Kronecker delta symbol. The orthogonality condition is equivalent to the statement that $U'U = I$, so a real symmetric matrix can be expressed as

$$A = U\Lambda U'$$

It is important in MATLAB that the symmetry condition $A' = A$ be satisfied perfectly. A matrix which is symmetric, except for roundoff, can be replaced by $(A + A')/2$ to guarantee perfect symmetry. The MATLAB function **eig** computes eigenvalues and eigenvectors. When a matrix is perfectly symmetric (not just to machine accuracy), **eig** generates real eigenvalues and orthonormalized eigenvectors.

An important property of symmetric matrices and the related orthonormal eigenvector set occurs in connection with quadratic forms expressed as

$$F(Y) = Y'AY$$

where Y is an arbitrary real vector and A is real symmetric. The function $F(Y)$ is a one-by-one matrix, hence it is a scalar function. The algebraic sign of the form for arbitrary nonzero choices of Y is important in physical applications. Let us use the eigenvector decomposition of A to write

$$F = Y'U'\Lambda UY = (UY)'\Lambda(UY)$$

Taking $X = UY$ and $Y = U'X$ gives

$$F = X'\Lambda X = \lambda_1 x_1^2 + \lambda_2 x_2^2 + \lambda_3 x_3^2 + \ldots + \lambda_n x_n^2$$

This diagonal form makes the algebraic character of F evident. If all λ_i are positive, then F is evidently positive whenever X has at least one nonzero component. Then the quadratic form is called positive definite. If the eigenvalues are all positive or zero, the form is called positive semidefinite since the form cannot assume a negative value but can equal zero without having $X = 0$. When both negative and positive eigenvalues are present, the form can change sign and is termed indefinite. When the eigenvalues are all negative, the form is classified as negative definite. Perhaps the most important of these properties is that a necessary and sufficient condition for the form to be positive definite is that all eigenvalues of A be positive.

An important generalization of the standard eigenvalue problem has the form

$$AX = \lambda BX$$

for arbitrary A and nonsingular B. If B is well conditioned, then it is computationally attractive to simply solve

$$B^{-1}AX = \lambda X$$

In general, it is safer, but much more time consuming, to call **eig** as

```
[EIGVECS,EIGVALS]=eig(A,B)
```

This returns the eigenvectors as columns of **EIGVECS** and also gives a diagonal matrix **EIGVALS** which contains the eigenvalues.

3.5.2 Application to Solution of Matrix Differential Equations

One of the most important applications of eigenvalues concerns the solution of the linear, constant-coefficient matrix differential equation

$$B\dot{Y}(t) = AY(t) \qquad Y(0) = Y_0$$

Component solutions can be written as

$$Y = Xe^{\lambda t} \qquad \dot{Y} = \lambda Xe^{\lambda t}$$

where X and λ are constant. Substitution into the differential equation gives

$$(A - \lambda B)Xe^{\lambda t} = 0$$

Since $e^{\lambda t}$ cannot vanish we need

$$AX = \lambda BX$$

After the eigenvalues and eigenvectors have been computed, a general solution is constructed as a linear combination of component solutions

$$Y = \sum_{j=1}^{n} X_j e^{\lambda_j t} c_j$$

The constants c_j are obtained by imposing the initial condition

$$Y(0) = [X_1, X_2, \ldots, X_n]c$$

Assuming that the eigenvectors are linearly independent we get

$$c = [X_1, \ldots, X_n]^{-1} Y_0$$

3.6   Column Space, Null Space, Orthonormal Bases, and SVD

A final topic discussed in this chapter is the factorization known as singular value decomposition, or SVD. We will briefly explain the structure of SVD and some of its applications. It is known that any real matrix having n rows, m columns, and rank r can be decomposed into the form

$$A = USV'$$

where

- U is an orthogonal n by n matrix such that $U'U = I$

- V is an orthogonal m by m matrix such that $V'V = I$

- S is an n by m diagonal matrix of the form

$$S = \begin{bmatrix} \sigma_1 & 0 & 0 & 0 & 0 & 0 \\ 0 & \sigma_2 & 0 & 0 & 0 & 0 \\ 0 & 0 & \ddots & 0 & 0 & 0 \\ 0 & 0 & 0 & \sigma_r & 0 & 0 \\ 0 & 0 & 0 & 0 & 0 & 0 \\ 0 & 0 & 0 & 0 & 0 & 0 \end{bmatrix}$$

where $\sigma_1, \ldots, \sigma_r$ are positive numbers on the main diagonal with $\sigma_i \geq \sigma_{i+1}$. Constants σ_j are called the singular values with the number of nonzero values being equal to the rank r.

To understand the structure of this decomposition, let us study the case where $n \geq m$. Direct multiplication gives

$$A'AV = V \mathbf{diag}([\sigma_1^2, \ldots, \sigma_r^2, \mathbf{zeros}(1, m - r)])$$

and

$$AA'U = U \mathbf{diag}([\sigma_1^2, \ldots, \sigma_r^2, \mathbf{zeros}(1, n - r)])$$

Consequently, the singular values are square roots of the eigenvalues of the symmetric matrix $A'A$. Matrix V contains the orthonormalized eigenvectors arranged so that $\sigma_\iota \geq \sigma_{\iota+1}$. Although the eigenvalues of $A'A$ are obviously real, it may appear that this matrix could have some negative eigenvalues leading to pure imaginary singular values. However, this cannot happen because $A'AY = \lambda Y$ implies $\lambda = (AY)'(AY)/(Y'Y)$, which clearly is non-negative. Once the eigenvectors and eigenvalues of $A'A$ are computed, columns of matrix U can be found as orthonormalized solutions of

$$[A'A - \sigma_\jmath I]U_\jmath = 0 \qquad \sigma_\jmath = 0 \qquad \jmath > r$$

The arguments just presented show that performing singular value decomposition involves solving a symmetric eigenvalue problem. However, SVD requires additional computation beyond solving a symmetric eigenvalue problem. It can be very time consuming for large matrices. The SVD has various uses, such as solving the normal equations. Suppose an n by m matrix A has $n > m$ and $r = m$. Substituting the SVD into

$$A'AX = A'B$$

gives

$$V \mathbf{diag}(\sigma_1^2, \ldots, \sigma_m^2)V'X = VS'U'B$$

Consequently, the solution of the normal equations is

$$X = V \mathbf{diag}(\sigma_1^{-1}, \ldots, \sigma_m^{-1})S'U'B$$

Another important application of the SVD concerns generation of orthonormal bases for the column space and the row space. The column space has dimension r and the null space has dimension $m - r$. Consider a consistent system

$$AX = B = U(SV'X)$$

Denote $SV'X$ as Y and observe that $y_\jmath = 0$ for $\jmath > r$ since $\sigma_\jmath = 0$. Because B can be any vector in the column space, it follows that the first r columns of U, which are also orthonormal, are a basis for the column space. Furthermore, the decomposition can be written as

$$AV = US$$

This implies

$$AV_j = U_j \sigma_j = 0 \qquad j > r$$

which shows that the final $m - r$ columns of V form an orthonormal basis for the null space. The reader can verify that bases for the row space and left null space follow analogously by considering $A' = V S' U'$, which simply interchanges the roles of U and V.

MATLAB provides numerous other useful matrix decompositions such as LU, QR, and Cholesky. Some of these are employed in other sections of this book. The reader will find it instructive to read the built-in help information for MATLAB functions which describe these decomposition methods. For instance, the command **help** \ gives extensive documentation on the operation for matrix inversion.

3.7   Program Comparing FLOP Counts for Various Matrix Operations

This chapter is concluded with a program to measure the number of floating point operations, FLOPs, needed to perform various algebraic operations on real square matrices. These operations are approximately proportional to n^3 for sufficiently high matrix order. The program output for orders $n = 100$ and $n = 200$ is included in the next section. Notice that the SVD takes more than 23 times as much computation as Gauss reduction, and solving $AX = \lambda BX$ by the QZ method takes approximately 6 times as much computation as that required to solve $B^{-1}AX = \lambda X$. It also appears that MATLAB produces an unreasonably low flop count for the general symmetric eigenvalue problem.

MATLAB Example

Output from Example

```
FLOP COUNT FOR VARIOUS MATRIX OPERATIONS


For n = 100


Gauss reduction                         =>    0.742 n^3
Cholesky decomposition                  =>    0.338 n^3
matrix multiplication                   =>    2.000 n^3
matrix inversion                        =>    2.050 n^3
solve A*X=(lambda)*X for symmetric A    =>    9.927 n^3
solve A*X=(lambda)*X for general A      =>   26.682 n^3
solve inv(B)*A*X=(lambda)*X             =>   25.493 n^3
solve A*X=(lambda)*B*X by QZ method     => 157.175 n^3
solve A*X=(lambda)*B*X for A,B symmetric=>    0.364 n^3
singular value decomposition  A=U*S*V'  =>   17.174 n^3


For n = 200


Gauss reduction                         =>    0.704 n^3
Cholesky decomposition                  =>    0.336 n^3
matrix multiplication                   =>    2.000 n^3
matrix inversion                        =>    2.025 n^3
solve A*X=(lambda)*X for symmetric A    =>    9.677 n^3
solve A*X=(lambda)*X for general A      =>   25.751 n^3
solve inv(B)*A*X=(lambda)*X             =>   24.794 n^3
solve A*X=(lambda)*B*X by QZ method     => 150.207 n^3
solve A*X=(lambda)*B*X for A,B symmetric=>    0.348 n^3
singular value decomposition  A=U*S*V'  =>   16.367 n^3
```

Script File flopex

```
 1: %   Example:  flopex
 2: %   ~~~~~~~~~~~~~~~~~
 3: %   This program tests the number of floating
 4: %   point operations required to perform several
 5: %   familiar matrix calculations. The operations
 6: %   tested are:
 7: %
 8: %     - solve A*X=B by Gauss reduction
 9: %     - solve A*X=B by Cholesky decomposition
10: %     - multiply square matrices
11: %     - invert a matrix
12: %     - eigenvectors and eigenvalues for a
13: %       symmetric matrix
14: %     - eigenvalues and eigenvectors of a real
15: %       matrix
16: %     - eigenvalues and eigenvectors of
17: %       A*X = Lambda*B*X  for A symmetric and B
18: %       symmetric positive definite
19: %     - eigenvalues of A*X=Lambda*B by the
20: %       QZ method
21: %     - singular value decomposition A=U*S*V'
22: %
23: %   User m functions required:
24: %      floptest, trisub
25: %-------------------------------------------------
26:
27: fprintf('\nThese linear algebra calculations\n');
28: fprintf('can take a while\n');
29: for j=100:100:200
30:   disp(' '); floptest(j); clear;
31: end
32: fprintf('\nAll Done\n');
```

Function floptest

```
 1: function [fcount,flist]=floptest(n)
 2: %
 3: % [fcount,flist]=floptest(n)
 4: % ~~~~~~~~~~~~~~~~~~~~~~~~~~~
 5: % This function determines the flop counts
```

```
 6: % needed to perform various matrix operations
 7: % which are proportional to n^3 for
 8: % sufficiently large n.
 9: %
10: %[fgus;fchol;fmlt;finv;fseig;feig;fsgeig;fgeig]
11: % are the flop count multipliers for various
12: % matrix operations
13: %
14: % User m functions called: none
15: %-------------------------------------------------
16:
17: a=rand(n,n)-.5; b=rand(n,n)-.5; c=a*ones(n,1);
18: as=a+a'; bs=b+b';
19: ev=sort(eig(bs));
20: bs=bs+(1+abs(ev(1)))*eye(n,n);
21: n3=n^3;
22:
23: % Solve A*X=C by Gauss reduction
24: fgus=flops; x=a\c; fgus=(flops-fgus)/n3; x=[];
25:
26: % Solve BS*X=C by Cholesky decomposition
27: fchol=flops; lo=chol(bs)';
28: x=trisub(lo,c); x=trisub(lo',x,1);
29: fchol=(flops-fchol)/n3; x=[];
30:
31: % Multiply two square matrices
32: fmlt=flops; d=a*b;
33: fmlt=(flops-fmlt)/n3; d=[];
34:
35: % Perform a matrix inversion
36: finv=flops; d=inv(a);
37: finv=(flops-finv)/n3; d=[];
38:
39: % Eigenvalues and eigenvectors for a real
40: % symmetric matrix
41: fseig=flops; [vecs,vals]=eig(bs);
42: fseig=(flops-fseig)/n3;
43: vecs=[]; vals=[];
44:
45: % Eigenvalues and eigenvectors for a general
46: % real matrix
47: feig=flops; [vecs,vals]=eig(a);
48: feig=(flops-feig)/n3;
49: vecs=[]; vals=[];
50:
```

```
51: % Eigenvalues and eigenvectors for
52: % A*X=lambda*B*X using inv(B)*A
53:
54: fgeig1=flops; [vecs,vals]=eig(b\a);
55: fgeig1=(flops-fgeig1)/n3;
56: vecs=[]; vals=[];
57:
58: % Eigenvalues and eigenvectors for
59: % A*X=lambda*B*X using the more
60: % conservative QZ algorithm
61:
62: fgeig2=flops; [vecs,vals]=eig(a,b);
63: fgeig2=(flops-fgeig2)/n3;
64: vecs=[]; vals=[];
65:
66: % Eigenvalues and eigenvectors for
67: % A*X=lambda*B*X where A is symmetric
68: % and B is symmetric positive definite
69:
70: fsgeig=flops; lo=chol(bs)'; aas=trisub(lo,as);
71: aas=trisub(lo,aas')'; aas=(aas+aas')/2;
72: [vecs,vals]=eig(aas); vecs=trisub(lo',vecs,1);
73: fsgeig=(flops-fsgeig)/n3; vecs=[]; vals=[];
74:
75: % Singular value decomposition of
76: % a square matrix
77:
78: fsvd=flops; [bu,bs,bv]=svd(b);
79: fsvd=(flops-fsvd)/n3;
80: s0 ='   =>  %7.3f n^3\n';
81: s1 =['Gauss reduction           ', ...
82:      '            ',s0];
83: s2 =['Cholesky decomposition    ', ...
84:      '            ',s0];
85: s3 =['matrix multiplication     ', ...
86:      '           ',s0];
87: s4 =['matrix inversion          ', ...
88:      '           ',s0];
89: s5 =['solve A*X=(lambda)*X for symmetric', ...
90:      ' A     ',s0];
91: s6 =['solve A*X=(lambda)*X for general A', ...
92:      '          ',s0];
93: s7 =['solve inv(B)*A*X=(lambda)*X   ', ...
94:      '          ',s0];
95: s8 =['solve A*X=(lambda)*B*X by QZ method', ...
```

```
96:          '         ',s0];
97: s9 =['solve A*X=(lambda)*B*X for A,B',  ...
98:        ' symmetric ',s0];
99: s10=['singular value decomposition  ',  ...
100:       'A=U*S*V''     ',s0];
101:
102: %          1        2        3        4        5
103: fcount=[fgus; fchol;  fmlt;    finv;    fseig;...
104:         feig; fgeig1; fgeig2; fsgeig; fsvd];
105: %          6        7        8        9       10
106:
107: disp(' ');
108: fprintf('FLOP COUNT FOR VARIOUS ');
109: fprintf('MATRIX OPERATIONS\n\n');
110: disp(['For n = ',int2str(n)]), disp(' ');
111:
112: fprintf(s1,fgus);    fprintf(s2,fchol);
113: fprintf(s3,fmlt);    fprintf(s4,finv);
114: fprintf(s5,fseig);   fprintf(s6,feig);
115: fprintf(s7,fgeig1);  fprintf(s8,fgeig2);
116: fprintf(s9,fsgeig);  fprintf(s10,fsvd);
```

Function trisub

```
1: function x=trisub(lowr,b,ifupper)
2: %
3: % x=trisub(lowr,b,ifupper)
4: % ~~~~~~~~~~~~~~~~~~~~~~~~~
5: % Solve LOWR*X = B, where LOWR is lower
6: % triangular.  When ifupper is present, then
7: % LOWR is assumed to be an upper triangular
8: % matrix. Note that the right side matrix B can
9: % have more than one column.
10: %
11: % User m functions called:  none
12: %-------------------------------------------------
13:
14: [n,m]=size(lowr); [nb,mb]=size(b);
15: x=zeros(n,mb);
16: if nargin==3
17:    nn=n:-1:1; lowr=lowr(nn,nn); b=b(nn,:);
18: end
19: x(1,:)=b(1,:)/lowr(1,1);
```

Chapter 4

Methods for Interpolation and Numerical Differentiation

4.1 Concepts of Interpolation

Interpolation is a process whereby a function is approximated using data known at a discrete set of points. Typically we have points (x_i, y_i), $1 \leq i \leq n$, arranged such that $x_{i+1} > x_i$. These points are to be connected by a continuous interpolation function influenced by smoothness requirements such as:

a) the function should not deviate greatly from the data at points lying between the data values.

b) the function should satisfy a differentiability condition such as continuity of first and second derivatives.

In the most simple form of interpolation, we connect successive points by straight lines. This method, known as piecewise linear interpolation, satisfies condition (a) but has the disadvantage of producing piecewise constant slope values which yield slope discontinuities. An obvious cure for slope discontinuity is to use a curve such as a polynomial of degree $n - 1$ (through n points) to produce an interpolation function having all derivatives continuous. However, it was seen in Section 2.3 that a polynomial passing exactly through the data points may be highly irregular at intermediate values. Frequently, using polynomial interpolations higher than order five or six produces disappointing results. An excellent alternative to allowing either slope discontinuities or demanding slope continuity of all orders is to use cubic spline interpolation. This method connects successive points by cubic curves joined such that function continuity as well as continuity of the first two function derivatives is achieved.

MATLAB's intrinsic function **interp1** performs piecewise linear interpolation and intrinsic function **spline** performs piecewise cubic interpolation with derivative continuity through order two. Another related function **linspace** is convenient for generating a set of equally spaced values between specified limits. For example, spline interpolating through 21 points on the sine curve and evaluating the spline at 101 points is accomplished by the statements

```
xd = linspace(0,2*pi,21);
x = linspace(0,2*pi,101);
y = spline(xd,sin(xd),x);
```

Function **interp1** is restricted to handling points within the data range and does not allow jump discontinuities as is necessary to describe a function such as a sawtooth wave. Consequently, we have provided the following function **lintrp** which allows jump discontinuities characterized by a condition such as $x_{j+1} = x_j$, $y_{j+1} \neq y_j$. Furthermore, function values outside the original data range are evaluated by extending the line segments for the outermost data pairs at each end. Function **lintrp** is employed repeatedly in subsequent chapters.

An auxiliary library of MATLAB functions called "The Spline Toolbox" [96] is available from The MathWorks, Inc. This library includes the spline function as well as various other capabilities. However, the standard spline function intrinsic to MATLAB does not include spline differentiation and integration. To provide this capability we have developed two cubic spline functions **spc** and **spltrp** which are discussed in detail in this chapter and are used extensively in geometric property calculations presented in Chapter 5.

Before proceeding with the mathematical formulation for cubic splines, let us mention briefly the problem of passing a polynomial of degree $n - 1$ through n data points. In principle, polynomial interpolation can be readily accomplished with existing MATLAB functions with the statement

```
yinterp=polyval(...
        polyfit(xdata,ydata,length(xdata)-1),xinterp);
```

However, the resulting computations sometime lead to poorly conditioned simultaneous equations. A viable alternative is to use Newton's algorithm employing divided difference formulas developed in Conte and de Boor [20]. The following functions **polyterp**, **dvdcof**, and **dvdtrp** provide polynomial interpolation analogous to those given by **spline**.

We will not discuss this method further because interpolation using high order polynomials usually yields less satisfactory results than is obtained with splines.

```
function y=lintrp(xd,yd,x)
%
% y=lintrp(xd,yd,x)
% ~~~~~~~~~~~~~~~~~
% This function performs piecewise linear
% interpolation through data values stored in
% xd, yd, where xd values are arranged in
% nondecreasing order. The function can handle
% discontinuous functions specified when two
% successive values in xd are equal. Then the
% repeated xd values are shifted by a tiny
% amount to remove the discontinuities.
% Interpolation for any points outside the range
% of xd is also performed by continuing the line
% segments through the outermost data pairs.
%
% xd,yd - data vectors defining the
%          interpolation
% x     - matrix of values where interpolated
%          values are required
%
% y     - matrix of interpolated values
%
% NOTE:  This routine is dependent on MATLAB
%        Version 5.x function interp1q.  A
%        Version 4.x solution can be created
%        by renaming routine lntrp.m to
%        lintrp.
%
%-------------------------------------------------
```

```
xd=xd(:); yd=yd(:); [nx,mx]=size(x); x=x(:);
xsml=min(x); xbig=max(x);
if xsml<xd(1)
 ydif=(yd(2)-yd(1))*(xsml-xd(1))/(xd(2)-xd(1));
 xd=[xsml;xd]; yd=[yd(1)+ydif;yd];
end
n=length(xd); n1=n-1;
if xbig>xd(n)
  ydif=(yd(n)-yd(n1))*(xbig-xd(n))/ ...
       (xd(n)-xd(n1));
  xd=[xd;xbig]; yd=[yd;yd(n)+ydif];
end
k=find(diff(xd)==0);
if length(k)~=0
  n=length(xd);
  xd(k+1)=xd(k+1)+(xd(n)-xd(1))*1e3*eps;
end
y=reshape(interp1q(xd,yd,x),nx,mx);
```

4.1.1   Example: Newton Polynomial Interpolation

Function polyterp

```
1: function yterp=polyterp(xdat,ydat,xterp)
2: %
3: % yterp=polyterp(xdat,ydat,xterp)
4: % ~~~~~~~~~~~~~~~~~~~~~~~~~~~~~~~~
5: % Interpolate through n=length(xdat) data
6: % points using a polynomial of degree n-1.
7: %
8: % User m functions called:  dvdtrp, dvdcof
9: %------------------------------------------------
10:
11: yterp=dvdtrp(xdat,dvdcof(xdat,ydat),xterp);
```

Function dvdcof

```
1: function c=dvdcof(xdat,ydat)
2: %
3: % c=dvdcof(xdat,ydat)
4: % ~~~~~~~~~~~~~~~~~~~
5: % This function uses divided differences to
6: % compute coefficients c needed to perform
7: % polynomial interpolation by the Newton form
8: % of the interpolating polynomial. The data
9: % values used are xdat(i),ydat(x), i=1:n.
10: %
11: % Reference:
12: %    "Elementary Numerical Analysis"
13: %    by S. Conte and C. de Boor
14: %
15: % User m functions called:  none.
16: %------------------------------------------------
17:
18: n=length(xdat);
19: for k=1:(n-1)
20:   for i=1:(n-k)
21:     ydat(i)=...
22:     (ydat(i+1)-ydat(i))./(xdat(i+k)-xdat(i));
23:   end
24: end
25: c=ydat;
```

Function dvdtrp

```
1: function yterp=dvdtrp(xdat,c,xterp)
2: %
3: % yterp=dvdtrp(xdat,c,xterp)
4: % ~~~~~~~~~~~~~~~~~~~~~~~~~~~
5: % This function performs polynomial
6: % interpolation using the Newton form of
7: % the interpolating polynomial. The
8: % coefficients c are first computed by a
9: % call to function dvdcof. The
10: % points xdat(i) are abscissas of the data
11: % values.  The vector yterp is returned as the
12: % interpolated function for argument values
13: % defined by vector xterp.
14: %
15: % Reference:
16: %    "Elementary Numerical Analysis"
17: %    by S. Conte and C. de Boor
18: %
19: % User m functions called:  none.
20: %-------------------------------------------------
21:
22: xdat=xdat(:); c=c(:);
23: xterp=xterp(:);
24: n=length(xterp); m=length(xdat);
25: yterp=c(1)*ones(n,1);
26: for i=2:m
27:   yterp=c(i)+(xterp-xdat(i)).*yterp;
28: end
```

4.2 Interpolation, Differentiation, and Integration by Cubic Splines

Cubic spline interpolation is a versatile method commonly employed to pass a smooth curve through a sequence of data points. The technique connects each successive pair of points using a cubic polynomial. Boundary conditions are imposed to make $y(x)$, $y'(x)$, and $y''(x)$ continuous whenever contiguous intervals join. Consequently, the resulting piecewise cubic curve has continuous derivatives through order two. Therefore, $y(x)$, $y'(x)$, $y''(x)$, and $y'''(x)$ are, respectively, piecewise cubic, piecewise parabolic, piecewise linear, and piecewise constant. Once the polynomials for different intervals have been calculated, the function, the two derivatives and the integral can be evaluated. Two functions are developed below to perform spline interpolation. These functions extend the intrinsic spline functions of MATLAB to include differentiation and integration. In the first calculation phase, the values of $y''(x)$ needed to assure slope continuity are determined. This set of second derivatives is subsequently used as coefficients of the cubic polynomial for each interval. The coefficients completely define the interpolation curve. They may also be employed to evaluate the spline at any number of points. Details of the procedure necessary to perform these operations are described next[1]. Readers who want further detail on spline theory will find the books by de Boor [26] and Ahlberg and Nilson [2] to be comprehensive references.

First, we discuss how to compute a cubic polynomial where $y(x)$ and $y''(x)$ are specified at $x = 0$ and $x = h_1$ (see Figure 4.1). The values at the left and right ends are

$$y(0) = y_1 \qquad y''(0) = T_1 \qquad \text{left end}$$

$$y(h_1) = y_2 \qquad y''(h_1) = T_2 \qquad \text{right end}$$

A Taylor series expansion implies that

$$y(x) = y_1 + s_1 x + \frac{1}{2}T_1 x^2 + \frac{1}{6}V_1 x^3$$

where

$$s_1 = y'(0) \qquad V_1 = y'''(0)$$

Since $y''(x)$ is a linear function with

$$y''(0) = T_1 \qquad y''(h_1) = T_2$$

then

$$y''(x) = T_1 + (T_2 - T_1)\frac{x}{h_1}$$

---

[1] The spline function in standard MATLAB is part of the more comprehensive *The Spline Toolbox* marketed by The MathWorks, Inc. [96]. Function **splined**, from Section 2.6.2, differentiates cubic splines using functions provided in basic MATLAB.

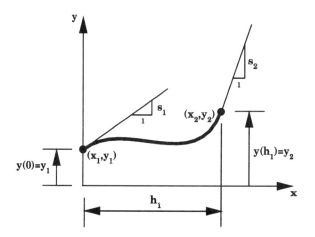

Figure 4.1. Cubic Segment

and

$$y'''(x) = (T_2 - T_1)\frac{1}{h_1}$$

Consequently,

$$y(x) = y_1 + s_1 x + \frac{1}{2}T_1 x^2 + \left(\frac{T_2 - T_1}{6h_1}\right) x^3$$

The imposed function value at $x = h_1$ requires

$$y_2 = y_1 + s_1 h_1 + \frac{(2T_1 + T_2)h_1}{6}$$

Therefore, the slope at $x = 0$ satisfies

$$y'(0) = s_1 = \frac{y_2 - y_1}{h_1} - \frac{(2T_1 + T_2)h_1}{6}$$

and, similarly, the slope at the right end is given by

$$y'(h_1) = s_2 = \frac{y_2 - y_1}{h_1} + \frac{(T_1 + 2T_2)h_1}{6}$$

The conditions necessary when contiguous cubic curves have the same slope and function values at the common interface point may now be developed. Suppose the left segment has length h_1 and satisfies $y = y_1$, $y'' = T_1$ at the left end, and $y = y_2$, $y'' = T_2$ at the right end. Similarly, the right segment of length h_2 satisfies $y = y_2$, $y'' = T_2$

at the left end, and $y = y_3$, $y'' = T_3$ at the right end. The equations developed above show that making the slope at the right end of the left interval match the slope at the left end of the right interval requires

$$\frac{y_2 - y_1}{h_1} + \frac{(T_1 + 2T_2)h_1}{6} = \frac{y_3 - y_2}{h_2} - \frac{(2T_2 + T_3)h_2}{6}$$

Consequently, the second derivatives T_1, T_2, and T_3 are connected by

$$h_1 T_1 + 2(h_1 + h_2)T_2 + h_2 T_3 = 6 \left[\frac{y_2 - y_1}{h_1} - \frac{y_3 - y_2}{h_2} \right]$$

The last relation can be generalized to handle a problem involving several intervals. If n data points are given as

$$x = x_j \qquad y = y_j \qquad x_j > x_{j-1} \qquad 1 \le j \le n$$

then $T_1, \ldots, T_n$ must satisfy

$$h_{j-1}T_{j-1} + 2(h_{j-1} + h_j)T_j + h_j T_{j+1} =$$

$$6 \left[\frac{y_j - y_{j-1}}{h_{j-1}} - \frac{y_{j+1} - y_j}{h_j} \right] \qquad 2 \le j \le n - 1$$

These $n - 2$ slope continuity conditions for interior points must be supplemented by end conditions at the left-most point x_1 and the right-most point x_n. The three types of conditions typically imposed are

a) slope, or

b) second derivative, or

c) continuity of $y'''(x)$ at the interior points nearest to the ends of the interval.

Boundary condition (c), pertaining to third order derivative continuity, has the effect of transferring information describing the function behavior at the interior points to the outside points. Since the value of $y'''(x)$ is constant for each subinterval, the requirement for $y'''(x_2)$ to be continuous is

$$\frac{T_2 - T_1}{h_1} = \frac{T_3 - T_2}{h_2}$$

which reduces to

$$h_2 T_1 - (h_1 + h_2)T_2 + h_3 T_3 = 0$$

The formulas developed above show that the types of boundary conditions applicable at the left end require

- for given slope:

$$2T_1 + T_2 = 6\left[\frac{y_2 - y_1}{h_1^2} - \frac{y'(x_1)}{h_1}\right]$$

- for given second derivative:

$$T_1 = y''(x_1)$$

- for continuous $y'''(x_2)$:

$$h_2 T_1 - (h_1 + h_2)T_2 + h_1 T_3 = 0$$

Similar boundary conditions imposed at $x = x_n$ yield

- for given slope:

$$T_{n-1} + 2T_n = 6\left[-\frac{y_n - y_{n-1}}{h_{n-1}^2} + \frac{y'(x_n)}{h_{n-1}}\right]$$

- for given second derivative:

$$T_n = y''(x_n)$$

- for continuous $y'''(x_{n-1})$:

$$h_{n-1}T_{n-2} - (h_{n-2} + h_{n-1})T_{n-1} + h_{n-2}T_n = 0$$

This set of relationships yields a sparse system of simultaneous equations solvable for $T_1, \ldots, T_n$. For example, when end conditions specifying slope or second derivative are used, the system has the tridiagonal form

$$a_j x_{j-1} + b_j x_j + c_j x_{j+1} = d_j$$

This system can be solved by Gauss reduction, or

```
for k = 1:n-1
  r = a(k+1) / b(k);
  b(k+1) = b(k+1) - r * c(k);
  d(k+1) = d(k+1) - r * d(k);
end
x(n) = d(n) / b(n);
for k = n-1:-1:1
  x(k) = ( d(k) - c(k) * x(k+1) ) / b(k);
end
```

The above code, which must be processed by the MATLAB interpreter, can be replaced with the more efficient intrinsic equation solver which has been compiled and therefore runs much faster than interpreted code.

Two MATLAB functions **spc** and **spltrp** are described below. Function **spc** accepts input data coordinates (x, y) and end conditions specified by vectors i and v. Constant $v(1)$ contains a value for the left end slope or deflection when $i(1)$ equals 1 or 2. The value of $v(1)$ is ignored when $i(1)$ equals 3. Parameters $i(2)$ and $v(2)$ impose right end conditions analogous to what $i(1)$ and $v(1)$ accomplish at the left end. One other argument is the vector $icrnr$ specifying indices of any interior points where the slope continuity condition is replaced by a condition requiring $y''(x)$ to be zero. This type of condition is comparable to placing a hinge in a deflected beam to create what is referred to here as a corner point (a point where the curve slope experiences a finite jump). No slope discontinuities occur if $icrnr$ has zero length or is omitted from the argument list. It is worth noting that piecewise linear interpolation is implied when a spline curve has $y''(x) = 0$ at both ends and all interior points are corner points. The final output from **spc** is matrix $splmat$ which contains all the data needed by function **spltrp** to interpolate, differentiate, or integrate the spline from x_1 to an arbitrary upper limit.

Function **spltrp** accepts input consisting of an array, mat, and a vector of arguments, x, where the spline interpolation is desired. Parameter $ideriv$ has values of 0, 1, 2, or 3. The value of $ideriv$ selects whether values of $y(x), y'(x), y''(x)$, or the integral from x_1 to x are requested. Another special characteristic of function **spltrp** is that function values outside of the original data range are based on the corresponding end tangents at x_1 and x_n. This choice seems preferable because evaluating the polynomials for the outer segments often produces very large function values at points outside the original data range.

Two examples illustrating spline interpolation are presented next. In the first example, a series of equally spaced points between 0 and 2π is used to approximate $y(x) = \sin(x)$. This function satisfies

$$y'(x) = \cos(x)$$

$$y''(x) = -\sin(x)$$

$$\int_0^x \sin(x)dx = 1 - \cos(x)$$

The approximations for the function, derivatives, and the integral shown in Figure 4.2 are quite satisfactory. The script file which produces this figure is listed as **sinetrp**.

In the second example, interpolation of a two-dimensional space curve where y cannot be interpolated as a single valued function of x is studied. In this case, a parameter t_j is employed which has a value equal to the index j for each (x_j, y_j) used. Then $x(t)$ and $y(t)$ are interpolated to produce a smooth curve through the data. A function **spcurv2d** is provided to compute points on a plane curve. Another script file named **matlbdat** produces data points spelling the word MATLAB

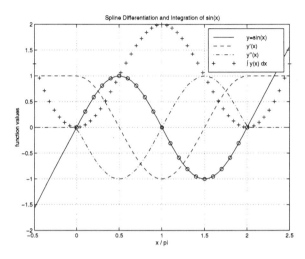

Figure 4.2. Spline Differentiation and Integration of $\sin(x)$

when connected. Results shown in Figure 4.3 illustrate the capability which splines provide to describe a complicated curved shape. The use of corner points is critical in this example to make the sharp turns needed to describe letters such as the "t".

It should be mentioned in conclusion that functions **spc** and **spltrp** are more general than the intrinsic spline function in MATLAB. That function does not provide for corner points, control of end conditions, or allow integration and differentiation.

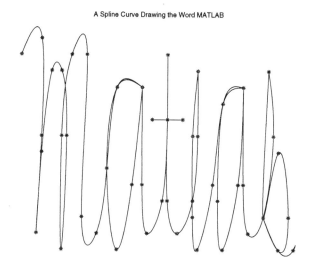

A Spline Curve Drawing the Word MATLAB

Figure 4.3. A Spline Curve Drawing the Word MATLAB

4.2.1   Example: Spline Interpolation Applied to $\sin(x)$

Script File sinetrp

```
 1: % Example: sinetrp
 2: % ~~~~~~~~~~~~~~~~~~
 3: % This example illustrates cubic spline
 4: % approximation of sin(x), its first two
 5: % derivatives, and its integral.
 6: %
 7: % User m functions required:
 8: %     spc, spltrp, genprint
 9: %-------------------------------------------------
10:
11: % Obtain data points on the spline curve
12: x=linspace(0,2*pi,21); y=sin(x);
13: xx=linspace(-pi/2,2.5*pi,51);
14: % Specify y' at first point and
15: % y'' at last point
16: i=[1,2]; v=[1,0];
17: % Get interpolation coefficients
18: splmat=spc(x,y,i,v);
19:
20: % Evaluate function values at a dense
21: % set of points
22: xx=linspace(-pi/2,2.5*pi,51);
23: z=xx/pi;
24: yy=spltrp(xx,splmat,0);
25: yyp=spltrp(xx,splmat,1);
26: yypp=spltrp(xx,splmat,2);
27: yyint=spltrp(xx,splmat,3);
28:
29: % Plot results
30: plot(z,yy,'-',z,yyp,'--',z,yypp,'-.', ...
31:      z,yyint,'+',x/pi,y,'o');
32: grid on;
33: title(['Spline Differentiation and ', ...
34:         'Integration of sin(x)']);
35: xlabel('x / pi'); ylabel('function values');
36: legend('y=sin(x)','y''(x)','y''''(x)', ...
37:         '\int y(x) dx');
38: figure(gcf); %pause; genprint('sintest');
```

Function spc

```
 1: function splmat=spc(x,y,i,v,icrnr)
 2: %
 3: % splmat=spc(x,y,i,v,icrnr)
 4: % ~~~~~~~~~~~~~~~~~~~~~~~~~~
 5: % This function computes matrix splmat
 6: % containing coefficients needed to perform a
 7: % piecewise cubic interpolation among data
 8: % values contained in vectors x and y. The
 9: % output from this function is used by function
10: % spltrp to evaluate the cubic spline function,
11: % its first two derivatives, or the function
12: % integral from x(1) to an arbitrary upper
13: % limit.
14: %
15: % x      - vector of abscissa values arranged
16: %          in increasing order. The number of
17: %          data values is denoted by n.
18: %
19: % y      - vector of ordinate values
20: %
21: % i      - a two component vector [i1,i2].
22: %          Parameters i1 and i2 refer to left
23: %          end and right end conditions,
24: %          respectively. These equal 1, 2,
25: %          or 3 as explained below.
26: %
27: % v      - a two component vector [v1,v2]
28: %          containing end values of y'(x)
29: %          or y''(x)
30: %
31: %          if i1=1: y'(x(1)) is set to v(1)
32: %             =2: y''(x(1)) is set to v(1)
33: %             =3: y''' is continuous at x(2)
34: %          if i2=1: y'(x(n)) is set to v(2)
35: %             =2: y''(x(n)) is set to v(2)
36: %             =3: y''' is continuous at x(n-1)
37: %
38: %          Note: When i1 or i2 equal 3 the
39: %                corresponding values of vector
40: %                v should be zero
41: %
42: % icrnr - A vector of indices identifying
```

```
43: %          interior points where slope
44: %          discontinuities are to be generated
45: %          by requiring y''(x) to equal zero.
46: %          Vector icrnr should have components
47: %          lying between 2 and n-1.  The vector
48: %          can be omitted from the argument list
49: %          if no slope discontinuities occur.
50: %
51: % User m functions called:  none
52: %-------------------------------------------------
53:
54: x=x(:); y=y(:);
55: n=length(x); a=zeros(n,1); b=a; c=a;
56: d=a; t=a; if nargin < 5, icrnr=[]; end
57: ncrnr=length(icrnr);
58: i1=i(1); i2=i(2);
59: v1=v(1); v2=v(2);
60:
61: % Form the tridiagonal system to solve for
62: % second derivative values at the data points
63:
64: n1=n-1; j=2:n1;
65: hj=x(j)-x(j-1); hj1=x(j+1)-x(j);
66: hjp=hj+hj1; a(j)=hj./hjp; wuns=ones(n-2,1);
67: b(j)=2*wuns; c(j)=wuns-a(j);
68: d(j)=6.*((y(j+1)-y(j))./hj1- ...
69:     (y(j)-y(j-1))./hj)./hjp;
70:
71: % Form  equations for the end conditions
72:
73: % slope specified at left end
74: if i1==1
75:    h2=x(2)-x(1); b(1)=2; c(1)=1;
76:    d(1)=6*((y(2)-y(1))/h2-v1)/h2;
77:
78: % second derivative specified at left end
79: elseif i1==2
80:    b(1)=1; c(1)=0; d(1)=v1;
81:
82: % not a knot condition
83: else
84:    b(1)=1;
85:    a(1)=hj(1)/hj(2);
86:    c(1)=-1-a(1);
87:    d(1)=0;
```

```
88: end
89:
90: % slope specified at right end
91: if i2==1
92:    hn=x(n)-x(n-1); a(n)=1; b(n)=2;
93:    d(n)=6.*(v2-(y(n)-y(n-1))/hn)/hn;
94:
95: % second derivative specified at right end
96: elseif i2==2
97:    a(n)=0; b(n)=1; d(n)=v2;
98:
99: % not a knot condition
100: else
101:    a(n)=1;
102:    b(n)=-(x(n-1)-x(n-2))/(x(n)-x(n-2));
103:    c(n)=-(x(n)-x(n-1))/(x(n)-x(n-2));
104: end
105:
106: % Adjust for slope discontinuity
107: % specified by lcrnr
108:
109: if ncrnr > 0
110:    zro=zeros(ncrnr,1);
111:    a(icrnr)=zro; c(icrnr)=zro;
112:    d(icrnr)=zro; b(icrnr)=ones(ncrnr,1);
113: end
114:
115: % Solve the tridiagonal system
116: % for t(1),...,t(n)
117:
118: bb=diag(a(2:n),-1)+diag(b)+diag(c(1:n-1),1);
119: if a(1)~=0, bb(1,3)=a(1); d(1)=0;    end
120: if c(n)~=0, bb(n,n-2)=c(n); d(n)=0; end
121: t=bb\d(:);
122:
123: % Save polynomial coefficients describing the
124: % cubics for each interval.
125:
126: j=1:n-1; k=2:n;
127: dx=x(k)-x(j); dy=y(k)-y(j); b=t(j)/2;
128: c=(t(k)-t(j))./(6*dx); a=dy./dx-(c.*dx+b).*dx;
129: int=(((c.*dx/4+b/3).*dx+a/2).*dx+y(j)).*dx;
130: int=[0;cumsum(int)]; n1=n-1;
131: ypn=dy(n1)/dx(n1)+dx(n1)*(2*t(n)+t(n1))/6;
132:
```

```
133: % The columns of splmat contain the
134: % following vectors
135: % [x,y,x_coef,x^2_coef,x^3_coef,integral_coef]
136:
137: splmat=[ ...
138:   [ x(1),  y(1),  a(1),   0,   0,   0      ];
139:   [ x(j),  y(j),  a,      b,   c,   int(j) ];
140:   [ x(n),  y(n),  ypn,    0,   0,   int(n) ] ];
```

Function spltrp

```
 1: function f=spltrp(x,mat,ideriv)
 2: %
 3: % f=spltrp(x,mat,ideriv)
 4: % ~~~~~~~~~~~~~~~~~~~~~~~
 5: % This function performs cubic spline
 6: % interpolation using data array mat obtained
 7: % by first calling function spc.
 8: %
 9: % x        - the vector of interpolation values
10: %            at which the spline is to be
11: %            evaluated.
12: % mat      - the matrix output from function spc.
13: %            This array contains the data points
14: %            and the polynomial coefficients
15: %            needed to define the piecewise
16: %            cubic curve connecting the data
17: %            points.
18: % ideriv   - a parameter specifying whether
19: %            function values, derivative values,
20: %            or integral values are computed.
21: %            Taking ideriv equal to 0, 1, 2, or
22: %            3, respectively, returns values of
23: %            y(x), y'(x), y''(x), or the integral
24: %            of y(x) from the first data point
25: %            defined in array mat to each point
26: %            in vector x specifying the chosen
27: %            interpolation points. For any points
28: %            outside the original data range,
29: %            the interpolation is performed by
30: %            extending the tangents at the first
31: %            and last data points.
32: %
```

```
33: % User m functions called:  none
34: %-----------------------------------------------
35:
36: % identify interpolation intervals
37: % for each point
38: [nrow,ncol]=size(mat);
39: xx=x(:)'; xd=mat(2:nrow);
40: xd=xd(:); np=length(x); nd=length(xd);
41: ik=sum(xd(:,ones(1,np)) < xx(ones(nd,1),:));
42: ik=1+ik(:);
43: xk=mat(ik,1); yk=mat(ik,2); dx=x(:)-xk;
44: ak=mat(ik,3); bk=mat(ik,4);
45: ck=mat(ik,5); intk=mat(ik,6);
46:
47: % obtain function values at x
48: if ideriv==0
49:   f=((ck.*dx+bk).*dx+ak).*dx+yk; return
50:
51: % obtain first derivatives at x
52: elseif ideriv==1
53:   f=(3*ck.*dx+2*bk).*dx+ak; return
54:
55: % obtain second derivatives at x
56: elseif ideriv==2
57:   f=6*ck.*dx+2*bk; return
58:
59: % obtain integral from xd(1) to x
60: elseif ideriv==3
61:   f=intk+(((ck.*dx/4+bk/3).*dx+ ...
62:     ak/2).*dx+yk).*dx;
63: end
```

4.2.2　Example: Plotting of General Plane Curves

Script File matlbdat

```
1: % Example: matlbdat
2: % ~~~~~~~~~~~~~~~~~~
3: % This example illustrates the use of splines
4: % to draw the word MATLAB.
5: %
6: % User m functions required:
7: %      spcurv2d, spc, spltrp, genprint
8: %-------------------------------------------------
9:
10: x=[13 17 17 16 17 19 21 22 21 21 23 26
11:     25 28 30 32 37 32 30 32 35 37 37 38
12:     41 42 42 42 45 39 42 42 44 47 48 48
13:     47 47 48 51 53 57 53 52 53 56 57 57
14:     58 61 63 62 61 64 66 64 61 64 67 67];
15: y=[63 64 58 52 57 62 62 58 51 58 63 63
16:     53 52 56 61 61 61 56 51 55 61 55 52
17:     54 59 63 59 59 59 59 54 52 54 58 62
18:     58 53 51 55 60 61 60 54 51 55 61 55
19:     52 53 58 62 53 57 53 51 53 51 51 51];
20: x=x'; x=x(:); y=y'; y=y(:);
21: ncrnr=[17 22 26 27 28 29 30 31 36 42 47 52];
22: clf; [xs,ys]=spcurv2d(x,y,10,ncrnr);
23: plot(xs,ys,'-',x,y,'*'), axis off;
24: title('A Spline Curve Drawing the Word MATLAB');
25: figure(gcf); %genprint('matlbdat');
```

Function spcurv2d

```
1: function [xout,yout,sout]= ...
2:          spcurv2d(xd,yd,nseg,ncrnr)
3: %
4: % [xout,yout]=spcurv2d(xd,yd,nseg,ncrnr)
5: % ~~~~~~~~~~~~~~~~~~~~~~~~~~~~~~~~~~~~~~~~
6: %
7: % This function tabulates points xout, yout on
8: % a spline curve connecting data points xd, yd
9: % on the cubic spline curve
10: %
11: %  xd,yd      - input data points
12: %  nseg       - number of tabulation intervals
13: %               used per spline segment
14: %  ncrnr      - point indices where corners are
15: %               required
16: %  xout,yout - output data points on the
17: %               spline curve. The number of
18: %               points returned equals
19: %
20: %               nout=(nd-1)*nseg+1.
21: %
22: % User m functions called:  spc, spltrp
23: %-----------------------------------------------
24:
25: nd=length(xd); sd=(1:nd)';
26: if nargin==2; nseg=10; end
27: if nargin<=3, ncrnr=[]; end;
28: nout=(nd-1)*nseg+1; sout=linspace(1,nd,nout);
29: if norm([xd(1)-xd(nd),yd(1)-yd(nd)]) < 100*eps
30:    yp=(yd(2)-yd(nd-1))/2;
31:    iend=[1;1]; vy=[yp;yp];
32:    xp=(xd(2)-xd(nd-1))/2; vx=[xp;xp];
33: else
34:    iend=[3;3]; vy=[0;0]; vx=vy;
35: end
36: matx=spc(sd,xd,iend,vy,ncrnr);
37: maty=spc(sd,yd,iend,vx,ncrnr);
38: xout=spltrp(sout,matx,0);
39: yout=spltrp(sout,maty,0);
```

4.3   Numerical Differentiation Using Finite Differences

Problems involving differential equations are often solved approximately by using difference formulas which approximate the derivatives in terms of function values at adjacent points. Deriving difference formulas by hand can be tedious, particularly when unequal point spacing is used. For this reason, we develop a numerical procedure allowing construction of formulas of arbitrary order and arbitrary truncation error. Of course, as the desired order of derivative and the order of truncation error increases, a larger number of points is needed to interpolate the derivative. We will show below that approximating a derivative of order k with a truncation error of order h^m generally requires $(k + m)$ points unless symmetric central differences are used.

Consider the Taylor series expansion

$$F(x + \alpha h) = \sum_{k=0}^{\infty} \frac{F^{(k)}(x)}{k!} (\alpha h)^k$$

where $F^{(k)}(x)$ means the k'th derivative of $F(x)$. This relation expresses values of F as linear combinations of the function derivatives at x. Conversely, the derivative values can be cast in terms of function values by solving a system of simultaneous equations. Let us take a series of points defined by

$$x_i = x + h\alpha_i \qquad 1 \le i \le n$$

where h is a fixed step-size and α_i are arbitrary parameters. Separating some leading terms in the series expansion gives

$$F(x_i) \;=\; \sum_{k=0}^{n-1} \frac{\alpha_i^k}{k!} \left[h^k F^{(k)}(x) \right] + \frac{\alpha_i^n}{n!} \left[h^n F^{(n)}(x) \right] +$$

$$\frac{\alpha_i^{n+1}}{(n+1)!} \left[h^{(n+1)} F^{(n+1)}(x) \right] + O(h^{n+2}) \qquad 1 \le i \le n$$

It is helpful to use the following notation:

α^k – a column vector with component i being equal to α_i^k

f – a column vector with component i being $F(x_i)$

fp – a column vector with component i being $h^i F^{(i)}(x)$

A – $[\alpha^0, \alpha^1, \ldots, \alpha^{n-1}]$, a square matrix with columns which are powers of α

Then the Taylor series expressed in matrix form is

$$f \;=\; A * fp + \frac{h^n F^{(n)}(x)}{n!} \alpha^n +$$

$$\frac{h^{n+1} F^{(n+1)}(x)}{(n+1)!} \alpha^{n+1} + O(h^{n+2})$$

Solving this system for the derivative matrix fp yields

$$fp = A^{-1}f - \frac{h^n F^{(n)}(x)}{n!}A^{-1}\alpha^n -$$

$$\frac{h^{n+1} F^{(n+1)}(x)}{(n+1)!}A^{-1}\alpha^{n+1} + O(h^{n+2})$$

In the last equation we have retained the first two remainder terms in explicit form to allow the magnitudes of these terms to be examined. Row $k+1$ of the previous equation implies

$$F^{(k)}(x) = h^{-k}(A^{-1}f)_{k+1} - \frac{h^{n-k}}{n!}F^{(n)}(x)(A^{-1}\alpha^n)_{k+1} -$$

$$\frac{h^{n-k+1}}{(n+1)!}F^{(n+1)}(x)(A^{-1}\alpha^{n+1})_{k+1} + O(h^{n-k+1})$$

Consequently, the rows of A^{-1} provide coefficients in formulas to interpolate derivatives. For a particular number of interpolation points, say N, the highest derivative approximated will be $F^{(N-1)}(x)$ and the truncation error will normally be of order k^1. Conversely, if we need to compute a derivative formula of order k with the truncation error being m, then it is necessary to use a number of points such that $n - k = m$; therefore $n = m + k$. For the case where interpolation points are symmetrically placed around the point where derivatives are desired, one higher power of accuracy order is achieved than might be expected. We can show, for example, that

$$\frac{d^4 F(x)}{dx^4} = \frac{1}{h^4}[F(x-2h) - 4F(x-h) + 6F(x) -$$

$$4F(x+h) + F(x+2h)] + O(h^2)$$

because the truncation error term associated with h^1 is found to be zero. At the same time, we can show that a forward difference formula for $f'''(x)$ employing equidistant point spacing is

$$\frac{d^3 F(x)}{dx^3} = \frac{1}{h^3}[-2.5F(x) - 9F(x+h) + 12F(x+2h) +$$

$$7F(x+3h) - 1.5F(x+4h)] + O(h^2)$$

Although the last two formulas contain arithmetically simple interpolation coefficients, due to equal point spacing, the method is certainly not restricted to equal spacing. The following program contains the function **derivtrp** which implements the ideas just developed. Since the program contains documentation which is output when it is executed, no additional example problem is included.

4.3.1   Example: Deriving General Difference Formulas

Output from Example

```
>> finidif

FINITE DIFFERENCE FORMULAS OF ARBITRARY ORDER

select the derivative order and
the truncation error order
(input 0,0 to stop)
 ? 3,2
give a 5 component vector defining the base points
 ? -3,-2,-1,0,1

derivative  interpolation coefficients
    order  c(1)      c(2)     c(3)     c(4)     c(5)
        0  0         0        0        1.0000   0
   1.0000 -0.0833    0.5000  -1.5000   0.8333   0.2500
   2.0000 -0.0833    0.3333   0.5000  -1.6667   0.9167

/---------------- Editorial Note ----------------\
|                                                 |
|     The following line gives the coefficients   |
|     to evaluate the third derivative using      |
|     five function values.                       |
|                                                 |
|    3.0000   0.5000 -3.0000   6.0000 -5.0000   1.5000|
|             ------  ------   ------   ------   ------|
\-------------------------------------------------/

       4.0000   1.0000 -4.0000   6.0000 -4.0000   1.0000

derivative   power of   error    power of    error
    order       h^k      coef.      h^k       coef.
        0     5.0000         0    6.0000          0
   1.0000     4.0000    0.0500    5.0000    -0.0417
   2.0000     3.0000    0.0833    4.0000    -0.0528


/---------------- Editorial Note ----------------\
|                                                 |
|     The truncation error coefficients of        |
|     order h^2 and h^3 are shown below           |
|                                                 |
|    3.0000    2.0000   -0.2500    3.0000    0.2500|
```

```
|                      ------              ------|
\------------------------------------------------/
```

```
    4.0000    1.0000   -1.0000    2.0000    0.6667
```

select the derivative order and
the truncation error order
(input 0,0 to stop)
 ? 4,1
give a 5 component vector defining the base points
 ? -2,-1,0,1,2

derivative order	interpolation coefficients				
	c(1)	c(2)	c(3)	c(4)	c(5)
0	0	0	1.0000	0	0
1.0000	0.0833	-0.6667	0	0.6667	-0.0833
2.0000	-0.0833	1.3333	-2.5000	1.3333	-0.0833
3.0000	-0.5000	1.0000	0	-1.0000	0.5000
4.0000	1.0000	-4.0000	6.0000	-4.0000	1.0000

derivative order	power of h^k	error coef.	power of h^k	error coef.
0	5.0000	0	6.0000	0
1.0000	4.0000	-0.0333	5.0000	0
2.0000	3.0000	0	4.0000	-0.0111
3.0000	2.0000	0.2500	3.0000	0
4.0000	1.0000	0	2.0000	0.1667

select the derivative order and
the truncation error order
(input 0,0 to stop)
 ? 0,0

Script file finidif

```
 1: % Example:  finidif
 2: % ~~~~~~~~~~~~~~~~~~
 3: % This program uses a truncated Taylor series
 4: % to compute finite difference formulas
 5: % approximating derivatives of arbitrary order
 6: % which are interpolated at an arbitrary set of
 7: % base points not necessarily evenly spaced.
 8: % The base point positions are expressed as
 9: % multiples of a stepsize variable h. For
10: % example, if function values at x1=x-h, x2=x,
11: % and x3=x+h are used to evaluate y''(x), then
12: % the base points would be defined by the
13: % vector [-1,0,1]. The corresponding
14: % approximation for y''(x) would be expressed
15: % as
16: %
17: %       ( c(1)*y1 + c(2)*y2 + c(3)*y3 )/h^2
18: %
19: % Furthermore, the truncation error
20: % coefficients for the next two terms in the
21: % Taylor series expansion are included in the
22: % table of printed output. Results are given
23: % in a tableau illustrated by the following
24: % examples pertaining to third and fourth
25: % derivative formulas.
26: %
27: % User m functions required:
28: %     derivtrp, readv, output
29: %---------------------------------------------------
30:
31: disp('');
32: fprintf('\nFINITE DIFFERENCE FORMULAS ');
33: fprintf('OF ARBITRARY ORDER\n\n'); disp('');
34: while 1
35:   disp('Select the derivative order and');
36:   disp('the truncation error order');
37:   disp('(input 0,0 to stop)');
38:   t=readv(2); n=sum(t); in=int2str(n);
39:   if n==0; break; end
40:   disp(' '); disp(['give a ',in, ...
41: ' component vector defining the base points']);
42:   basepts=readv(n);
```

```
43:     [cof,ct,cb,e1,e2]=derivtrp(basepts);
44:     output(cof,e1,e2);
45: end
```

Function derivtrp

```
 1: function [c,ctop,cbot,erc1,erc2]=...
 2:              derivtrp(baspts)
 3: %
 4: % [c,ctop,abot,erc1,erc2]=derivtrp(baspts)
 5: % ~~~~~~~~~~~~~~~~~~~~~~~~~~~~~~~~~~~~~~~~~~
 6: % This function computes coefficients to
 7: % interpolate derivatives by finite
 8: % differences.
 9: %
10: % baspts - The matrix specifying the base
11: %          points for which function values
12: %          are employed.  For example, using
13: %          function values of f(x-2h),
14: %          f(x-h), f(x), f(x), f(x+2h) and
15: %          f(x+3h) would imply baspts=[-2:3].
16: % c       - Matrix in which row k+2 contains
17: %           coefficients which properly weight
18: %           the function values corresponding
19: %           to base points defined by vector
20: %           baspts. The first row of c contains
21: %           the components of alp to help
22: %           identify which coefficients go with
23: %           which function values.
24: % ctop,   - These two matrices contain integers
25: % cbot      such that c is closely approximated
26: %           by ctop./cbot. When the difference
27: %           coefficients turn out to be exactly
28: %           expressible as rational functions,
29: %           then matrices ctop and cbot usually
30: %           give the desired coefficients
31: %           exactly.
32: % erc1,   - Vectors characterizing the
33: %           truncation error terms
34: % erc2      associated with various derivatives.
35: %
36: % User m functions called:  none.
37: %----------------------------------------------
```

```
38:
39: x=baspts(:); n=length(x); a=ones(n,n+2);
40: for k=2:n+2;
41:   a(:,k)=a(:,k-1).*x/(k-1);
42: end
43: c=inv(a(:,1:n));
44: erc1=c*a(:,n+1); erc2=c*a(:,n+2);
45: [ctop,cbot]=rat(c);
```

Function output

```
1: function output(a,ercof1,ercof2)
2: %
3: % output(a,ercof1,ercof2)
4: % ~~~~~~~~~~~~~~~~~~~~~~~~~
5: % This function prints the results.
6: %
7: % User m functions called: none.
8: %------------------------------------------------
9:
10: n=max(size(a)); mat1=[(0:n-1)',a];
11: mat2=[(0:n-1)',(n:-1:1)',ercof1(:), ...
12:       (n+1:-1:2)',ercof2];
13: s='        order';
14: for j=1:n;
15:   s=[s,'        c(',int2str(j),')'];
16: end
17: fprintf('\nderivative      interpolation ');
18: fprintf('coefficients\n');
19: disp(s); disp(mat1);
20: if n>6, pause, end
21: fprintf('\nderivative    power of  error');
22: fprintf('      power of    error   ');
23: fprintf('\n     order      h^k        coef.');
24: fprintf('        h^k       coef. \n');
25: disp(mat2);
```

Function readv

```
1: function [v,l]=readv(n)
2: %
3: % v=readv(n)
```

```
 4: %  ~~~~~~~~~~
 5: % This function inputs a vector of length n.
 6: % If fewer than n values are given, then the
 7: % vector is padded with zeros.
 8: %
 9: %   n  - number of values to be input
10: %   v  - a vector of length n. If fewer than n
11: %          values were input, the final values
12: %          of v are set to zero
13: %   l  - the number of values actually read
14: %
15: % User m functions called:  none
16: %-----------------------------------------------
17:
18: str=input('? > ','s');
19: str=['[',str,']'];
20: v=eval(str); l=length(v);
21: if l>n, v=v(1:n); end
22: if l<n, v=[v,zeros(1,n-l)]; end
```

4.3.2   Example: Deriving Adams-Type Integration Formulas

The same ideas employed to derive finite difference relations can be applied to develop Adams-type integration formulas used to integrate differential equations. Integrating the differential equation

$$y'(x) = f(x, y(x))$$

over a time step gives

$$y(x + h) = y(x) + h \sum_{k=0}^{\infty} \left[\frac{f^{(k)}(x, y(x))}{(k + 1)!} \right] h^k$$

Expressing the derivatives $f^{(k)}$ in terms of f values in a manner similiar to what was done earlier leads to formulas of the form

$$y_{i+1} = y_i + h \sum_j c_{ij} f_j$$

involving values of the derivative function $f(x, y(x))$ at selected points. The resulting formulas are either explicit if y_{i+1} is not involved in the points where f is evaluated or implicit if $f(x_{i+1}, y_{i+1})$ is present on the right side. The following short program computes Adams integration formulas. Two examples involving explicit and implicit formulas are given. Output produced by the program is also provided. This example concludes the current chapter on interpolation and differentiation.

MATLAB Example

Output from Example

```
*** Adams Integration Formulas for DE Solution ***

        Adams-Bashforth  order six explicit
          Index       C-top      C-bottom
            0           199          67
           -1         -1315         239
           -2          1324         191
           -3          -968         191
           -4           959         480
           -5           -95         288


        Adams-Moulton  order six implicit
          Index       C-top      C-bottom
            1            95          288
            0           439          443
           -1          -133         240
           -2           241         720
           -3          -173        1440
           -4             3         160
```

Script file adamsex

```
 1: % Example:  adamsex
 2: % ~~~~~~~~~~~~~~~~~~
 3: % This program illustrates use of function
 4: % adams2 for determining coefficients in the
 5: % explicit or implicit Adams-type formulas
 6: % used for differential equation solution.
 7: % These integration formulas have the form
 8: %
 9: %     ynplus1 = yn + h*Sum(cof(j)*f(j))
10: %
11: % where f(j) denotes the value of f for
12: %
13: %     x(j) = xn + h*alpha(j)
14: %
15: % The weighting coefficient cof(j) equals
16: % the quotient c(j,1)/c(j,2).
17: %
18: % User m functions required: adams2
19: %-------------------------------------------------
20:
21: fprintf('\n*** Adams Integration Formulas');
22: fprintf(' for DE Solution ***\n');
23: % order6_explicit=[(0:-1:-5)',adams2(0:-1:-5)]
24: % order6_implicit=[(1:-1:-4)',adams2(1:-1:-4)]
25: fprintf(...
26:   '\n     Adams-Bashforth  order six explicit');
27: fprintf(...
28:   '\n          Index     C-top     C-bottom \n');
29: disp([(0:-1:-5)',adams2(0:-1:-5)]);
30: fprintf(...
31:   '\n     Adams-Moulton  order six implicit');
32: fprintf(...
33:   '\n          Index     C-top     C-bottom \n');
34: disp([(1:-1:-4)',adams2(1:-1:-4)]);
```

Function adams2

```
 1: function c = adams2(alpha)
 2: %
 3: % c = adams2(alpha)
```

```
4: % ~~~~~~~~~~~~~~~~~
5: % This function determines coefficients in the
6: % Adams-type formulas used to solve
7: % y'(x)=f(x,y).  These integration formulas
8: % have the form
9: %
10: %      ynplus1 = yn + h*Sum(cof(j)*f(j))
11: %
12: % where f(j) denotes the value of f for
13: %
14: %      x(j) = xn + h*alpha(j)
15: %
16: % The weighting coefficient cof(j) equals
17: % the quotient c(j,1)/c(j,2). For example,
18: % adams2(0:-1:-4) determines the Adams-
19: % Bashforth formula of order 5. Similarly,
20: % adams2(1:-1:-3) determines the Adams-Moulton
21: % formula of order 5.
22: %
23: % alpha - vector of coefficients defining the
24: %         base points used for the integration
25: % c     - weighting coefficients for the
26: %         integration formula
27: %
28: % User m functions called:  none.
29: %-------------------------------------------------
30:
31: alpha=alpha(:)'; n=length(alpha);
32: a=(ones(n,1)*alpha).^((0:n-1)'*ones(1,n));
33: [coftop,cofbot]=rat(a\(1 ./(1:n)'));
34: c=[coftop,cofbot];
```

Chapter 5

Gaussian Integration with Applications to Geometric Properties

5.1 Fundamental Concepts and Intrinsic Integration Tools Provided in MATLAB

Numerical integration methods approximate a definite integral by evaluating the integrand at several points and taking a weighted combination of those integrand values. The weight factors can be obtained by interpolating the integrand at selected points and integrating the interpolating function exactly. For example, the Newton-Cotes formulas result from polynomial interpolation through equidistant base points. This chapter presents concepts of numerical integration adequate to handle many practical applications.

Let us assume that an integral over limits a to b is to be evaluated. We can write

$$\int_a^b f(x)dx = \sum_{i=1}^n W_i f(x_i) + E$$

where E represents the error due to replacement of the integral by a sum. This is called an n-point quadrature formula. The points x_i where the integrand is evaluated are base points and the constants W_i are weight factors. Most integration formulas depend on approximating the integrand by a polynomial. Consequently, they give exact results when the integrand is a polynomial of sufficiently low order. Different choices of x_i and W_i will be discussed below.

It is helpful to express an integral over general limits in terms of some fixed limits, say -1 to 1. This is accomplished by introducing a linear change of variables

$$x = \alpha + \beta t$$

Requiring that $x = a$ corresponds to $t = -1$ and $x = b$ corresponds to

$t = 1$ gives $\alpha = (a + b)/2$ and $\beta = (b - a)/2$, so that one obtains the identity

$$\int_a^b f(x)dx = \frac{1}{2}(b - a) \int_{-1}^1 f\left[\frac{a + b}{2} + \frac{b - a}{2}t\right] dt = \int_{-1}^1 F(t)dt$$

where $F(t) = f[(a+b)/2+(b-a)t/2](b-a)/2$. Thus, the dependence of the integral on the integration limits can be represented parametrically by modifying the integrand. Consequently, if an integration formula is known for limits -1 to 1, we can write

$$\int_a^b f(x)dx = \beta \sum_{i=1}^n W_i f(\alpha + \beta x_i) + E$$

The idea of shifting integration limits can be exploited further by dividing the interval a to b into several parts and using the same numerical integration formula to evaluate the contribution from each interval. Employing m intervals of length $\ell = (b - a)/m$, we get

$$\int_a^b f(x)dx = \sum_{j=1}^m \int_{a+(j-1)\ell}^{a+j\ell} f(x)dx$$

Each of the integrals in the summation can be transformed to have limits -1 to 1 by taking

$$x = \alpha_j + \beta t$$

with

$$\alpha_j = a + (j - .5)\ell \qquad \beta = .5\ell$$

Therefore we obtain the identity

$$\int_a^b f(x)dx = \sum_{j=1}^m .5\ell \int_{-1}^1 f(\alpha_j + \beta t)dt$$

Applying the same n-point quadrature formula in each of m equal intervals gives what is termed a composite formula

$$\int_a^b f(x)dx = .5\ell \sum_{j=1}^m \sum_{i=1}^n W_i f(\alpha_j + \beta x_i) + E$$

By interchanging the summation order in the previous equation we get

$$\int_a^b f(x)dx = .5\ell \sum_{i=1}^n W_i \sum_{j=1}^m f(\alpha_j + \beta x_i) + E$$

Let us now turn to certain choices of weight factors and base points. Two of the most widely used approximations assume that the integrand is represented satisfactorily as either piecewise linear or piecewise cubic. Approximating the integrand by a straight line through the integrand end points gives the following formula

$$\int_{-1}^{1} f(x)dx = f(-1) + f(1) + E$$

A much more accurate formula results by using a cubic approximation matching the integrand at $x = -1, 0, 1$. Let us write

$$f(x) = c_1 + c_2 x + c_3 x^2 + c_4 x^3$$

then

$$\int_{-1}^{1} f(x)dx = 2c_1 + \frac{2}{3}c_3$$

Evidently the linear and cubic terms do not influence the integral value. Also, $c_1 = f(0)$ and $f(-1) + f(1) = 2 c_1 + 2 c_3$ so that

$$\int_{-1}^{1} f(x)dx = \frac{1}{3}\left[f(-1) + 4f(0) + f(1)\right] + E$$

The error E in this formula is zero when the integrand is any polynomial of order 3 or lower. Expressed in terms of more general limits this result is

$$\int_{a}^{b} f(x)dx = \frac{(b - a)}{6}\left[f(a) + 4f(.5\,(a + b)) + f(b)\right] + E$$

which is called Simpson's rule.

Analyzing the integration error for a particular choice of integrand and quadrature formula can be complex. In practice, the usual procedure taken is to apply a composite formula with m chosen large enough so the integration error is expected to be negligibly small. The value for m is then increased until no further significant change in the integral approximation results. Although this procedure involves some risk of error, adequate results can be obtained in most practical situations.

In the subsequent discussions the integration error which results by replacing an integral by a weighted sum of integrand values will be neglected. It must nevertheless be kept in mind that this error depends on the base points, weight factors, and the particular integrand. Most importantly, the error typically decreases as the number of function values is increased.

It is convenient to summarize the composite formulas obtained by employing a piecewise linear or piecewise cubic integrand approximation. Using m intervals and letting $\ell = (b - a)/m$, it is easy to obtain the composite trapezoidal formula which is

$$\int_{a}^{b} f(x)dx = \ell \left[\frac{f(a) + f(b)}{2} + \sum_{j=1}^{m-1} f(a + j\ell)\right]$$

This formula assumes that the integrand is satisfactorily approximated as piecewise linear. A similar but much more accurate result is obtained for the composite integration formula based on cubic approximation. For this case, taking m intervals implies $2m + 1$ function evaluations. If we let $g = (b - a)/(2m)$ and $h = 2g$, then

$$x_j = a + gj \qquad f_j = f(x_j) \qquad j = 0, 1, 2, \ldots, 2m$$

where $f(x_0) = f(a)$ and $f(x_{2m}) = f(b)$. Combining results for all intervals gives

$$\int_a^b f(x)dx \;=\; \frac{h}{6}\left[(f_0 + 4f_1 + f_2) + (f_2 + 4f_3 + f_4) + \ldots +\right.$$

$$\left.(f_{2m-2} + 4f_{2m-1} + f_{2m})\right]$$

The terms in the previous formula can be rearranged into a form more convenient for computation. We get

$$\int_a^b f(x)dx = \frac{h}{6}\left[f(a) + 4f_1 + f(b) + \sum_{i=1}^{m-1}(4f_{2i+1} + 2f_{2i})\right]$$

This formula, known as the composite Simpson rule, is one of the most commonly used numerical integration methods. A MATLAB implementation of Simpson's rule is listed below.

An important goal in numerical integration is to achieve accurate results with only a few function evaluations. It was shown for Simpson's rule that three function evaluations are enough to exactly integrate a cubic polynomial. By choosing the base point locations properly, a much higher accuracy can be achieved for a given number of function evaluations than would be obtained by using evenly spaced base points. Results from orthogonal function theory lead to the following conclusions. If the base points are located at the zeros of the Legendre polynomials (all these zeros are between -1 and 1) and the weight factors are computed as certain functions of the base points, then the formula

$$\int_{-1}^1 f(x)dx = \sum_{i=1}^n W_i f(x_i)$$

```
function ansr=simpson(funct,a,b,neven)
%
% ansr=simpson(funct,a,b,neven)
% ----------------------------
%
% This function integrates "funct" from
% "a" to "b" by Simpson's rule using
% "neven+1" function values.  Parameter
% "neven" should be an even integer.
%
% Example use:  ansr=simpson('sin',0,pi/2,4)
%
% funct    -  character string name of
%             function integrated
% a,b      -  integration limits
% neven    -  an even integer defining the
%             number of integration intervals
% ansr     -  Simpson rule estimate of the
%             integral
%
% User m functions called: argument funct
%-------------------------------------------------

ne=max(2,2*round(.1+neven/2)); d=(b-a)/ne;
x=a+d*(0:ne); y=feval(funct,x);
ansr=(d/3)*(y(1)+y(ne+1)+4*sum(y(2:2:ne))+...
     2*sum(y(3:2:ne-1)));
```

is exact for a polynomial integrand of degree $2n - 1$. Although the theory proving this property is not elementary, the final results are quite simple. The base points and weight factors for a particular order can be computed once and stored as data statements in the program. Formulas which use the Legendre polynomial roots as base points are called Gauss quadrature formulas. In a typical application, Gauss integration gives much more accurate results than Simpson's rule for an equivalent number of function evaluations. Since it is equally easy to use, the Gauss formula is preferable to the well known Simpson's rule.

MATLAB also provides two functions **quad** and **quad8** which perform numerical integration by adaptive methods. These methods repeatedly modify estimates of an integral until

$$abs \left(\frac{\text{error\_estimate}}{\text{integral\_value}} \right)$$

is smaller than some error tolerance. Presently, such procedures have not yet been suitably refined to produce accuracy levels comparable to what is routinely taken for granted (for example, in the evaluation of elementary functions). Since the adaptive calculation process is vulnerable to convergence failure, error estimates are usually made quite conservatively in order to produce results at least as good as what the chosen error tolerances would imply. The MATLAB functions **quad** and **quad8** usually work well. However, these integrators should be used with some degree of caution. These functions employ conservative error measures, so warning messages sometimes result when simple functions such as $\sqrt{x}$ are integrated. Furthermore, choosing too crude an error tolerance may yield inaccurate results, whereas using too stringent a tolerance can require excessive computation. If a user is sufficiently informed to choose a nonadaptive integrator known to be accurate enough for a particular application, that formula may frequently run much faster than a corresponding adaptive formula. For example, evaluating $\int_0^1 e^{10x} \cos(10\pi x) \, dx$ with **quad8** and a tolerance of 10^{-3} takes 14 times as much computer time and produces an integration error 127 times as large as that resulting when **gquad10** (a tenth-order Gauss integrator provided by the authors) is used with 40 function evaluations.

5.2   Concepts of Gauss Integration

This section summarizes properties of Gauss integration which, for the same number of function evaluations, are typically much more accurate than comparable Newton-Cotes formulas. It can be shown for Gauss integration [20] that

$$\int_{-1}^{1} f(x) \, dx = \sum_{j=1}^{n} w_j f(x_j) + E(f)$$

where the integration error term is representable as

$$E = \frac{-n(n-1)^3 2^{2n-1}[(n-2)!]^4}{(2n-1)[(2n-2)!]^3} f^{(2n-2)}(\xi) \qquad -1 < \xi < 1$$

The base points in the Gauss formula of order n are the roots of the Legendre polynomial of order n and the weight factors are expressible concisely in terms of the base points. The quadrature error term for an n-point formula involves the integrand derivative of order $2n - 2$, which implies a zero error for any polynomial of order $2n - 1$ or lower. The coefficient of the derivative term in E decreases very rapidly with increasing n. For example, $n = 10$ gives a coefficient of $2.03E-21$. Thus, a function having well behaved high order derivatives can be integrated accurately with a formula of fairly low order.

The base points x_j are all distinct, lie between -1 and 1, and are the eigenvalues of a symmetric tridiagonal matrix [25] which can be analyzed efficiently with function **eigen**. Furthermore, the weight factors are simply twice the squares of the first components of the orthonormalized eigenvectors. Because **eigen** returns orthonormalized eigenvectors for symmetric matrices having distinct eigenvalues, only lines 33-35 in function **gaussint** shown below are needed to compute base points and weight factors. The computing time to evaluate x and w on an Intel-based microcomputer is essentially negligible for $n = 10$ and $n = 20$. Function **gaussint** returns base points and weight factors as well as an integral value if a function is specified.

Using **gaussint** to integrate $\exp(x)$ from 0 to 1 with $n = 10$ yields 15 digit accuracy. Even when **gaussint** is employed to integrate the singular function $1/\sqrt{x}$ from 0 to 1 with $n = 100$, the integration error is less than 0.5%. This example involving a singular integrand does not imply that Gauss integration should be applied casually to evaluate singular integrals. In such cases considerable caution is needed. However, MATLAB functions **quad** and **quad8** give infinite results for this case.

Although the computation times cited above to obtain **bp** and **wf** are small, when a formula of a particular order is used many times, obtaining these values once and using them repeatedly can improve efficiency. The composite integration formula **gquad10** employed below in the computation comparisons uses this idea by including the base points and weight factors as data values. A similar six-point formula, **gquad6**, is provided later for use in geometry problems.

```
function [bp,wf,val]=gaussint(n,a,b,fun)
%
% [bp,wf,val]=gaussint(n,a,b,fun)
% -------------------------------
% This function generates Gaussian base points
% and weight factors of arbitrary order. It also
% integrates a function over arbitrary limits if
% appropriate input values are given. A Gauss
% formula of order n integrates any polynomial
% having degree up to 2*n-1 exactly. The theory
% employed here appears on page 93 of
% 'Methods of Numerical Integration' by
% Philip Davis and Philip Rabinowitz.
%
% n     - order of the formula (default is 20)
% a,b   - integration limits applicable when a
%         function name fun is input
% fun   - a character string giving the name of
%         the function to be integrated. This
%         argument is omitted if only base
%         points and weight factors are needed.
%
% bp,wf - Gauss base points and weight factors
% val   - integral of fun over limits from a
%         to b if a function name and limits
%         are given
%
% User m functions called:  none
%-------------------------------------------------

if nargin < 4, fun=[]; end;
if nargin ==0, n=20; end
u=(1:n-1)./sqrt((2*(1:n-1)).^2-1);
[vc,bp]=eig(diag(u,-1)+diag(u,1));
[bp,k]=sort(diag(bp)); wf=2*vc(1,k)'.^2;
if isempty(fun)
  val=[];
else
  x=(a+b)/2+(b-a)/2*bp; f=feval(fun,x)*(b-a)/2;
  val=wf(:)'*f(:);
end
```

Function	Method	# of function values	cpu seconds	percent error	t/tgquad10	Notes
sqrt(x)	quad	80	0.196	6E-4	30.2	1
	quad8	177	0.213	4E-6	32.8	
	gquad10	100	0.007	4E-4	1	
log(x)	quad	∞ (16)	-	∞	-	2
	quad8	∞ (17)	-	∞	-	2
	gquad10	100	0.007	6E-2	1	
humps(x)	quad	120	0.226	4E-5	26.6	
	quad8	65	0.058	5E-5	6.8	
	gquad10	100	0.009	4E-12	1	
exp(10x) cos(10πx)	quad	492	0.923	2E-4	102	
	quad8	129	0.109	8E-8	12.1	
	gquad10	200	0.009	3E-13	1	
cos[20πx) - 20 sin(πx)]	quad	1100	2.230	2E-5	262	1
	quad8	289	0.247	6E-6	29.1	
	gquad10	200	0.009	3E-13	1	
Note 1: warnings						
Note 2: infinite						

Table 5.1. Comparative Results from Three Integrators

5.3 Examples Comparing Different Integration Methods

We conclude this section by comparing answers obtained when **quad**, **quad8**, and **gquad10** are used to integrate several difficult integrands over limits of zero to one. The examples include a) $\sqrt{x}$, which has infinite slope at $x = 0$; b) $\log(x)$, which is infinite at $x = 0$ but is still integrable; c) **humps**(x), which is a test function provided in MATLAB and has complex poles near $x = 0.3$ and $x = 0.9$, d) $e^{10x} \cos(10\pi x)$, and e) is a highly oscillatory function, the integral of which gives the Bessel function $J_{20}(20)$. Table 5.1 summarizes results obtained and shows function value counts, cpu times, error percentages, and relative computation times (compared with **gquad10**).

Table 5.1 indicates that **quad** gave numerous singularity warnings for **sqrt(x)**. Both **quad** and **quad8** failed on **log(x)**. The integrator **gquad10** gave good results on all problems and was between 6 and 262 times faster than the other integrators. The authors have found high order Gauss quadrature to be a useful tool in many applications. Readers may wish to consider Gauss quadrature as a useful alternative to **quad** and **quad8**.

5.3.1   Example: Comparison of Integration Methods

Script File runcases

```
 1: % Example:  runcases
 2: % ~~~~~~~~~~~~~~~~~~
 3: % Comparison of numerical integration results
 4: % produced by quad, quad8, and gquad10. The
 5: % functions integrated over limits zero to
 6: % one are: sqrt(x), log(x), humps(x),
 7: % exp(10*x)*cos(10*x), and
 8: % cos(20*pi*x+20*sin(pi*x)). Successive rows
 9: % in matrix r contain the following values.
10: %
11: % row           result
12: %  1      exact integral values
13: %  2      cputime using quad
14: %  3      cputime using quad8
15: %  4      cputime using gquad10
16: %  5      time_quad./time_gquad10
17: %  6      time_quad8./time_gquad10
18: %  7      integrals using quad
19: %  8      integrals using quad8
20: %  9      integrals using gquad10
21: % 10      number of function values by quad
22: % 11      number of function values by quad8
23: % 12      number of function values by gquad10
24: % 13      percent error using quad
25: % 14      percent error using quad8
26: % 15      percent error using gquad10
27: %
28: % User m functions called:
29: %    besf, expc, hmpf, logf, sqtf,
30: %    gquad10
31: %------------------------------------------------
32:
33: clear
34: global nval_
35:
36: % results are saved in matrix r
37: r=zeros(15,5); ng10=[10,10,10,20,20]; m=20;
38:
39: % names of functions integrated for 0 to 1
40: a=['sqtf';'logf';'hmpf';'expc';'besf'];
```

```
41:
42: % exact values of the integrals
43: r(1,:)=[2/3, -1, gquad10('hmpf',0,1,100), ...
44:       real((exp(10+10*pi*i)-1)/(10+10*pi*i)),...
45:       bessel(20,20)];
46:
47: % integrate using quad, quad8 and gquad10
48: for k=1:5
49:    ak=['''',a(k,:),'''']; nk=num2str(ng10(k));
50:    str1=['r(7,k)=quad(',ak,',0,1);'];
51:    str2=['r(8,k)=quad8(',ak,',0,1);'];
52:    str3=['r(9,k)=gquad10(',ak,',0,1,',nk,');'];
53:
54: % integrate using quad
55:    disp('*** USING QUAD ***')
56:    nval_=0; r(2,k)=cputime;
57:       for j=1:m, eval(str1); end
58:    r(2,k)=cputime-r(2,k); r(10,k)=nval_;
59:
60: % integrate using quad8
61:    disp('*** USING QUAD8 ***')
62:    nval_=0; r(3,k)=cputime;
63:       for j=1:m, eval(str2); end
64:    r(3,k)=cputime-r(3,k); r(11,k)=nval_;
65:
66: % integrate using gquad10
67:    disp('*** USING GQUAD10 ***')
68:    nval_=0; r(4,k)=cputime;
69:       for j=1:m, eval(str3); end
70:    r(4,k)=cputime-r(4,k); r(12,k)=nval_;
71: end
72:
73: % time ratios relative to gquad10
74: r(5,:)=r(2,:)./r(4,:); r(6,:)=r(3,:)./r(4,:);
75:
76: % percent errors of different methods
77: r(13,:)=100*abs(1-r(7,:)./r(1,:));
78: r(14,:)=100*abs(1-r(8,:)./r(1,:));
79: r(15,:)=100*abs(1-r(9,:)./r(1,:));
80:
81: exaval=r(1,:);tquad=r(2,:)/m;tquad8=r(3,:)/m;
82: tgqd10=r(4,:)/m;quadrat=r(5,:);quad8rat=r(6,:);
83: qdval=r(7,:);qd8val=r(8,:);gqd10vl=r(9,:);
84: qdnf=r(10,:)/m;qd8nf=r(11,:)/m;gqd10nf=r(12,:)/m;
85: erqd=r(13,:);erqd8=r(14,:);ergqd10=r(15,:);
```

```
86:
87: % Print results obtained from the test cases
88:
89: format long
90: exaval,
91: qdval,    qd8val,    gqd10vl
92: erqd,     erqd8,     ergqd10
93: tquad,    tquad8,    tgqd10
94: qdnf,     qd8nf,     gqd10nf
95: quadrat, quad8rat
96: format short
97:
98: disp('All Done')
```

Function besf

```
1: function y=besf(x)
2: %
3: % y=besf(x)
4: % ~~~~~~~~~~
5: %
6: % Companion function for script runcases.
7: %
8: % User m functions called:  none
9: %-----------------------------------------------
10:
11: global nval_
12: nval_=nval_+length(x);
13: % integrand defining bessel(20,20)
14: y=cos(20*pi*x-20*sin(pi*x));
```

Function expc

```
1: function y=expc(x)
2: %
3: % y=expc(x)
4: % ~~~~~~~~~~
5: %
6: % Companion function for script runcases.
7: %
8: % User m functions called:  none
9: %-----------------------------------------------
```

```
10:
11: global nval_
12: nval_=nval_+length(x);
13: y=exp(10*x).*cos(10*pi*x);
```

Function hmpf

```
1: function y = hmpf(x)
2: %
3: % y = hmpf(x)
4: % ~~~~~~~~~~~~
5: %
6: % Companion function for script runcases.
7: %
8: % User m functions called:  none
9: %------------------------------------------------
10:
11: global nval_
12: nval_=nval_+length(x);
13: y=1 ./((x-.3).^2+.01)+1 ./((x-.9).^2+.04)-6;
```

Function logf

```
1: function y=logf(x)
2: %
3: % y=logf(x)
4: % ~~~~~~~~~~
5: %
6: % Companion function for script runcases.
7: %
8: % User m functions called:  none
9: %------------------------------------------------
10:
11: global nval_
12: nval_=nval_+length(x);
13: y=log(x);
```

Function sqtf

```
1: function y=sqtf(x)
```

```
2: %
3: % y=sqtf(x)
4: % ~~~~~~~~~
5: % Companion function for script runcases.
6: %
7: % User m functions called:  none
8: %---------------------------------------------------
9:
10: global nval_
11: nval_=nval_+length(x);
12: y=sqrt(x);
```

Function gquad

```
1:
2: function area= ...
3:          gquad(fun,xlow,xhigh,mparts,bp,wf)
4: %
5: % area = gquad (fun,xlow,xhigh,mparts,bp,wf)
6: % ~~~~~~~~~~~~~~~~~~~~~~~~~~~~~~~~~~~~~~~~~~~
7: % This function evaluates the integral of an
8: % externally defined function fun(x) between
9: % limits xlow and xhigh. The numerical
10: % integration is performed using a composite
11: % Gauss integration rule.  The whole interval
12: % is divided into mparts subintervals and the
13: % integration over each subinterval is done
14: % with an nquad point Gauss formula which
15: % involves base points bp and weight factors
16: % wf.  The normalized interval of integration
17: % for the bp and wf constants is -1 to +1.
18: % The algorithm is described by the summation
19: % relation
20: %
21: % x=b
22: % integral(f(x)*dx)=
23: % x=a
24: %
25: %             j=n k=m
26: %         d1*sum sum(wf(j)*fun(a1+d*k+d1*bp(j)))
27: %             j=1 k=1
28: %
29: % where bp are base points, wf are
```

```
30: % weight factors
31: %         m = mparts, and n = length(bp) and
32: %         d = (b-a)/m, d1 = d/2, a1 = a-d1
33: %
34: % The base points and weight factors must
35: % first be generated by a call to gaussint of
36: % the form
37: %
38: % [bp,wf] = gaussint(nquad)
39: %
40: % fun          - function to be integrated
41: % xlow,xhigh   - integration limits
42: % mparts       - number of integration intervals
43: %                used
44: % bp,wf        - base points and weight factors
45: %                obtained by an initial call
46: %                to grule
47: % area         - numerically approximated
48: %                integral value
49: %
50: % User m functions called:  argument fun
51: %-----------------------------------------------
52:
53: bp=bp(:); wf=wf(:);
54: d=(xhigh-xlow)/mparts; d2=d/2;
55: nquad=length(bp);
56: x=(d2*bp)*ones(1,mparts)+ ...
57:   (d*ones(nquad,1))*(1:mparts);
58: x=x(:)+(xlow-d2);
59: fv=feval(fun,x);
60: wv=wf*ones(1,mparts);
61: area=d2*(wv(:)'*fv(:));
62:
```

Function gquad10

```
1: function area=gquad10(fun,xlow,xhigh,mparts)
2: %
3: % area = gquad10(fun,xlow,xhigh,mparts)
4: % ~~~~~~~~~~~~~~~~~~~~~~~~~~~~~~~~~~~~~~~
5: % This function determines the area under an
6: % externally defined function fun(x) between
7: % limits xlow and xhigh. The numerical
```

```
 8: % integration is performed using a composite
 9: % gauss integration rule.  The whole interval
10: % is divided into mparts subintervals and the
11: % integration over each subinterval is done
12: % with a ten point Gauss formula which
13: % involves base points bp and weight factors
14: % wf.  The normalized interval of integration
15: % for the bp and wf constants is -1 to +1.
16: % The algorithm is structured in terms of a
17: % parameter mquad = 6 which can be changed
18: % along with bp and wf to accommodate a
19: % different order formula.  The composite
20: % algorithm is described by the following
21: % summation relation:
22: %
23: % x=b
24: % integral( f(x)*dx ) =
25: % x=a
26: %
27: %          j=n k=m
28: %      d1*sum sum( wf(j)*fun(a1+d*k+d1*bp(j)) )
29: %          j=1 k=1
30: %
31: % where d = (b-a)/m, d1 = d/2, a1 = a-d1,
32: %        m = mparts, and n = nquad.
33: %
34: % User m functions called:  argument fun
35: %-------------------------------------------------
36:
37: if nargin ==3, mparts=1; end
38:
39: % The weight factors are
40: wf=[  6.66713443086879e-02; ...
41:      14.94513491505806e-02; ...
42:      21.90863625159821e-02; ...
43:      26.92667193099964e-02; ...
44:      29.55242247147528e-02 ];
45: wf=[wf; wf([5:-1:1])];
46:
47: % The base points are
48: bp=[-97.39065285171718e-02; ...
49:     -86.50633666889846e-02; ...
50:     -67.94095682990245e-02; ...
51:     -43.33953941292472e-02; ...
52:     -14.88743389816312e-02 ];
```

```
53: bp=[bp;-bp([5:-1:1])];
54:
55: d=(xhigh-xlow)/mparts;
56: d2=d/2; nquad=length(bp);
57: x=(d2*bp)*ones(1,mparts)+ ...
58:    (d*ones(nquad,1))*(1:mparts);
59: x=x(:)+(xlow-d2);
60: fv=feval(fun,x); wv=wf*ones(1,mparts);
61:
62: area=d2*(wv(:)'*fv(:));
```

5.4   Evaluating a Multiple Integral

Gauss integration can be used effectively to evaluate multiple integrals having variable limits. For instance, let us consider the instance typified by the following triple integral

$$I = \int_{c_1}^{c_2} \int_{b_1(z)}^{b_2(z)} \int_{a_1(y,z)}^{a_2(y,z)} F(x, y, z)\, dx\, dy\, dz$$

This integral can be changed into one with constant limits by the substitutions

$$z = c_p + c_m u \qquad\qquad -1 \le u \le 1$$

$$c_p = \frac{c_2 + c_1}{2} \qquad\qquad c_m = \frac{c_2 - c_1}{2}$$

$$y = b_p + b_m t \qquad\qquad -1 \le t \le 1$$

$$b_p = \frac{b_2 + b_1}{2} \qquad\qquad b_m = \frac{b_2 - b_1}{2}$$

$$x = a_p + a_m s \qquad\qquad -1 \le s \le 1$$

$$a_p = \frac{a_2 + a_1}{2} \qquad\qquad a_m = \frac{a_2 - a_1}{2}$$

and yields

$$I = \int_{-1}^{1} \int_{-1}^{1} \int_{-1}^{1} c_m b_m a_m f(s, t, u)\, ds\, dt\, du$$

where

$$f(s, t, u) = F(a_p + a_m s, b_p + b_m t, c_p + c_m u)$$

$$a_m = a_m(y, z) = a_m(b_p + b_m t, c_p + c_m u)$$

$$b_m = b_m(z) = b_m(c_p + c_m u)$$

Thus, the integral has the form

$$I = \int_{-1}^{1} \int_{-1}^{1} \int_{-1}^{1} G(s, t, u)\, ds\, dt\, du$$

where

$$G = c_m b_m a_m f$$

Performing the integration over each limit using an n-point quadrature formula with weight factors w_i and base points x_i yields

$$I = \sum_{k=1}^{n} \sum_{j=1}^{n} \sum_{i=1}^{n} w_k w_j w_i G(x_i, x_j, x_k)$$

A function allowing an integrand and integration limits of general form was developed. Let us consider an example where the inertial moment of a sphere having unit radius, unit mass density, and centered at $(0, 0, 0)$ is to be obtained about an axis through $x = 2, y = 0$, parallel to the z-axis. The related integral

$$I = \int_{-1}^{1} \int_{-\sqrt{1-z^2}}^{\sqrt{1-z^2}} \int_{-\sqrt{1-y^2-z^2}}^{\sqrt{1-y^2-z^2}} \left[(x - 2)^2 + y^2 \right] dx\, dy\, dz$$

has a value of $88\pi/5$. Shown below is a function **quadit3d** and related limit and integrand functions. The function **triplint(n)** computes the ratio of the numerically integrated function to the exact result. The function specification **triplint(20)** yields a value of 1.000067. Even though the triple integration procedure is not computationally very fast, it is nevertheless robust enough to produce accurate results when a sufficiently high integration order is chosen.

5.4.1   Example: Evaluating a Multiple Integral

Function triplint

```
1: function val=triplint(n)
2: %
3: % val=triplint(n)
4: % ~~~~~~~~~~~~~~~~
5: % Triple integration example on inertial
6: % moment of a sphere.
7: %
8: % User m functions called:  fsphere, bs1, bs2,
9: %                           as1, as2
10: %-----------------------------------------------
11:
12: val=quadit3d('fsphere',[-1,1],'bs1','bs2',...
13:              'as1','as2',n)/(88*pi/15);
```

Function fsphere

```
1: function v=fsphere(x,y,z)
2: %
3: % v=fsphere(x,y,z)
4: % ~~~~~~~~~~~~~~~~
5: % Integrand.
6: %
7: % User m functions called:  none
8: %-----------------------------------------------
9:
10: v=(x-2).^2+y.^2;
```

Function as1

```
1: function x=as1(y,z)
2: %
3: % x=as1(y,z)
4: % ~~~~~~~~~~
5: % Lower x integration limit.
6: %
7: % User m functions called:  none
8: %-----------------------------------------------
```

```
 9:
10: x=-sqrt(1-y.^2-z.^2);
```

Function as2

```
 1: function x=as2(y,z)
 2: %
 3: % x=as2(y,z)
 4: % ~~~~~~~~~~
 5: % Upper x integration limit.
 6: %
 7: % User m functions called:  none
 8: %--------------------------------------------
 9:
10: x=sqrt(1-y.^2-z.^2);
```

Function bs1

```
 1: function y=bs1(z)
 2: %
 3: % y=bs1(z)
 4: % ~~~~~~~~
 5: % Lower y integration limit.
 6: %
 7: % User m functions called:  none
 8: %--------------------------------------------
 9:
10: y=-sqrt(1-z.^2);
```

Function bs2

```
 1: function y=bs2(z)
 2: %
 3: % y=bs2(z)
 4: % ~~~~~~~~~~
 5: % Upper y integration limit.
 6: %
 7: % User m functions called:  none
 8: %--------------------------------------------
 9:
```

```
10: y=sqrt(1-z.^2);
```

Function quadit3d

```
1: function s = quadit3d(f,c,b1,b2,a1,a2,w)
2: %
3: % s = quadit3d(f,c,b1,b2,a1,a2,w)
4: % ~~~~~~~~~~~~~~~~~~~~~~~~~~~~~~~~~
5: % This function computes the iterated integral
6: %
7: % s = integral(...
8: %       f(x,y,z), x=a1..a2, y=b1..b2, z=c1..c2)
9: %
10: % where a1 and a2 are functions of y and z, b1
11: % and b2 are functions of z, and c is a vector
12: % containing constant limits on the z variable.
13: % Hence, as many as five external functions may
14: % be involved in the call list. For example,
15: % when the integrand and limits are:
16: %
17: % f  = x.^2+y^2+z^2
18: % a2 = sqrt(4-y^2-z^2)
19: % a1 = -a2
20: % b2 = sqrt(4-z^2)
21: % b1 = -b2
22: % c = [-2,2]
23: %
24: % then the exact value is 128*pi/5.
25: % The approximation produced from a 12 point
26: % Gauss formula is accurate within .05 percent.
27: %
28: % f      - a function f(x,y,z) which must return
29: %          a vector value when x is a vector,
30: %          and y and z are scalar.
31: % a1,a2 - integration limits on the x variable
32: %          which may specify names of functions
33: %          or have constant values. If a1 is a
34: %          function it should have a call list
35: %          of the form a1(y,z). A similar form
36: %          applies to a2.
37: % b1,b2 - integration limits on the y variable
38: %          which may specify functions of z or
39: %          have constant values.
```

```
40: % c       - a vector defined by c=[c1,c2] where
41: %           c1 and c2 are fixed integration
42: %           limits for the z direction.
43: % w       - this argument defines the quadrature
44: %           formula used. It has the following
45: %           three possible forms. If w is omitted,
46: %           a Gauss formula of order 12 is used.
47: %           If w is a positive integer n, a Gauss
48: %           formula of order n is used. If w is an
49: %           n by 2 matrix, w(:,1) contains the base
50: %           points and w(:,2) contains the weight
51: %           factors for a quadrature formula over
52: %           limits -1 to 1.
53: %
54: % s       - the numerically evaluated integral
55: %
56: % User m functions called:  gaussint
57: %-------------------------------------------------
58:
59: if nargin<7
60: % function gaussint generates base points
61: % and weight factors
62:    n=12; [x,W]=gaussint(n);
63: elseif size(w,1)==1 & size(w,2)==1
64:    n=w; [x,W]=gaussint(n);
65: else
66:    n=size(w,1); x=w(:,1); W=w(:,2);
67: end
68: s=0; cp=(c(1)+c(2))/2; cm=(c(2)-c(1))/2;
69: for k=1:n
70:    zk=cp+cm*x(k);
71:    if ischar(b1)
72:       B1=feval(b1,zk);
73:    else
74:       B1=b1;
75:    end
76:    if ischar(b2)
77:       B2=feval(b2,zk);
78:    else
79:       B2=b2;
80:    end
81:    Bp=(B2+B1)/2; Bm=(B2-B1)/2; sj=0;
82:    for j=1:n
83:       yj=Bp+Bm*x(j);
84:       if ischar(a1)
```

```
85:        A1=feval(a1,yj,zk);
86:     else
87:        A1=a1;
88:     end
89:     if ischar(a2)
90:        A2=feval(a2,yj,zk);
91:     else
92:        A2=a2;
93:     end
94:     Ap=(A2+A1)/2; Am=(A2-A1)/2;
95:     fval=feval(f, Ap+Am*x, yj, zk);
96:     si=fval(:).'*W(:); sj=sj+W(j)*Am*si;
97:   end
98:   s=s+W(k)*Bm*sj;
99: end
100: s=cm*s;
```

5.5   Line Integrals for Geometric Properties of Plane Areas

The next article applies Gauss quadrature and spline interpolation to analyze a practical problem. Engineering applications often require computation of geometrical properties of plane areas. The main properties needed are the area,

$$A = \int \int dx\, dy$$

the first moments of area,

$$A_x = \int \int x\, dx\, dy \qquad A_y = \int \int y\, dx\, dy$$

and the three inertial moments,

$$A_{xx} = \int \int x^2\, dx\, dy \qquad A_{xy} = \int \int xy\, dx\, dy$$

$$A_{yy} = \int \int y^2\, dx\, dy$$

This problem is important enough that we will analyze it for regions having one or several parts which can be solid or can contain holes. Two types of solutions are presented. The first handles any region bounded by straight lines and circular arcs. The second solution deals with more general shapes where the boundary is approximated by a spline curve.

The desired area properties all have the form

$$I_{nm} = \int \int x^n y^m\, dx\, dy$$

where n and m are integers. This two-dimensional integral can be converted into a one-dimensional integral evaluated over the boundary curve.[1] The one-dimensional integrals can be either computed exactly or numerically. Consider the general area in Figure 5.1 bounded by a curve L described in parametric form as

$$(x(t), y(t)) \qquad a \le t \le b$$

The area integral can be transformed into a line integral by using polar coordinates in the form

$$x = \rho \cos \theta \qquad y = \rho \sin \theta$$

$$\rho = (x^2 + y^2)^{\frac{1}{2}} \qquad \theta = \tan^{-1}\left(\frac{y}{x}\right)$$

---

[1] Wilson and Farrior [112] published an early paper on this topic.

$$dA = \rho \, d\rho \, d\theta \qquad d\theta = \frac{x \, dy - y \, dx}{x^2 + y^2}$$

These relations yield

$$I_{nm} = \int_\theta \int_0^{r(\theta)} (\rho \cos \theta)^n (\rho \sin \theta)^m \rho \, d\rho \, d\theta$$

$$I_{nm} = \int_\theta (\cos \theta)^n (\sin \theta)^m \int_0^r \rho^{n+m+1} d\rho \, d\theta$$

$$I_{nm} = \frac{1}{n+m+2} \int_\theta [\rho \cos \theta]^n [\rho \sin \theta]^m \rho^2 d\theta$$

Returning to cartesian coordinates gives

$$\int \int x^n y^m dx \, dy = \frac{1}{n+m+2} \int_L x^n y^m (x \, dy - y \, dx)$$

This formula is good for non-negative integer values of n and m. The formula even works for negative indices provided $x = 0$ is outside L for negative n, and $y = 0$ is outside L for negative m. The important cases include:

- $n = m = 0$ for area,

- $n = 1$, $m = 0$ or $n = 0$, $m = 1$ for first moments, and

- $n = 2$, $m = 0$, or $n = m = 1$, or $n = 0$, $m = 2$ for inertial moments.

A general boundary shape may have several parts each of which can be represented by a parametric equation. For example, a typical line segment from (x_1, y_1) to (x_2, y_2) can be expressed as

$$x = x_1 + (x_2 - x_1)t \qquad y = y_1 + (y_2 - y_1)t \qquad 0 \le t \le 1$$

Since

$$x \, dy - y \, dx = (x_1 y_2 - x_2 y_1)dt = g_{12}dt$$

the integral contributions from the line segment are

$$\frac{1}{2} \int x \, dy - y \, dx = \frac{1}{2}(x_1 y_1 - y_1 x_2) = \frac{1}{2}g_{12}$$

$$\frac{1}{3} \int x(x \, dy - y \, dx) = \frac{1}{6}(x_1 + x_2)g_{12}$$

$$\frac{1}{3} \int y(x \, dy - y \, dx) = \frac{1}{6}(y_1 + y_2)g_{12}$$

$$\frac{1}{4} \int x^2(x \, dy - y \, dx) = \frac{1}{12}(x_1^2 + x_1 x_2 + x_2^2)g_{12}$$

$$\frac{1}{4}\int xy(x\ dy - y\ dx) = \frac{1}{24}(2x_1y_1 + x_1y_2 + x_2y_1 + 2x_2y_2)g_{12}$$

$$\frac{1}{4}\int y^2(x\ dy - y\ dx) = \frac{1}{12}(y_1^2 + y_1y_2 + y_2^2)g_{12}$$

Similar formulas can be developed for a circular arc described as

$$x = x_0 + r\cos t \qquad y = y_0 + r\sin t \qquad \theta_1 \le t \le \theta_2$$

Then

$$x\ dy - y\ dx = r_0\left[r_0 + x_0\cos t - y_0\sin t\right]dt$$

and the integral pertaining to area is

$$\frac{1}{2}\int (x\ dy - y\ dx) = r_0\left[r_0\langle\theta_2 - \theta_1\rangle + x_0\langle s_1 - s_2\rangle + y_0\langle c_1 - c_2\rangle\right]$$

where

$$s_1 = \sin\theta_1 \qquad c_1 = \cos\theta_1 \qquad s_2 = \sin\theta_2 \qquad c_2 = \cos\theta_2$$

The other integrals for first and second area moments are similar but algebraically tedious. Those expressions are not repeated here since they are included in the MATLAB functions which compute exactly the properties of any region bounded by straight lines and circular arcs. The realistic geometry described by circular arcs and straight lines in Figure 5.2 involves half of an annulus placed over a hollow square. It will be used later to compare exact results with those obtained when the boundary is approximated by a spline. However, it will be helpful to first consider a simple example to demonstrate some ideas.

5.5.1 Geometry Example Using a Simple Spline Interpolated Boundary

Let us consider a simple case before proceeding with the treatment of general spline interpolated boundaries. Suppose a set of data points (x_i, y_i), $1 \le i \le n$, are to be connected by a smooth closed curve having parametric form $x(t), y(t), a \le t \le b$. The area, centroidal coordinates, and inertial moments resulting from evaluation of the six basic integrals are also sought. To get a smooth curve through the data we choose points $[x_n, x_1, x_2, \cdots, x_n, x_1, x_2]$ and $[y_n, y_1, y_2, \cdots, y_n, y_1, y_2]$ which are spline interpolated (using **spline** from MATLAB) as functions of $t = [0, 1, 2, \cdots, n + 2]$. The end points added to the data make the curve close and give a smoothly turning tangent. The remaining calculations needed to get area properties are to evaluate $x'(t)$ and $y'(t)$ by a method such as finite differences, and then evaluate the integrals by using an algorithm such as Simpson's rule. The following short program is adequate to handle smooth shapes. An example was run to approximate a rotated ellipse using 20 data points as shown in Figure 5.3. The maximum error in computed geometrical properties was about 0.4%. Although this simple program illustrates the general procedure for computing area properties, dealing with complex shapes having sharp corners takes additional effort. More careful attention is needed regarding boundary slope representation, derivative evaluation, and numerical integration. The general spline approximation problem is discussed next.

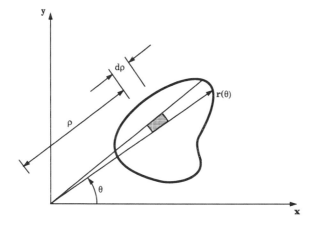

Figure 5.1. General Two-Dimensional Area

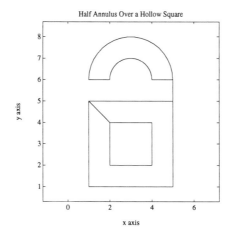

Figure 5.2. Half Annulus Over a Hollow Square

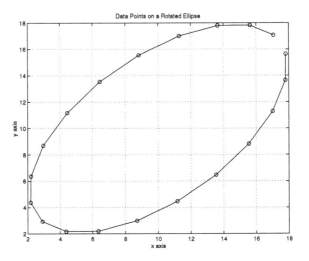

Figure 5.3. Rotated Ellipse

MATLAB Example

Script File elipprop

```
 1: % Example: elipprop
 2: % ~~~~~~~~~~~~~~~~~~~~
 3: % Compute the area, centroidal coordinates,
 4: % and inertial moments of a rotated ellipse
 5: % centered at (cx,cy) and having diameters
 6: % of 2*rx and 2*ry
 7: %
 8: % User m functions required:
 9: %    allprop, dife, simpsum, pauz, genprint
10: %-----------------------------------------------
11:
12: cx=10; cy=10; rx=10; ry=5;
13: np=20; th=2*pi/np*(0:np-1);
14: z=(rx*cos(th)+i*ry*sin(th))* ...
15:    exp(i*pi/4)+(cx+i*cy);
16: xdat=real(z); ydat=imag(z);
17:
18: clf; plot(xdat,ydat,'-',xdat,ydat,'o');
19: xlabel('x axis'); ylabel('y axis'); grid;
20: title('Data Points on a Rotated Ellipse');
21: figure(gcf); disp(' ');
22: disp('Press [Enter] to continue');
23: disp(' '); pauz; %genprint('ellipse');
24:
25: [area,xb,yb,axx,axy,ayy]= ...
26:    allprop(xdat,ydat,101);
27:
28: fprintf('area   = %7.4f\n',area);
29: fprintf('xcentr = %7.4f\n',xb);
30: fprintf('ycentr = %7.4f\n',yb);
31: fprintf('axx    = %7.4f\n',axx);
32: fprintf('axy    = %7.4f\n',axy);
33: fprintf('ayy    = %7.4f\n',ayy);
34:
35: % Exact values would give unity
36: % for aa, xx and yy
37: aa=area/(pi*rx*ry); xx=xb/cx; yy=yb/cy;
38:
39: fprintf(...
40:    '\napprox_area / exact_area   = %7.4f\n',aa);
```

```
41: fprintf(...
42:   '\napprox_xcentr / exact_xcntr = %7.4f\n',xx);
43: fprintf(...
44:   '\napprox_ycentr / exact_ycntr = %7.4f\n',yy);
45: disp(' '); disp('All Done');
```

Function dife

```
 1: function ydif=dife(y,h)
 2: %
 3: % ydif=dife(y,h)
 4: % ~~~~~~~~~~~~~~~
 5: % This function differentiates data evenly
 6: % spaced with the argument increment being h.
 7: %
 8: % User m functions called:  none
 9: %-----------------------------------------------
10:
11: % Use central differences for interior points
12: y=y(:); n=length(y);
13: ydif=( y(3:n)-y(1:n-2) )/2;
14:
15: % Use forward and backward differences at
16: % the end points
17: ydif=[(-3*y(1)+4*y(2)-y(3))/2; ydif;
18:       (y(n-2)-4*y(n-1)+3*y(n))/2];
19:
20: % Default value of h is one
21: if nargin ==2, ydif=ydif/h; end
```

Function simpsum

```
 1: function asmp=simpsum(y,a,b)
 2: %
 3: % asmp=simpsum(y,a,b)
 4: % ~~~~~~~~~~~~~~~~~~~~
 5: % Simpson's rule for a matrix
 6: % having an odd number of rows corresponding to
 7: % argument values evenly spaced from a to b.
 8: %
 9: % User m functions called:  none
10: %-----------------------------------------------
```

```
11:
12: [n,nc]=size(y); h=(b-a)/(n-1);
13:
14: % If the number of function values is even,
15: % then pad with a zero at the bottom. This
16: % can affect accuracy adversely, so taking n
17: % odd is best.
18: if n==2*fix(n/2), n=n+1; y=[y;0]; b=b+h; end
19:
20: % Accumulate the summation
21: asmp=(b-a)/(3*(n-1))*(y(1,:)+y(n,:)+...
22:         4*sum(y(2:2:n-1,:))+2*sum(y(3:2:n-2,:)));
```

Function allprop

```
 1: function [area,xcentr,ycentr,axx,axy,ayy]=...
 2:               allprop(xdat,ydat,ntrp)
 3: %
 4: % [area,xcentr,ycentr,axx,axy,ayy]=...
 5: %               allprop(xdat,ydat,ntrp)
 6: % ~~~~~~~~~~~~~~~~~~~~~~~~~~~~~~~~~~~~~~~~~
 7: % This function computes area, centroidal
 8: % coordinates, and second moments of area
 9: % (inertial moments) for a spline curve passed
10: % through points (xdat,ydat).
11: %
12: % area            - area enclosed by the curve
13: % xcentr,ycentr   - centroidal coordinates
14: % axx             - integral( x^2*d(area) )
15: % axy             - integral( y*x*d(area) )
16: % ayy             - integral( y^2*d(area) )
17: %
18: % User m functions called:  dife, simpsum
19: %-------------------------------------------------
20:
21: if nargin < 3, ntrp=101; end; nd=length(xdat);
22: xdat=xdat(:); ydat=ydat(:);
23:
24: % Make curve close and extend the end values
25: xdat=[xdat(nd);xdat;xdat(1:2)];
26: ydat=[ydat(nd);ydat;ydat(1:2)];
27:
28: % Be sure to use an odd number
```

```
29:  % of quadrature points
30:  if ntrp==2*fix(ntrp/2), ntrp=ntrp+1; end
31:  h=(nd-1)/ntrp; t=1:h:nd+1; tdat=0:nd+2;
32:
33:  % Spline interpolate to get integrand values
34:  x=spline(tdat,xdat,t); x=x(:);
35:  y=spline(tdat,ydat,t); y=y(:);
36:
37:  % Approximate derivatives by
38:  % finite differences
39:  xd=dife(x,h); yd=dife(y,h);
40:
41:  % Compute area by Simpson's rule
42:  a=x.*yd-y.*xd;
43:  am=zeros(length(a),6); am(:,1)=a;
44:  am(:,2)=a.*x; am(:,3)=a.*y;
45:  am(:,4)=am(:,2).*x;
46:  am(:,5)=am(:,2).*y; am(:,6)=am(:,3).*y;
47:
48:  % Calculate centroidal coordinates
49:  v=simpsum(am,1,nd+1); area=v(1)/2;
50:  xcentr=v(2)/(3*area); ycentr=v(3)/(3*area);
51:  axx=v(4)/4; axy=v(5)/4; ayy=v(6)/4;
```

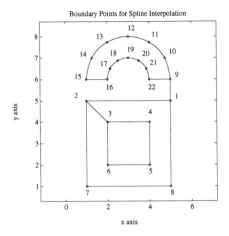

Figure 5.4. Boundary Points for Spline Interpolation

5.6 Spline Approximation of General Boundary Shapes

In Chapter 4 spline interpolation was used to represent a plane curve by piecewise cubic polynomials. We will briefly review some ideas about splines. Suppose a closed curve is defined by connecting points (x_1, y_1), $\cdots, (x_n, y_n)$ where $x_1 = x_n$, $y_1 = y_n$ is required for the curve to close. At each point (x_j, y_j), a parameter $t_j = j$ is assigned. Functions $x(t)$ and $y(t)$ are determined by spline interpolation using functions **spc** and **spltrp**. Boundary points where slope discontinuities occur, such as the corners of a rectangle, are defined as corner points. The example in Figure 5.4 involving two arcs and ten straight line segments can be approximated well by a spline. Seven points are needed to describe the semicircle accurately. Any spline segment with a corner at each end reduces exactly to a straight line. The boundary curve chosen for spline interpolation has 22 points specified by the point sequence 1-6, 3, 2, 7, 8, 1, 9-22, 9, 1. Furthermore, corners are used at 1-9, 15, 16, and 22. Figure 5.6 displays the merits of corner points by showing the unsatisfactory results produced without the use of corner points. The curve employing corner points, shown in Figure 5.5, approximates well the original geometry involving straight lines and circular arcs.

Let us examine the integrands occurring in the line integrals for area properties of the spline interpolated boundary. Any boundary segment has the form

$$x(t) = a_1t^3 + a_2t^2 + a_3t + a_4 \qquad y(t) = b_1t^3 + b_2t^2 + b_3t + b_4$$

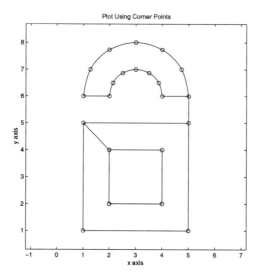

Figure 5.5. Plot Using Corner Points

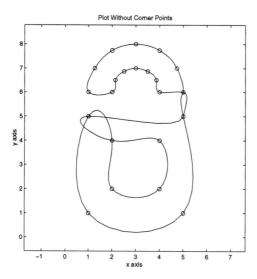

Figure 5.6. Plot Without Corner Points

Evidently, $x(t)y'(t) - y(t)x'(t)$ is a polynomial of degree four. For a typical property such as

$$a_{xy} = \frac{1}{4} \int_L xy(x\,dy - y\,dx) = \frac{1}{4} \int_1^n x(t)y(t)[x(t)y'(t) - y(t)x'(t)]dt$$

the integrand has piecewise polynomial form of degree ten. Because an M-point Gauss formula integrates exactly any polynomial of degree $2M - 1$, taking $M = 6$ is adequate to exactly evaluate all the geometrical properties associated with the spline boundary. Furthermore, a boundary with n segments requires a spline interpolation using $n + 1$ points. Thus, $24n$ spline evaluations are needed to compute x, x', y, and y' values. Using these values to evaluate integrands and accumulate numerical integration sums only requires an additional $120n$ floating point calculation to compute the six area properties of interest.

This chapter concludes with an explanation of the structure of the two geometric property programs. The first program, for geometries defined by circular arcs and straight line segments, evaluates all integrals exactly. The second program, based on spline interpolation, is approximate in the sense that most geometries will not have boundaries exactly defined by a spline. However, no numerical integration error is encountered in the geometric property evaluations because six-point composite Gauss quadrature works exactly for the piecewise polynomial integrands.

Sometimes it is necessary to compute the six area properties for new axes translated and rotated relative to the initial frame. This type of transformation can be described conveniently using complex arithmetic. Let a new frame $\hat{z}$ be obtained by first translating the origin to Z_0 and then rotating the axes by an angle θ counterclockwise. The new axes are related to the initial axes according to

$$\hat{z} = (Z - Z_0)e^{-i\theta} \qquad Z = X + iY \qquad \hat{z} = \hat{x} + i\hat{y}$$

To understand how the area properties transform, start with

$$A = \int\int dx\,dy \qquad AZ = \int\int Z\,dx\,dy \qquad Z_c = \frac{AZ}{A}$$

and introduce

$$P_1 = \frac{1}{2} \int\int Z^2\,dx\,dy \qquad P_2 = \frac{1}{2} \int\int |Z|^2\,dx\,dy$$

which yields the following inertial moment terms.

$$\begin{aligned} A_{xx} &= \mathbf{real}(P_1 + P_2) \\ A_{yy} &= \mathbf{real}(P_1 - P_2) \\ A_{xy} &= \mathbf{imag}(P_1) \end{aligned}$$

Since P_1 and P_2 concisely describe the inertial properties, let us demonstrate how P_1 and P_2 transform under a translation defined by $z = Z - Z_0$, or

$$z_c = Z_c - Z_0$$

$$p_1 = \frac{1}{2} \int \int z^2 \, dx \, dy = P_1 - Z_0 Z_c A + \frac{1}{2} Z_0^2 A$$

$$p_2 = \frac{1}{2} \int \int |z|^2 \, dx \, dy = P_2 - \mathbf{real}(\bar{Z}_0 Z_c) A + \frac{1}{2} |Z|_0^2 A$$

The new inertial moments are

$$\int \int x^2 \, dx \, dy = \mathbf{real}(p_1 + p_2) \qquad \int \int y^2 \, dx \, dy = \mathbf{real}(p_1 - p_2)$$

$$\int \int xy \, dx \, dy = \mathbf{imag}(p_1)$$

After the translation is completed, the axes can be rotated counterclockwise to a new frame $\hat{z}$ defined by $\hat{z} = z e^{-\imath \theta}$. The new properties are $\hat{z}_c = z_c e^{-\imath \theta}$, $\hat{p}_1 = e^{-2\imath \theta} p_1$, and $\hat{p}_2 = p_1$. Consequently,

$$\int \int \hat{x}^2 \, d\hat{x} \, d\hat{y} = \mathbf{real}(\hat{p}_1 + \hat{p}_2)$$

$$\int \int \hat{y}^2 \, d\hat{x} \, d\hat{y} = \mathbf{real}(\hat{p}_1 + \hat{p}_2)$$

$$\int \int \hat{x}\hat{y} \, d\hat{x} \, d\hat{y} = \mathbf{imag}(\hat{p}_1)$$

This completes the presentation of the formulas used in function **shft-prop** to determine area properties produced by translation and rotation of axes.

5.6.1 Program for Exact Properties of Any Area Bounded by Straight Lines and Circular Arcs

Two programs were written to compute geometrical properties of plane areas. The first program, **arearun**, reads a geometry file specifying the straight lines and circular arcs encountered in a counterclockwise circuit around the boundary. Function **crclovsq** defines the general shape plotted in Figure 5.2. The first data item is a title. Then a matrix named *dat* specifies point data. Nonzero values in the first column of *dat* identify the start of a polyline (connecting several successive points) or an arc. For example, the value 11 indicates a line connecting eleven points with (x, y) coordinates contained in the second and third columns of *dat*. Lines such as 24 and 25 in **crclovsq** specify a single line segment. A value of -2 in column one identifies two data lines characterizing an arc. The center coordinates of the arc are contained in the second two items after the -2. The last two items of the line immediately following this line contain the initial angle and the subtended angle (in radians measured positive counterclockwise). For example, the inner semicircle (lines 34 and 35) starts at pi and has a subtended angle of $-pi$ because it turns clockwise. Notice that symbolic expressions such as $pi/2$ may be used rather than rounded values such as 1.57. Furthermore, it is very important that the figure must close. Otherwise, incorrect results will be produced.

Functions **arcprop**, **lineprop**, and **shftprop** perform geometrical property calculations. Function **areaprop** decodes the data from matrix *dat* and shifts the points by (xs, ys) if required. The data is broken into polylines and arcs. Function **lineprop** evaluates contributions of a polyline. Function **arcprop** produces the contribution from an arc. (Note that the various analytical expressions in this routine are algebraically complex.) Function **areaprop** accumulates the properties and returns a single vector containing the area, the centroidal coordinates, and the three inertial moments. A set of boundary points suitable for plotting is also returned. Function **shftprop** takes area properties for one axis system and produces values for a new system derived by shifting and rotating the reference axes. Output produced by the program for the data specified by **crclovsq** is used to compute properties for the original axes and for axes shifted to the centroid.

MATLAB Example

Output from Example arearun

```
>> arearun

                GEOMETRICAL PROPERTIES OF PLANE AREAS

Give the script file name which defines the data
(omit the .m extension)
? > crclovsq

                    SUMMARY OF AREA PROPERTIES
    area     xbar     ybar     axx       axy       ayy
  16.7124   3.0000   4.1251  176.3020  206.8230  359.5365


To compute properties relative to shifted axes input
x,y coordinates of the shifted origin and a rotation
angle measured + counterclockwise.
(Press [Enter] to stop)
? > 3.0,4.1251,0

                    PROPERTIES FOR SHIFTED AXES
    area     xbar     ybar     axx       axy       ayy
  16.7124   0.0000   0.0000   25.8905  -0.0000   75.1450


To compute properties relative to shifted axes input
x,y coordinates of the shifted origin and a rotation
angle measured + counterclockwise.
(Press [Enter] to stop)
? > 0,0,0

All done
```

Script File crclovsq

```
 1: % Example: crclovsq
 2: % ~~~~~~~~~~~~~~~~~~
 3: % Script file defining a half annulus above a
 4: % square which contains a square hole.  Used
 5: % as input for script file arearun.
 6: %
 7: % User m functions required:
 8: %    none
 9: %-------------------------------------------------
10:
11: titl='Half Annulus Over a Hollow Square';
12:
13: % Be sure to traverse the boundary
14: % counterclockwise
15:
16: dat=...
17: [11 ,5  ,5  % <-- polyline traversing hollow
18:  0  ,1  ,5  %     square
19:  0  ,2  ,4
20:  0  ,4  ,4
21:  0  ,4  ,2
22:  0  ,2  ,2
23:  0  ,2  ,4
24:  0  ,1  ,5
25:  0  ,1  ,1
26:  0  ,5  ,1
27:  0  ,5  ,5
28:  2  ,5  ,5  % <-- line joining the square
29:  0  ,5  ,6  %     and half annulus
30: -2  ,3  ,6  % <-- outer semicircle of
31:  0  ,0  ,pi %     the half annulus
32:  2  ,1  ,6  % <-- left side horizontal line
33:  0  ,2  ,6
34: -1  ,3  ,6  % <-- inner semicircle of
35:  0  ,pi ,-pi%     the half annulus
36:  2  ,4  ,6  % <-- right side horizontal line
37:  0  ,5  ,6
38:  2  ,5  ,6  % <-- final line joining
39:  0  ,5  ,5];%     the annulus and square
```

Script File arearun

```
1:  % Example: arearun
2:  % ~~~~~~~~~~~~~~~~~
3:  % This program computes the area, centroidal
4:  % coordinates, moments of inertia, and product
5:  % of inertia for any area bounded by straight
6:  % lines and circular arcs. The boundary
7:  % geometry is also depicted graphically. Each
8:  % piecewise linear boundary part is defined by
9:  % a sequence of x,y coordinates. Each arc is
10: % specified by a radius, center coordinates,
11: % initial polar angle, and a subtended angle.
12: % Computation of properties for new reference
13: % axes translated and rotated relative to the
14: % original reference frame is also provided.
15: %
16: % User m functions required:
17: %     areaprop, shftprop, arcprop, lineprop,
18: %     crclovsq, cubrange, read, pauz
19: %-------------------------------------------------
20:
21: fprintf('\n           GEOMETRICAL ');
22: fprintf('PROPERTIES OF PLANE AREAS');
23: fprintf('\n\nGive the script file name');
24: fprintf(' which defines the data\n');
25: fprintf('(omit the .m extension)\n');
26: fn=input('? > ','s'); eval(fn);
27: [a,x,y]=areaprop(dat);
28: rng=cubrange([x(:),y(:)],1.2);
29: plot(x,y); axis(rng); axis('square');
30: xlabel('x axis'); ylabel('y axis');
31: title(titl); figure(gcf);
32: disp(' ');
33: disp('Press [Enter] to continue'); pauz;
34: fprintf('\n                  SUMMARY ');
35: fprintf('OF AREA PROPERTIES');
36: fprintf('\n   area         xbar          ');
37: fprintf('ybar         axx');
38: fprintf('         axy          ayy \n');
39: disp(a');
40: while 1
41:   fprintf('\nTo compute properties ');
42:   fprintf('relative to shifted axes input');
```

```
43:    fprintf('\nx,y coordinates of shifted ');
44:    fprintf('origin and a rotation angle');
45:    fprintf('\nmeasured + counterclockwise.');
46:    fprintf(' (Press [Enter] to stop)\n');
47:    [xshf, yshf, theta] = read;
48:    if norm([xshf,yshf,theta])==0, break, end
49:    theta=pi*theta/180;
50:    ashf=shftprop(a,xshf,yshf,theta);
51:    fprintf('\n                 PROPERTIES ');
52:    fprintf('FOR SHIFTED AXES');
53:    fprintf('\n    area        xbar        ');
54:    fprintf('ybar        axx');
55:    fprintf('          axy        ayy \n');
56:    disp(ashf');
57: end
58: fprintf('\n\nAll Done\n');
```

Function areaprop

```
1: function [aprops,xb,yb]=areaprop(dat,xs,ys)
2: %
3: % [aprops,xb,yb]=areaprop(dat,xs,ys)
4: % ~~~~~~~~~~~~~~~~~~~~~~~~~~~~~~~~~~~~~~
5: % This function determines geometrical
6: % properties of any area bounded by straight
7: % lines and circular arcs. Coordinates are
8: % shifted by (xs,ys) if these parameters are
9: % included in the argument list.
10: %
11: % dat      - array containing data defining
12: %            arcs and lines
13: % (xs,ys) - coordinates to which origin is
14: %            shifted
15: % aprops  - [area; xbar; ybar; axx; axy; ayy]
16: % xb,yb   - boundary point table suitable for
17: %            plotting
18: %
19: % User m functions called:  arcprop, lineprop
20: %-------------------------------------------------
21:
22: if nargin==1, xs=0; ys=0; end
23: aprops=zeros(6,1); ndx=find(dat(:,1)~=0);
24: jline=1; nparts=length(ndx); xb=[]; yb=[];
```

```
25: for k=1:nparts
26:   nk=ndx(k); n=dat(nk,1);
27:   if n>0
28:     jnext=jline+n; kk=jline:jnext-1;
29:     x=dat(kk,2); y=dat(kk,3); x=x-xs; y=y-ys;
30:     aprops=aprops+lineprop(x,y); jline=jnext;
31:     if nargout >1, xb=[xb;x]; yb=[yb;y]; end
32:   else
33:     rad=-n; r=dat(jline,2:3)-[xs,ys];
34:     th1=dat(jline+1,2);
35:     th12=dat(jline+1,3); jline=jline+2;
36:     aprops=aprops+ ...
37:             arcprop(r(1),r(2),rad,th1,th12);
38:     if nargout > 1
39:       nplot=max(10,round(abs(th12*20/pi)));
40:       theta=th1+(0:nplot-1)'* ...
41:             (th12/(nplot-1));
42:       x=r(1)+rad*cos(theta);
43:       y=r(2)+rad*sin(theta);
44:       xb=[xb;x]; yb=[yb;y];
45:     end
46:   end
47: end
48: aprops(2)=aprops(2)/aprops(1);
49: aprops(3)=aprops(3)/aprops(1);
```

Function arcprop

```
1:  function arcprp=arcprop(x0,y0,r0,t1,t12)
2:  %
3:  % arcprp = arcprop(x0,y0,r0,t1,t12)
4:  % ~~~~~~~~~~~~~~~~~~~~~~~~~~~~~~~~~~~
5:  % This function computes area properties
6:  % of a circular arc.
7:  %
8:  % (x0,y0)        - center coordinates
9:  % r0             - arc radius
10: % t1             - polar angle of the
11: %                  starting point
12: % t12            - angle subtended at the
13: %                  center (positive direction
14: %                  is counterclockwise)
15: % arcprp         -[area;axbar;aybar;axx;axy;ayy]
```

```
16: % P(n,m)            - integral of
17: %                     (x^n y^m (x y'(t)-y x'(t)) dt;
18: %                     [t=t1 => t=t1+t12] )/(n+m+2)
19: % area              - P(0,0)
20: % [axbar,aybar]     - [ P(1,0),P(0,1) ]
21: % [axx,axy,ayy]     - [ P(2,0),P(1,1),P(0,2) ]
22: %
23: % User m functions called:  none
24: %-----------------------------------------------
25:
26: t2=t1+t12; s1=sin(t1); c1=cos(t1);
27: s2=sin(t2); c2=cos(t2);
28: fs=c1-c2; fc=s2-s1;
29: fss=.5*(t12-s2*c2+s1*c1); fcc=t12-fss;
30: fcs=.5*(s2+s1)*(s2-s1); fcss=(s2^3-s1^3)/3;
31: fccs=(c1^3-c2^3)/3; fccc=fc-fcss;
32: fsss=fs-fccs;
33: aa=x0*fc+y0*fs+r0*t12; area=aa*r0/2;
34: e1=(x0*fcc+y0*fcs+r0*fc)*r0;
35: e2=(x0*fcs+y0*fss+r0*fs)*r0;
36: axbar=(x0*aa+e1)*r0/3; aybar=(y0*aa+e2)*r0/3;
37: r2=r0/4; r4=(r0^3)/4;
38: axx=r2*x0*(x0*aa+2*e1)+ ...
39:     r4*(x0*fccc+y0*fccs+r0*fcc);
40: axy=r2*(x0*y0*aa+e1*y0+e2*x0)+ ...
41:     r4*(x0*fccs+y0*fcss+r0*fcs);
42: ayy=r2*y0*(y0*aa+2*e2)+ ...
43:     r4*(x0*fcss+y0*fsss+r0*fss);
44: arcprp=[area;axbar;aybar;axx;axy;ayy];
```

Function lineprop

```
 1: function [Lineprp]=lineprop(x,y)
 2: %
 3: % [Lineprp]=lineprop(x,y)
 4: % ~~~~~~~~~~~~~~~~~~~~~~~
 5: % This function computes the area property
 6: % contributions associated with a polyline.
 7: %
 8: % x,y      - vectors containing data
 9: %            coordinates
10: % Lineprp - vector of geometrical properties
11: %            the components of which are
```

```
12: %              [area; axbar; aybar; axx; axy; ayy]
13: %
14: % User m functions called:  none
15: %--------------------------------------------------
16:
17: n=length(x); x=x(:)'; y=y(:)'; r=[x;y];
18: if n==2 % Case for a single line element
19:   area=det(r)/2;
20:   r1=r(:,1); r2=r(:,2); rs=r1+r2;
21:   arbar=(r1+r2)*area/3;
22:   arr=(r1*r1'+rs*rs'+r2*r2')*area;
23: else    % Case for a sequence of line segments
24:   J=[1:n-1]; J1=J+1;
25:   s=x(J).*y(J1)-y(J).*x(J1);
26:   area=sum(s)/2;
27:   arbar=(r(:,J)+r(:,J1)).*s([1 1],:);
28:   arbar=sum(arbar')/6;
29:   arr=zeros(2,2); rj=r(:,1); rrj=rj*rj';
30:   for j=1:n-1
31:     rj1=r(:,j+1); rrj1=rj1*rj1'; t=rj*rj1';
32:     arr=arr+(rrj+rrj1+(t+t')/2)*s(j);
33:     rj=rj1; rrj=rrj1;
34:   end
35: end
36: Lineprp=[area; arbar(:); ...
37:          [arr(1,1); arr(1,2); arr(2,2)]]/12;
```

Function shftprop

```
1: function AP = shftprop(ap,xctr,yctr,theta);
2: %
3: % AP = shftprop(ap,xctr,yctr,theta)
4: % ~~~~~~~~~~~~~~~~~~~~~~~~~~~~~~~~~~
5: % This function computes area properties for a
6: % set of axes centered at (xctr,yctr) and
7: % rotated counterclockwise through an angle
8: % theta relative to the original axes.
9: %
10: % ap - [ area; xbar; ybar; axx; axy; ayy ]
11: % AP - Transformed property vector for the
12: %       new axes
13: %
14: % User m functions called:  none
```

```
15: %-----------------------------------------------
16:
17: AP=ap(:); a=ap(1);
18: x=ap(2); y=ap(3); i=sqrt(-1);
19: if nargin==4
20:   X=x-xctr; Y=y-yctr; AP(2)=X; AP(3)=Y;
21:   AP(4)=AP(4)+a*(X*X-x*x);
22:   AP(5)=AP(5)+a*(X*Y-x*y);
23:   AP(6)=AP(6)+a*(Y*Y-y*y);
24: end
25: z=(AP(2)+i*AP(3))*exp(-i*theta);
26: AP(2)=real(z); AP(3)=imag(z);
27: p1=(AP(4)+AP(6))/2;
28: p2=(AP(4)-AP(6))/2+i*AP(5);
29: p2=p2*exp(-2*i*theta); AP(4)=real(p1+p2);
30: AP(5)=imag(p2); AP(6)=real(p1-p2);
```

Property	Exact Results	Spline Results	Percent Error
area	16.7124	16.6899	0.1343
xbar	3.0000	3.0000	0.0000
ybar	4.1251	4.1221	0.0733
Axx	176.3020	175.9907	0.1766
Axy	206.8230	206.3937	0.2076
Ayy	359.5365	358.6261	0.2532

Table 5.2. Comparison of Area Properties

5.6.2   Program to Analyze Spline Interpolated Boundaries

The second program produces properties for a spline interpolated geometry. Function **makcrcsq** has been written to generate spline data suitable to approximate the exact geometry discussed above. Function **spcurv2d** is used to graph the spline interpolated geometry for both cases, with and without use of corner points. Results appear in Figures 5.5 and 5.6. The spline geometry in Figure 5.5 is used to reproduce Figure 5.2. The segments with corners at each end give perfect straight lines. The spline curve in Figure 5.6, which was obtained without use of corner points, is clearly unsatisfactory. Function **splaprop** computes geometrical properties for the spline interpolated boundary. Functions **spc** and **spltrp** are called to evaluate the necessary integrands. The numerical integration is performed with function **cbpwf6** utilizing composite base points and weight factors for a six-point Gauss integration formula.

A comparison between the exact results calculated by **areaprop** and the spline results produced by **splaprop** is quite interesting. Table 5.2 indicates good agreement between both methods. The error from the spline method is entirely attributable to imperfect representation of the circular arcs rather than to approximate quadrature. That error can be reduced by employing additional interpolation points to refine the arc definitions.

The programs developed in this chapter are adequate to analyze a variety of practical geometries. Furthermore, the results clearly show the power of cubic spline interpolation to describe general boundary shapes.

MATLAB Example

Script File crnrtest

```
 1: % Example: crnrtest
 2: % ~~~~~~~~~~~~~~~~~~
 3: % MATLAB example showing effect of corner
 4: % points on spline curve approximation of
 5: % a general geometry.
 6: %
 7: % User m functions required:
 8: %     makcrcsq, spcurv2d, splaprop, cbpwf6,
 9: %     spc, cubrange, spltrp, genprint, pauz
10: %------------------------------------------------
11:
12: fprintf('\nAREA PROPERTIES OF A SPLINE ');
13: fprintf('INTERPOLATED GEOMETRY\n\n');
14: [xd,yd,icrnr]=makcrcsq;
15: [xxc,yyc]=spcurv2d(xd,yd,8,icrnr);
16: [xxnc,yync]=spcurv2d(xd,yd,8);
17:
18: clf; plot(xxnc,yync,xd,yd,'o');
19: axis(cubrange([xxnc(:),yync(:)],1.2));
20: xlabel('x axis'); ylabel('y axis');
21: title('Plot Without Corner Points');
22: axis('square'); figure(gcf);
23: %genprint('nocorner');
24: disp('Press [Enter] to continue'); pauz;
25:
26: clf; plot(xxc,yyc,xd,yd,'o');
27: axis(cubrange([xxc(:),yyc(:)],1.2));
28: xlabel('x axis'); ylabel('y axis');
29: title('Plot Using Corner Points');
30: axis('square'); figure(gcf);
31: %genprint('crnrtst');
32: fprintf('Press [Enter] to continue\n'); pauz;
33:
34: aprop=splaprop(xd,yd,icrnr);
35: fprintf('\n    area          xbar        ybar');
36: fprintf('         axx         axy         ayy\n');
37: disp(aprop(:)');
```

Function splaprop

```
 1: function aprop=splaprop(xd,yd,ic)
 2: %
 3: % aprop=splaprop(xd,yd,ic)
 4: % ~~~~~~~~~~~~~~~~~~~~~~~~~
 5: % This function computes geometrical properties
 6: % of an area bounded by a spline curve passed
 7: % through data points xd, yd. Vector aprop
 8: % contains the following quantities
 9: %
10: %    aprop - [ area; xbar; ybar; axx; axy; ayy]
11: %
12: % where (xbar,ybar) are centroidal coordinates,
13: % and
14: %
15: %    axx - integral of x*x*d(area)
16: %    axy - integral of x*y*d(area)
17: %    ayy - integral of y*y*d(area)
18: %
19: % User m functions called:  cbpwf6, spc, spltrp
20: %-----------------------------------------------
21:
22: n=length(xd); xd=xd(:); yd=yd(:);
23: if nargin==2, ic=[]; end
24: xp=[xd(2)-xd(n-1)]/2;
25: yp=[yd(2)-yd(n-1)]/2; id=[1;1];
26: s=(1:n)'; [bp,wf]=cbpwf6(s(1),s(n),n-1);
27: matx=spc(s,xd,id,xp*id,ic);
28: maty=spc(s,yd,id,yp*id,ic);
29: x=spltrp(bp,matx,0); xp=spltrp(bp,matx,1);
30: y=spltrp(bp,maty,0); yp=spltrp(bp,maty,1);
31: mat=zeros(length(x),6);
32: mat(:,1)=x.*yp-y.*xp; mat(:,2)=mat(:,1).*x;
33: mat(:,3)=mat(:,1).*y; mat(:,4)=mat(:,2).*x;
34: mat(:,5)=mat(:,2).*y; mat(:,6)=mat(:,3).*y;
35: aprop=wf'*mat;
36: aprop=aprop(:)./[2; 3; 3; 4; 4; 4];
37: aprop(2)=aprop(2)/aprop(1);
38: aprop(3)=aprop(3)/aprop(1);
```

Function makcrcsq

```
1:  function [x,y,icrnr]=makcrcsq
2:  %
3:  % [x,y,icrnr]=makcrcsq
4:  % ~~~~~~~~~~~~~~~~~~~~
5:  % This function creates data for a geometry
6:  % involving half of an annulus placed above a
7:  % square containing a square hole.
8:  %
9:  % x,y   - data points characterizing the data
10: % icrnr - index vector defining corner points
11: %
12: % User m functions called:  none
13: %-----------------------------------------------
14:
15: xshift=3.0; yshift=3.0;
16: a=2; b=1; narc=7; x0=0; y0=2*a-b;
17: xy=[ a,   a
18:     -a,   a
19:     -b,   b
20:      b,   b
21:      b,  -b
22:     -b,  -b
23:     -b,   b
24:     -a,   a
25:     -a,  -a
26:      a,  -a
27:      a,   a];
28: theta=linspace(0,pi,narc);
29: c=cos(theta); s=sin(theta);
30: c=c(:); s=s(:);
31: xy=[xy;[x0+a*c,y0+a*s]];
32: c=flipud(c); s=flipud(s);
33: xy=[xy;[x0+b*c,y0+b*s];[a,y0];[a,a]];
34: x=xy(:,1)+xshift; y=xy(:,2)+yshift;
35: icrnr=[(1:12)';11+narc;12+narc; ...
36:        11+2*narc;12+2*narc;13+2*narc];
```

Function cbpwf6

```
1:  function [cbp,cwf]=cbpwf6(xlow,xhigh,mparts)
```

```
 2: %
 3: % [cbp,cwf]=cbpwf6(xlow,xhigh,mparts)
 4: % ~~~~~~~~~~~~~~~~~~~~~~~~~~~~~~~~~~~~~~
 5: % This function computes base points, cbp,
 6: % and weight factors, cwf, in a composite
 7: % quadrature formula which integrates an
 8: % arbitrary function from xlow to xhigh by
 9: % dividing the interval of integration into
10: % mparts equal parts and integrating over
11: % each part using a Gauss formula requiring
12: % six function values.
13: %
14: % xlow, xhigh - integration limits
15: % mparts      - number of subintervals used
16: %                 for composite integration
17: % cbp, cwf    - vectors of length 6*mparts
18: %                 which contain the composite
19: %                 base points and weight factors
20: %
21: % User m functions called:  none
22: %------------------------------------------------
23:
24: wf=[ 1.71324492379170d-01; ...
25:      3.60761573048139d-01; ...
26:      4.67913934572691d-01];
27: wf=[wf;wf([3 2 1])];
28: bp=[-9.32469514203152d-01; ...
29:     -6.61209386466265d-01;...
30:     -2.38619186083197d-01];
31: bp=[bp;-bp([3 2 1])];
32:
33: d=(xhigh-xlow)/mparts; d1=d/2; nquad=6;
34: cbp=(d1*bp)*ones(1,mparts)+ ...
35:     (d*ones(nquad,1))*(1:mparts);
36: cbp=cbp(:)+(xlow-d1);
37: cwf=(d1*wf)*ones(1,mparts);
38: cwf=cwf(:);
```

Function spcurv2d

```
1: function [xout,yout,sout]= ...
2:          spcurv2d(xd,yd,nseg,ncrnr)
3: %
```

```
 4: % [xout,yout]=spcurv2d(xd,yd,nseg,ncrnr)
 5: % ~~~~~~~~~~~~~~~~~~~~~~~~~~~~~~~~~~~~~~~~~~~
 6: %
 7: % This function tabulates points xout, yout on
 8: % a spline curve connecting data points xd, yd
 9: % on the cubic spline curve
10: %
11: % xd,yd      - input data points
12: % nseg       - number of tabulation intervals
13: %              used per spline segment
14: % ncrnr      - point indices where corners are
15: %              required
16: % xout,yout  - output data points on the
17: %              spline curve. The number of
18: %              points returned equals
19: %
20: %              nout=(nd-1)*nseg+1.
21: %
22: % User m functions called:  spc, spltrp
23: %-------------------------------------------------
24:
25: nd=length(xd); sd=(1:nd)';
26: if nargin==2; nseg=10; end
27: if nargin<=3, ncrnr=[]; end;
28: nout=(nd-1)*nseg+1; sout=linspace(1,nd,nout);
29: if norm([xd(1)-xd(nd),yd(1)-yd(nd)]) < 100*eps
30:    yp=(yd(2)-yd(nd-1))/2;
31:    iend=[1;1]; vy=[yp;yp];
32:    xp=(xd(2)-xd(nd-1))/2; vx=[xp;xp];
33: else
34:    iend=[3;3]; vy=[0;0]; vx=vy;
35: end
36: matx=spc(sd,xd,iend,vy,ncrnr);
37: maty=spc(sd,yd,iend,vx,ncrnr);
38: xout=spltrp(sout,matx,0);
39: yout=spltrp(sout,maty,0);
```

Function spc

```
1: function splmat=spc(x,y,i,v,icrnr)
2: %
3: % splmat=spc(x,y,i,v,icrnr)
4: % ~~~~~~~~~~~~~~~~~~~~~~~~~~~
```

```
5:  % This function computes matrix splmat
6:  % containing coefficients needed to perform a
7:  % piecewise cubic interpolation among data
8:  % values contained in vectors x and y. The
9:  % output from this function is used by function
10: % spltrp to evaluate the cubic spline function,
11: % its first two derivatives, or the function
12: % integral from x(1) to an arbitrary upper
13: % limit.
14: %
15: %  x     - vector of abscissa values arranged
16: %            in increasing order. The number of
17: %            data values is denoted by n.
18: %
19: %  y     - vector of ordinate values
20: %
21: %  i     - a two component vector [i1,i2].
22: %            Parameters i1 and i2 refer to left
23: %            end and right end conditions,
24: %            respectively. These equal 1, 2,
25: %            or 3 as explained below.
26: %
27: %  v     - a two component vector [v1,v2]
28: %            containing end values of y'(x)
29: %            or y''(x)
30: %
31: %            if i1=1: y'(x(1)) is set to v(1)
32: %                =2: y''(x(1)) is set to v(1)
33: %                =3: y''' is continuous at x(2)
34: %            if i2=1: y'(x(n)) is set to v(2)
35: %                =2: y''(x(n)) is set to v(2)
36: %                =3: y''' is continuous at x(n-1)
37: %
38: %            Note: When i1 or i2 equal 3 the
39: %                  corresponding values of vector
40: %                  v should be zero
41: %
42: % icrnr - A vector of indices identifying
43: %            interior points where slope
44: %            discontinuities are to be generated
45: %            by requiring y''(x) to equal zero.
46: %            Vector icrnr should have components
47: %            lying between 2 and n-1. The vector
48: %            can be omitted from the argument list
49: %            if no slope discontinuities occur.
```

```
50: %
51: % User m functions called:  none
52: %------------------------------------------------
53:
54: x=x(:); y=y(:);
55: n=length(x); a=zeros(n,1); b=a; c=a;
56: d=a; t=a; if nargin < 5, icrnr=[]; end
57: ncrnr=length(icrnr);
58: i1=i(1); i2=i(2);
59: v1=v(1); v2=v(2);
60:
61: % Form the tridiagonal system to solve for
62: % second derivative values at the data points
63:
64: n1=n-1; j=2:n1;
65: hj=x(j)-x(j-1); hj1=x(j+1)-x(j);
66: hjp=hj+hj1; a(j)=hj./hjp; wuns=ones(n-2,1);
67: b(j)=2*wuns; c(j)=wuns-a(j);
68: d(j)=6.*((y(j+1)-y(j))./hj1- ...
69:     (y(j)-y(j-1))./hj)./hjp;
70:
71: % Form  equations for the end conditions
72:
73: % slope specified at left end
74: if i1==1
75:   h2=x(2)-x(1); b(1)=2; c(1)=1;
76:   d(1)=6*((y(2)-y(1))/h2-v1)/h2;
77:
78: % second derivative specified at left end
79: elseif i1==2
80:   b(1)=1; c(1)=0; d(1)=v1;
81:
82: % not a knot condition
83: else
84:   b(1)=1;
85:   a(1)=hj(1)/hj(2);
86:   c(1)=-1-a(1);
87:   d(1)=0;
88: end
89:
90: % slope specified at right end
91: if i2==1
92:   hn=x(n)-x(n-1); a(n)=1; b(n)=2;
93:   d(n)=6.*(v2-(y(n)-y(n-1))/hn)/hn;
94:
```

```
95: % second derivative specified at right end
96: elseif i2==2
97:   a(n)=0; b(n)=1; d(n)=v2;
98:
99: % not a knot condition
100: else
101:   a(n)=1;
102:   b(n)=-(x(n-1)-x(n-2))/(x(n)-x(n-2));
103:   c(n)=-(x(n)-x(n-1))/(x(n)-x(n-2));
104: end
105:
106: % Adjust for slope discontinuity
107: % specified by lcrnr
108:
109: if ncrnr > 0
110:   zro=zeros(ncrnr,1);
111:   a(icrnr)=zro; c(icrnr)=zro;
112:   d(icrnr)=zro; b(icrnr)=ones(ncrnr,1);
113: end
114:
115: % Solve the tridiagonal system
116: % for t(1),...,t(n)
117:
118: bb=diag(a(2:n),-1)+diag(b)+diag(c(1:n-1),1);
119: if a(1)~=0, bb(1,3)=a(1); d(1)=0;   end
120: if c(n)~=0, bb(n,n-2)=c(n); d(n)=0; end
121: t=bb\d(:);
122:
123: % Save polynomial coefficients describing the
124: % cubics for each interval.
125:
126: j=1:n-1; k=2:n;
127: dx=x(k)-x(j); dy=y(k)-y(j); b=t(j)/2;
128: c=(t(k)-t(j))./(6*dx); a=dy./dx-(c.*dx+b).*dx;
129: int=(((c.*dx/4+b/3).*dx+a/2).*dx+y(j)).*dx;
130: int=[0;cumsum(int)]; n1=n-1;
131: ypn=dy(n1)/dx(n1)+dx(n1)*(2*t(n)+t(n1))/6;
132:
133: % The columns of splmat contain the
134: % following vectors
135: % [x,y,x_coef,x^2_coef,x^3_coef,integral_coef]
136:
137: splmat=[ ...
138:   [ x(1),  y(1),  a(1),   0,   0,   0     ];
139:   [ x(j),  y(j),  a,      b,   c,   int(j) ];
```

```
140:    [ x(n),   y(n),   ypn,     0,    0,    int(n) ] ];
```

Function spltrp

```
 1: function f=spltrp(x,mat,ideriv)
 2: %
 3: % f=spltrp(x,mat,ideriv)
 4: % ~~~~~~~~~~~~~~~~~~~~~~~
 5: % This function performs cubic spline
 6: % interpolation using data array mat obtained
 7: % by first calling function spc.
 8: %
 9: % x      - the vector of interpolation values
10: %           at which the spline is to be
11: %           evaluated.
12: % mat    - the matrix output from function spc.
13: %           This array contains the data points
14: %           and the polynomial coefficients
15: %           needed to define the piecewise
16: %           cubic curve connecting the data
17: %           points.
18: % ideriv - a parameter specifying whether
19: %           function values, derivative values,
20: %           or integral values are computed.
21: %           Taking ideriv equal to 0, 1, 2, or
22: %           3, respectively, returns values of
23: %           y(x), y'(x), y''(x), or the integral
24: %           of y(x) from the first data point
25: %           defined in array mat to each point
26: %           in vector x specifying the chosen
27: %           interpolation points. For any points
28: %           outside the original data range,
29: %           the interpolation is performed by
30: %           extending the tangents at the first
31: %           and last data points.
32: %
33: % User m functions called:  none
34: %-------------------------------------------------
35:
36: % identify interpolation intervals
37: % for each point
38: [nrow,ncol]=size(mat);
39: xx=x(:)'; xd=mat(2:nrow);
```

```
40: xd=xd(:); np=length(x); nd=length(xd);
41: ik=sum(xd(:,ones(1,np)) < xx(ones(nd,1),:));
42: ik=1+ik(:);
43: xk=mat(ik,1); yk=mat(ik,2); dx=x(:)-xk;
44: ak=mat(ik,3); bk=mat(ik,4);
45: ck=mat(ik,5); intk=mat(ik,6);
46:
47: % obtain function values at x
48: if ideriv==0
49:    f=((ck.*dx+bk).*dx+ak).*dx+yk; return
50:
51: % obtain first derivatives at x
52: elseif ideriv==1
53:    f=(3*ck.*dx+2*bk).*dx+ak; return
54:
55: % obtain second derivatives at x
56: elseif ideriv==2
57:    f=6*ck.*dx+2*bk; return
58:
59: % obtain integral from xd(1) to x
60: elseif ideriv==3
61:    f=intk+(((ck.*dx/4+bk/3).*dx+ ...
62:      ak/2).*dx+yk).*dx;
63: end
```

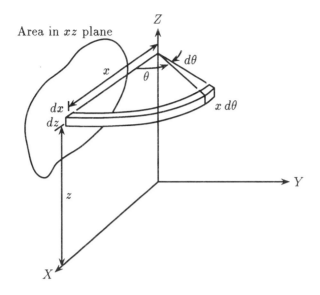

Figure 5.7. Generation of a Volume of Revolution

5.7 Geometrical Properties for a Volume of Revolution

The concepts developed using line integrals to compute properties for plane areas can be extended to partial or full volumes of revolution [110]. We will analyze volumes generated when an area in the xz-plane is rotated about the z-axis through limits $\theta_1 \leq \theta \leq \theta_2$, where the positive direction for θ is counterclockwise according to the right-hand rule. As illustrated in Figure 5.7, an area element $dx\,dz$ corresponds to a volume element

$$dV = x\,d\theta\,dx\,dz$$

and a generic point in three dimensions is described by a column vector

```
R=[c; s; 1].*[x; x; z];   % R=[X; Y; Z];
```

where $c = \cos(\theta)$, $s = \sin(\theta)$, and .* indicates component by component matrix multiplication. Integration gives

$$V = \int\int\int x\,dx\,dz\,d\theta = (\theta_2 - \theta_1)\int\int x\,dx\,dz$$

$$V_r = \int R\,dV = \int_{\theta_1}^{\theta_2} [\texttt{c};\ \texttt{s};\ 1]\,d\theta\ .*\ \int\int [\texttt{x};\ \texttt{x};\ \texttt{z}]\,x\,dx\,dz$$

$$V_{rr} = \int RR^T \, dV$$

$$= \int_{\theta_1}^{\theta_2} [\text{c; s; 1}] \, [\text{c, s, 1}] \, d\theta \, .*$$

$$\int\int [\text{x; x; z}] \, [\text{x, x, z}] \, x \, dx \, dz$$

where V is the volume, V_r is the first moment of the volume, and V_{rr} is the second moment of the volume. These integrals lead to the following area properties:

$$\int\int \left[1, x, z, x^2, xz, z^2, x^3, x^2z, xz^2\right] \, dx \, dy$$

Even though the integrals involving

$$[1, z, z^2] \, dx \, dy$$

do not occur in the volume calculations, they are included as area properties of practical importance. The area integrals are classified in terms of the property I_{nm} from Section 5.5 according to

$$I_{nm} = \int\int x^n z^m \, dx \, dz = \frac{1}{n+m+2} \int_L x^n y^m \, (x \, dy - y \, dx)$$

with $\int_L$ denoting a line integral around the boundary of the related area. Then we get

$$P = \int\int [\text{x; x; z}] \, [\text{x, x, z}] \, x \, dx \, dz$$

$$= \begin{bmatrix} I_{30} & I_{30} & I_{21} \\ I_{30} & I_{30} & I_{21} \\ I_{21} & I_{21} & I_{12} \end{bmatrix}$$

The second order moment quantities are expressed as

VRR=P.*G

where

$$G = \int_{\theta_1}^{\theta_2} [\text{c; s; 1}] \, [\text{c, s, 1}] \, d\theta$$

$$= \begin{bmatrix} CC & CS & C \\ CS & SS & S \\ C & S & T \end{bmatrix}$$

and

$$T = \int_{\theta_1}^{\theta_2} d\theta = \theta_2 - \theta_1$$

$$C = \int_{\theta_1}^{\theta_2} c \, d\theta = \sin(\theta_2) - \sin(\theta_1)$$

$$S = \int_{\theta_1}^{\theta_2} s \, d\theta = -\cos(\theta_2) + \cos(\theta_1)$$

$$CC = \int_{\theta_1}^{\theta_2} c^2 \, d\theta = \frac{\sin(2\theta_2) - \sin(2\theta_1)}{4} + \frac{T}{2}$$

$$SS = \int_{\theta_1}^{\theta_2} s^2 \, d\theta = \frac{\sin(2\theta_2) - \sin(2\theta_1)}{4} - \frac{T}{2}$$

$$CS = \int_{\theta_1}^{\theta_2} cs \, d\theta = \frac{\sin^2(\theta_2) - \sin^2(\theta_1)}{2}$$

These properties involving area moments through third order can be evaluated for curved boundary shapes using spline procedures developed previously. However, attention will be limited to an exact solution for an arbitrary n-sided polygon. The parametric equation for a typical line segment from (x_i, y_i) to (x_j, y_j) is

$$x = x_i + x_{ji}t \qquad z = z_i + z_{ji}t \qquad 0 \le t \le 1$$

where

$$x_{ji} = x_j - x_i \qquad z_{ji} = z_j - z_i$$

The area property for the segment is

$$\int^{L_i} x^n z^m (x \, dz - z \, dx) = (x_i z_j - z_i x_j) \int_0^1 (x_i + x_{ji}t)^n (z_i + z_{ji}t)^m \, dt$$

The highest polynomial order occurring in our application is $n + m = 3$. A two-point Gauss formula produces exact results for this case. The two-point formula [20] can be expressed as

$$\int_0^1 g(x) \, dx = \frac{1}{2}[g(b_1) + g(b_2)]$$

where

$$b_1 = \frac{1 - \frac{1}{\sqrt{3}}}{2} \qquad b_2 = \frac{1 + \frac{1}{\sqrt{3}}}{2}$$

The function **polgnvol** computes properties for the rotated polygon. Data input includes corner coordinates of the polygon traversed in counterclockwise order, and θ limits for the rotation angle. Function **rotapoly** is used to plot the geometry. When **polgnvol** is executed with no input arguments, properties are generated for a quarter cylinder of radius 2 with its axis lying on the z-axis and $1 \leq z \leq 3$. The geometrical properties for this solid are given exactly by

$$V = \frac{\pi r^2}{4} \qquad V_1 = V_2 = \frac{r^3 \ell}{3} \qquad V_3 = \frac{\pi r^2 (z_1 + z_2)}{8}$$

$$V_{11} = V_{22} = \frac{\pi r^4 \ell}{16} \qquad V_{33} = \frac{\pi r^2 (z_2^3 - z_1^3)}{12}$$

$$V_{12} = \frac{r^4 \ell}{8} \qquad V_{13} = V_{23} = \frac{r^3 (z_2^2 - z_1^2)}{6}$$

where

$$r = 2 \quad z_1 = 1 \quad z_2 = 3 \quad \ell = 2$$

The reader can verify that the program computes these values correctly. When the function **rotapoly** is executed with no input, polygon data is automatically generated which represents the lower half of a circle of unit radius on top of which is placed a square of unit side length which is capped by a semicircle of radius $\frac{1}{2}$. The cross section is rotated through $270°$ to produce the volume shown in Figure 5.8. The example program **polgnrun** demonstrates the use functions **polgnvol** and **rotapoly**. The output is included in Section 5.7.1. This example illustrates how a curve such as a circle can be handled satisfactorily provided enough chord segments are used to describe the geometry well. Readers may find it interesting to compute the approximate properties of a sphere by rotating a semicircle constructed from chord segments. Even though results for polygonal cross sections are computed exactly, a considerable number of chords is needed on the circle to produce sphere properties accurate to several digits.

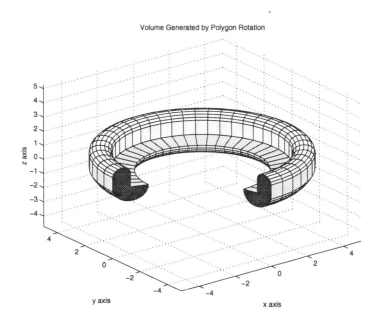

Figure 5.8. Volume Generated by Polygon Rotation

5.7.1   Program Output and Code

Output from Example polgnrun

Geometrical Properties for a Partial
       Volume of Revolution

Main area properties are:
        area   xcent   zcent
       2.8177   4.2229   0.1141

Main volume properties are:
        volume   xcent   ycent   zcent
       56.0735   0.9076   0.9076   0.1544

Inertia tensor defined by
Integral([X;Y;Z]*[X,Y,Z]*d(vol)) is:
     518.8605 -110.1056   9.7313
    -110.1056   518.8605   9.7313
        9.7313   9.7313   25.9992

Script File polgnrun

```
 1: % Example: polgnrun
 2: % ~~~~~~~~~~~~~~~~~
 3: % This program illustrates the computation of
 4: % geometrical properties of a volume generated
 5: % by revolving a polygon partially or
 6: % completely about the z axis.
 7: %
 8: % User m functions called: polgnvol, rotapoly,
 9: %                           cubrange, genprint
10: %-----------------------------------------------
11:
12: a1=linspace(0,pi,9); a2=linspace(pi,2*pi,17);
13: c1=cos(a1); s1=sin(a1); c2=cos(a2); s2=sin(a2);
14: x=4+[1,(c1+1)/2,0,c2]; z=[0,1+s1/2,s2,0];
15: t1=-90; t2=180; nt=fix(24/180*abs(t2-t1));
16: disp(' ');
17: disp('Geometrical Properties for a Partial');
18: disp('        Volume of Revolution'); disp(' ');
19: [a,v]=polgnvol(x,z,[t1,t2]);
20: disp('Main area properties are:');
21: disp('      area       xcent       zcent');
22: disp([a(1);a(2:3)/a(1)]');
23: disp('Main volume properties are:');
24: disp( ...
25:   '     volume      xcent       ycent       zcent');
26: disp([v(1);v([2:4])/v(1)]')
27: disp('Inertia tensor defined by')
28: disp('Integral([X;Y;Z]*[X,Y,Z]*d(vol)) is:')
29: disp(reshape(v([5 8 9 8 6 10 9 10 7]),3,3));
30: rotapoly(x,z,t1,t2,nt); %genprint('threeqrt');
```

Function polgnvol

```
 1: function [aprop,vprop]=polgnvol(x,z,th)
 2: %
 3: % [aprop,vprop]=polgnvol(x,z,th)
 4: % ~~~~~~~~~~~~~~~~~~~~~~~~~~~~~~~~
 5: %
 6: % This function computes area and volume
 7: % properties for a polygonal area in the xz
 8: % plane which is rotated partially or completely
 9: % about the z axis to generate a volume of
10: % revolution.
11: %
12: % x,z    - vectors of the polygon corner
13: %          coordinates. The polygon should not
14: %          have the first and last points equal.
15: %          The program performs the closure
16: %          calculation.  The positive direction
17: %          for traversing the boundary is
18: %          counterclockwise.
19: % th     - vector [theta1,theta2] containing the
20: %          rotation angles in degrees measured
21: %          positive counterclockwise about the z
22: %          axis (the positive direction follows
23: %          the right hand rule).
24: %
25: % aprop - vector of area property integrals
26: %          defined by
27: %            integral( [1,x,z,x^2,xz,z^2,
28: %                       x^3,x^2*z,x*z^2] ) dxdz
29: % vprop - vector of volume property integrals
30: %          defined by
31: %            integral( [1,X,Y,Z,X^2,Y^2,
32: %                       Z^2,XY,XZ,YZ] ) d(vol)
33: %
34: % User m functions called:  none
35: %------------------------------------------------
36:
37: if nargin<3, th=[0,360]; end;
38: % Default data generates a quarter cylinder of
39: % radius 2 with its axis lying between (0,0,1)
40: % and (0,0,3).
41: if nargin==0
42:   x=[0 2 2 0]; z=[1 1 3 3]; th=[0,90]; end
```

```
43:
44: % Evaluate quantities relating to the
45: % rotation angle
46: th=pi/180*th; n=length(x);
47: s1=sin(th(1)); s2=sin(th(2));
48: c1=cos(th(1)); c2=cos(th(2));
49: t=th(2)-th(1); s=-c2+c1;
50: c=s2-s1; cs=(s2^2-s1^2)/2;
51: cc=(s2*c2-s1*c1+t)/2; ss=t-cc;
52: g=[t c s cc cs ss];
53:
54: % Form integrands for the line integrals
55: % defining various area properties
56: x=x(:); z=z(:); X=x([2:n,1]);
57: Z=z([2:n,1]); dx=X-x; dz=Z-z;
58: f=(x.*dz-z.*dx)/2; f=[f;f];
59: r=1/sqrt(3); b=(1-r)/2;
60: B=(1+r)/2; xx=[x+b*dx;x+B*dx];
61: zz=[z+b*dz;z+B*dz];
62: u=zeros(2*n,9); u(:,1)=f;
63: u(:,2)=f.*xx; u(:,3)=f.*zz; u(:,4)=u(:,2).*xx;
64: u(:,5)=u(:,2).*zz; u(:,6)=u(:,3).*zz;
65: u(:,7)=u(:,4).*xx; u(:,8)=u(:,4).*zz;
66: u(:,9)=u(:,5).*zz;
67:
68: % Evaluate the line integrals by
69: % summing columns
70: p=sum(u)./[2 3 3 4 4 4 5 5 5];
71: vprop=p([2 4 4 5 7 7 9 7 8 8]).* ...
72:       g([1 2 3 1 4 6 1 5 2 3]);
73: vprop=vprop(:); aprop=p(:);
```

Function rotapoly

```
1: function [X,Y,Z]=rotapoly(x,z,t1,t2,nt)
2: %
3: % [X,Y,Z]=rotapoly(x,z,t1,t2,nt)
4: % ~~~~~~~~~~~~~~~~~~~~~~~~~~~~~~~~
5: %
6: % This function plots a partial surface of
7: % revolution produced by rotating a polygon.
8: %
9: % x,z   - vectors of points defining a polygon
```

```
10: %              in the x,z plane
11: % t1,t2 - rotation angles in degrees. The
12: %          polygon is rotated clockwise about
13: %          the z-axis from t1 to t2 to generate
14: %          a volume of revolution.
15: % nt     - the number of values between t1 and
16: %          t2 used to plot the surface
17: % X,Y,Z - coordinate values of points on the
18: %          surface of the volume
19: %
20: % User m functions called: cubrange
21: %-------------------------------------------------
22:
23: % Generate a default data case
24: if nargin==0
25:   a1=linspace(0,pi,9); a2=linspace(pi,2*pi,17);
26:   c1=cos(a1); s1=sin(a1);
27:   c2=cos(a2); s2=sin(a2);
28:   x=4+[1,(c1+1)/2,0,c2]; z=[0,1+s1/2,s2,0];
29:   t1=-90; t2=180; nt=fix(24/180*abs(t2-t1));
30: end
31:
32: % Make sure the polygon closes.
33: x=x(:)'; z=z(:)'; np=length(x);
34: if norm([x(1)-x(np),z(1)-z(np)])~=0
35:   x=[x,x(1)]; z=[z,z(1)];
36: end
37:
38: % Create the points defining the surface
39: t=linspace(t1,t2,nt)'*pi/180;
40: X=cos(t)*x; Y=sin(t)*x; Z=z(ones(nt,1),:);
41:
42: % Plot the surface of revolution
43: hold off; clf; surf(X,Y,Z); xlabel('x axis');
44: ylabel('y axis'); zlabel('z axis');
45: colr=[1 0 1]; colormap(colr);
46: title('Volume Generated by Polygon Rotation');
47: axis(cubrange([X(:),Y(:),Z(:)])); m=size(X,1);
48:
49: % Use polygon fill to cap the ends of
50: % the solid
51: hold on;
52: fill3(X(1,:),Y(1,:),Z(1,:),colr);
53: fill3(X(m,:),Y(m,:),Z(m,:),colr);
54: grid on; figure(gcf); hold off;
```

5.8 Geometrical Properties of a Polyhedron

A polyhedron is a solid covered by polygonal faces. Since polyhedra with sufficiently many faces can approximate volumes of complex shape, computing the volume, centroidal position, and inertia tensor of a polyhedron has useful applications. For mathematical analysis, a polyhedron can be treated as the combination of a number of pyramids with bases which are the polyhedron faces and apexes located at the coordinate origin. Once the geometrical properties of a pyramid are known, results for a polyhedron are found by combining results for all faces [109].

Consider a general volume V covered by surface S. It follows from the divergence theorem of Gauss [58] that

$$\int\int_V\int X^n Y^m Z^\ell \, dX \, dY \, dZ = \frac{1}{n+m+\ell+3} \int_S\int X^n Y^m Z^\ell \, \hat{\eta} \cdot R \, dS$$

where $\hat{\eta}$ is the outward directed surface normal and R is the column vector $[X;Y;Z]$. This formula implies

$$V = \int\int_V\int dX \, dY \, dZ = \frac{1}{3} \int_S\int \hat{\eta} \cdot R \, dS$$

$$V_R = \int\int_V\int R \, dX \, dY \, dZ = \frac{1}{4} \int_S\int R\hat{\eta} \cdot R \, dS$$

and

$$V_{RR} = \int\int_V\int RR^T \, dX \, dY \, dZ = \frac{1}{5} \int_S\int RR^T \hat{\eta} \cdot R \, dS$$

where R^T means the transpose of R. Let us apply these formulas to a pyramid with the apex at $R = 0$ and the base being a planar region S_b of area A. For points on the side of the pyramid of height h we find that $\hat{\eta} \cdot R = 0$ and for points on the base $\hat{\eta} \cdot R = h$. Consequently

$$V = \frac{1}{3} \int_{S_b}\int h \, dS = \frac{h}{3} A$$

$$V_R = \frac{1}{4} \int_{S_b}\int Rh \, dS = \frac{h}{4} \int_{S_b}\int R \, dS$$

$$V_{RR} = \frac{1}{5} \int_{S_b}\int RR^T h \, dS = \frac{h}{5} \int_{S_b}\int RR^T \, dS$$

The volume is seen to equal one third of the height times the base area, regardless of the base shape. If $\bar{R}_b$ and $\bar{R}_p$ signify the centroidal radii of the base and the pyramid volume, respectively, we get

$$\bar{R}_p = \frac{V_R}{V} = \frac{\frac{h}{4}\bar{R}_b A}{\frac{h}{3} A} = \frac{3}{4}\bar{R}_b$$

Therefore, the centroid of the volume lies $\frac{3}{4}$ of the way along a line from the apex to the centroid of the base. For any planar area it is not hard to show that the area A and unit surface normal $\hat{\eta}$ can be computed using the line integral

$$\hat{\eta}A = \frac{1}{2}\int_L R \times \, dR$$

The last formula simplifies for a polygon having corners at $R_1, R_2, \ldots, R_n$ to yield

$$\hat{\eta}A = \frac{1}{2}\sum_{j=1}^{n} R_j \times R_{j+1} \qquad R_{n+1} = R_1$$

To compute the first and second area moments for a general planar area, it is helpful to introduce coordinates centered anywhere in the plane containing the base. We let

$$R = R_0 + \hat{\imath}x + \hat{\jmath}y$$

where R_0 is a vector to a point in the plane of the base, and $\hat{\imath}$ and $\hat{\jmath}$ are orthonormal unit vectors which are tangent to the plane and are chosen such that $\hat{\imath}$, $\hat{\jmath}$, $\hat{n}$ form a right-handed system. The local coordinates (x, y) can be computed using matrix products as

$$x = (R - R_0)^T \hat{\imath} \qquad y = (R - R_0)^T \hat{\jmath}$$

Then we get

$$
\begin{aligned}
V_R &= \frac{h}{4}\int_{S_b}\int (R_0 + \hat{\imath}x + \hat{\jmath}y)\, dx\, dy \\
&= \frac{h}{4}(R_0 + \hat{\imath}\bar{x} + \hat{\jmath}\bar{y})A \\
&= \frac{h}{4}\bar{R}_b A
\end{aligned}
$$

where $(\bar{x}, \bar{y})$ are the centroidal coordinates of the area measured relative to the local axes. Similarly we have

$$
\begin{aligned}
V_{RR} &= \frac{h}{5}\int_{S_b}\int \left[R_0 R_0^T + (R_0\hat{\imath}^T + \hat{\imath}R_0^T)x + (R_0\hat{\jmath}^T + \hat{\jmath}R_0^T)y + \right. \\
&\qquad \left. (\hat{\imath}\hat{\jmath}^T + \hat{\jmath}\hat{\imath}^T)xy + \hat{\imath}\hat{\imath}^T x^2 + \hat{\jmath}\hat{\jmath}^T y^2\right] dx\, dy \\
&= \frac{h}{5}\left[R_0 R_0^T + (R_0\hat{\imath}^T + \hat{\imath}R_0^T)\bar{x} + (R_0\hat{\jmath}^T + \hat{\jmath}R_0^T)\bar{y}\right] A + \\
&\qquad \frac{h}{5}\left[\hat{\imath}\hat{\imath}^T A_{xx} + \hat{\jmath}\hat{\jmath}^T A_{yy} + (\hat{\imath}\hat{\jmath}^T + \hat{\jmath}\hat{\imath}^T)A_{xy}\right]
\end{aligned}
$$

where

$$A_{xx} = \int \int x^2 \, dx \, dy \qquad A_{xy} = \int \int xy \, dx \, dy \qquad A_{yy} = \int \int y^2 \, dx \, dy$$

The formula for V_{RR} simplifies when R_0 is chosen as the centroidal radius $\bar{R}_b$. Then $\bar{x} = \bar{y} = 0$ and gives

$$V_{RR} = \frac{h}{5} \left[\bar{R}_b \bar{R}_b^T A + \hat{\imath}\hat{\imath}^T \bar{A}_{xx} + \hat{\jmath}\hat{\jmath}^T \bar{A}_{yy} + (\hat{\imath}\hat{\jmath}^T + \hat{\jmath}\hat{\imath}^T)\bar{A}_{xy} \right]$$

with the quantities $\bar{A}_{xx}, \bar{A}_{xy}, \bar{A}_{yy}$ denoting reference to the centroidal axes.

The analysis to compute polyhedron properties can now be completed using vector algebra along with area property calculations of the type introduced earlier. To define data for a particular polyhedron we provide vectors x, y, z containing global coordinates of all corners. We also employ a matrix named idface having a row dimension equal to the number of faces on the polyhedron and a column dimension equal to the largest number of corners on any face. Row $\imath$ of idface consists of corner indices of the $\imath$'th face with the row being padded with zeros on the right if necessary. Each face is traversed in the counterclockwise sense relative to the outward normal. To fix ideas consider a figure showing a triangular block with a hole. It has twelve corners and eight faces as shown in Figure 5.9. The required geometry descriptions are defined in example **polhdrun**. The results produced for this example are

$$V = 15 \qquad \bar{r}^T = [0, \ 2.6667, \ 1.3333]$$

$$V_{RR} = \begin{bmatrix} 5.0 & 0.0 & 0.0 \\ 0.0 & 120.8333 & 60.4167 \\ 0.0 & 60.4167 & 40.8333 \end{bmatrix}$$

These can be easily verified by manual calculations.

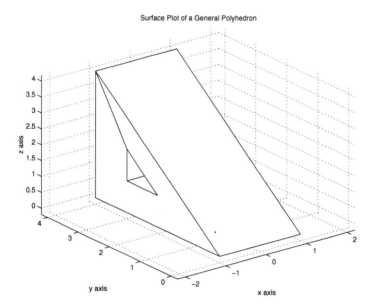

Figure 5.9. Surface Plot of a General Polygon

5.8.1   Program Output and Code

Script File polhdrun

```
 1:  % Example: polhdrun
 2:  % ~~~~~~~~~~~~~~~~~~
 3:  %
 4:  % This program illustrates the use of routine
 5:  % polhedrn to calculate the geometrical
 6:  % properties of a polyhedron.
 7:  %
 8:  % User m functions called:
 9:  %      crosmat, polyxy, cubrange, pyramid,
10:  %      polhdplt, polhedrn, genprint
11:  %------------------------------------------------
12:
13:  x=[2 2 2 2 2 2 0 0 0 0 0 0]-1;
14:  y=[0 4 4 2 3 3 0 4 4 2 3 3];
15:  z=[0 0 4 1 1 2 0 0 4 1 1 2];
16:  idface=[1  2  3  6  5  4  6  3; ...
17:          1  3  9  7  0  0  0  0; ...
18:          1  7  8  2  0  0  0  0; ...
19:          2  8  9  3  0  0  0  0; ...
20:          7  9 12 10 11 12  9  8; ...
21:          4 10 12  6  0  0  0  0; ...
22:          4  5 11 10  0  0  0  0; ...
23:          5  6 12 11  0  0  0  0];
24:  polhdplt(x,y,z,idface); %genprint('polhedrn');
25:  [v,rc,vrr]=polhedrn(x,y,z,idface)
```

Function polhedrn

```
1: function [v,rc,vrr]=polhedrn(x,y,z,idface)
2: %
3: % [v,rc,vrr]=polhedrn(x,y,z,idface)
4: % ~~~~~~~~~~~~~~~~~~~~~~~~~~~~~~~~~~~
5: %
6: % This function determines the volume,
7: % centroidal coordinates and inertial moment
8: % for an arbitrary polyhedron.
9: %
10: % x,y,z  - vectors containing the corner
11: %           indices of the polyhedron
12: % idface - a matrix in which row j defines the
13: %           corner indices of the j'th face.
14: %           Each face is traversed in a
15: %           counterclockwise sense relative to
16: %           the outward normal. The column
17: %           dimension equals the largest number
18: %           of indices needed to define a face.
19: %           Rows requiring fewer than the
20: %           maximum number of corner indices are
21: %           padded with zeros on the right.
22: %
23: % User m functions called: pyramid
24: %-------------------------------------------------
25:
26: r=[x(:),y(:),z(:)]; nf=size(idface,1);
27: v=0; vr=0; vrr=0;
28: for k=1:nf
29:   i=idface(k,:); i=i(find(i>0));
30:   [u,ur,urr]=pyramid(r(i,:));
31:   v=v+u; vr=vr+ur; vrr=vrr+urr;
32: end
33: rc=vr/v;
```

Function polyxy

```
1: function [area,xbar,ybar,axx,axy,ayy]=polyxy(x,y)
2: %
3: % [area,xbar,ybar,axx,axy,ayy]=polyxy(x,y)
4: % ~~~~~~~~~~~~~~~~~~~~~~~~~~~~~~~~~~~~~~~~~~
```

```
 5: %
 6: % This function computes the area, centroidal
 7: % coordinates, and inertial moments of an
 8: % arbitrary polygon.
 9: %
10: % x,y        - vectors containing the corner
11: %               coordinates. The boundary is
12: %               traversed in a counterclockwise
13: %               direction
14: %
15: % area       - the polygon area
16: % xbar,ybar - the centroidal coordinates
17: % axx        - integral of x^2*dxdy
18: % axy        - integral of xy*dxdy
19: % ayy        - integral of y^2*dxdy
20: %
21: % User m functions called: none
22: %-------------------------------------------------
23:
24: n=1:length(x); n1=n+1;
25: x=[x(:);x(1)]; y=[y(:);y(1)];
26: a=(x(n).*y(n1)-y(n).*x(n1))';
27: area=sum(a)/2; a6=6*area;
28: xbar=a*(x(n)+x(n1))/a6; ybar=a*(y(n)+y(n1))/a6;
29: ayy=a*(y(n).^2+y(n).*y(n1)+y(n1).^2)/12;
30: axy=a*(x(n).*(2*y(n)+y(n1))+x(n1).* ...
31:     (2*y(n1)+y(n)))/24;
32: axx=a*(x(n).^2+x(n).*x(n1)+x(n1).^2)/12;
```

Function pyramid

```
 1: function [v,vr,vrr,h,area,n]=pyramid(r)
 2: %
 3: % [v,vr,vrr,h,area,n]=pyramid(r)
 4: % ~~~~~~~~~~~~~~~~~~~~~~~~~~~~~~~~~~~~
 5: %
 6: % This function determines geometrical
 7: % properties of a pyramid with the apex at the
 8: % origin and corner coordinates of the base
 9: % stored in the rows of r.
10: %
11: % r    - matrix containing the corner
12: %          coordinates of a polygonal base stored
```

```
13: %          in the rows of matrix r.
14: %
15: % v    - the volume of the pyramid
16: % vr   - the first moment of volume relative to
17: %          the origin
18: % vrr  - the second moment of volume relative
19: %          to the origin
20: % h    - the pyramid height
21: % area - the base area
22: % n    - the ourward directed unit normal to
23: %          the base
24: %
25: % User m functions called: crosmat, polyxy
26: %------------------------------------------------
27:
28: ns=size(r,1);
29: na=sum(crosmat(r,r([2:ns,1],:)))'/2;
30: area=norm(na); n=na/area; p=null(n');
31: i=p(:,1); j=p(:,2);
32: if det([p,n])<0, j=-j; end;
33: r1=r(1,:); rr=r-r1(ones(ns,1),:);
34: x=rr*i; y=rr*j;
35: [areat,xc,yc,axx,axy,ayy]=polyxy(x,y);
36: rc=r1'+xc*i+yc*j; h=r1*n;
37: v=h*area/3; vr=v*3/4*rc;
38: axx=axx-area*xc^2; ayy=ayy-area*yc^2;
39: axy=axy-area*xc*yc;
40: vrr=h/5*(area*rc*rc'+axx*i*i'+ayy*j*j'+ ...
41:     axy*(i*j'+j*i'));
```

Function polhdplt

```
1: function polhdplt(x,y,z,idface,colr)
2: %
3: % polhdplt(x,y,z,idface,colr)
4: % ~~~~~~~~~~~~~~~~~~~~~~~~~~~~
5: %
6: % This function makes a surface plot of an
7: % arbitrary polyhedron.
8: %
9: % x,y,z  - vectors containing the corner
10: %          indices of the polyhedron
11: % idface - a matrix in which row j defines the
```

```
12: %              corner indices of the j'th face.
13: %              Each face is traversed in a
14: %              counterclockwise sense relative to
15: %              the outward normal. The column
16: %              dimension equals the largest number
17: %              of indices needed to define a face.
18: %              Rows requiring fewer than the
19: %              maximum number of corner indices are
20: %              padded with zeros on the right.
21: % colr    - character string or a vector
22: %              defining the surface color
23: %
24: % User m functions called: cubrange
25: %-------------------------------------------------
26:
27: if nargin<5, colr=[1 0 1]; end
28: hold off, clf, close; nf=size(idface,1);
29: v=cubrange([x(:),y(:),z(:)],1.1);
30: for k=1:nf
31:    i=idface(k,:); i=i(find(i>0));
32:    xi=x(i); yi=y(i); zi=z(i);
33:    fill3(xi,yi,zi,colr); hold on;
34: end
35: axis(v); grid on;
36: xlabel('x axis'); ylabel('y axis');
37: zlabel('z axis');
38: title('Surface Plot of a General Polyhedron');
39: figure(gcf); hold off;
```

Function crosmat

```
1: function c=crosmat(a,b)
2: %
3: % c=crosmat(a,b)
4: % ~~~~~~~~~~~~~~~
5: %
6: % This function computes the vector cross
7: % product for vectors stored in the rows
8: % of matrices a and b, and returns the
9: % results in the rows of c.
10: %
11: % User m functions called: none
12: %-------------------------------------------------
```

```
13:
14: c=[a(:,2).*b(:,3)-a(:,3).*b(:,2),...
15:     a(:,3).*b(:,1)-a(:,1).*b(:,3),...
16:     a(:,1).*b(:,2)-a(:,2).*b(:,1)];
```

Chapter 6

Fourier Series and the FFT

6.1 Definitions and Computation of Fourier Coefficients

Trigonometric series are useful to represent periodic functions. A function defined for $-\infty < x < \infty$ has a period of 2π if $f(x + 2\pi) = f(x)$ for all x. In most practical situations, such a function can be expressed as a complex Fourier series

$$f(x) = \sum_{j=-\infty}^{\infty} c_j e^{ijx} \qquad i = \sqrt{-1}$$

The numbers c_j, called complex Fourier coefficients, are computed by integration as

$$c_j = \frac{1}{2\pi} \int_0^{2\pi} f(x) e^{-ijx} dx$$

The Fourier series can also be rewritten using sines and cosines as

$$f(x) = c_0 + \sum_{j=1}^{\infty} (c_j + c_{-j}) \cos(jx) + i(c_j - c_{-j}) \sin(jx)$$

Denoting

$$a_j = c_j + c_{-j} \qquad b_j = i(c_j - c_{-j})$$

yields

$$f(x) = \frac{1}{2} a_0 + \sum_{j=1}^{\infty} a_j \cos(jx) + b_j \sin(jx)$$

which is called a Fourier sine-cosine expansion. This series is most appealing when $f(x)$ is real valued. For that case $c_{-j} = \bar{c}_j$ for all j, which implies that c_0 must be real and

$$a_j = 2\,\mathbf{real}(c_j) \qquad b_j = -2\,\mathbf{imag}(c_j) \qquad j > 0$$

Suppose we want a Fourier series expansion for a more general function $f(x)$ having period p instead of 2π. If we introduce a new function

$g(x)$ defined by

$$g(x) = f(\frac{px}{2\pi})$$

then $g(x)$ has a period of 2π. Consequently, $g(x)$ can be represented as

$$g(x) = \sum_{j=-\infty}^{\infty} c_j e^{ijx}$$

From the fact that $f(x) = g(2\pi x/p)$ we deduce that

$$f(x) = \sum_{j=-\infty}^{\infty} c_j e^{2\pi ijx/p}$$

A need sometimes occurs to expand a function as a series of sine terms only, or as a series of cosine terms only. If the function is originally defined for $0 < x < \frac{p}{2}$, then making $f(x) = -f(p - x)$ for $\frac{p}{2} < x < p$ gives a series involving only sine terms. Similarly, if $f(x) = +f(p - x)$ for $\frac{p}{2} < x < p$, only cosine terms arise. Thus we get

$$f(x) = c_0 + \sum_{j=1}^{\infty}(c_j + c_{-j})\cos(2\pi jx/p) \qquad \text{if:} \quad f(x) = f(p - x)$$

or

$$f(x) = \sum_{j=1}^{\infty} i(c_j - c_{-j})\sin(2\pi jx/p) \qquad \text{if:} \quad f(x) = -f(p - x)$$

When the Fourier series of a function is approximated using a finite number of terms, the resulting approximating function may oscillate in regions where the actual function is discontinuous or changes rapidly. This undesirable behavior can be reduced by using a smoothing procedure described by Lanczos [59]. Use is made of Fourier series of a closely related function $\hat{f}(x)$ defined by a local averaging process according to

$$\hat{f}(x) = \frac{1}{\Delta} \int_{x-\frac{\Delta}{2}}^{x+\frac{\Delta}{2}} f(\zeta)d\zeta$$

where the averaging interval Δ should be a small fraction of the period p. Hence we write $\Delta = \alpha p$ with $\alpha < 1$. The functions $\hat{f}(x)$ and $f(x)$ are identical as $\alpha \to 0$. Even for $\alpha > 0$, these functions also match exactly at any point x where $f(x)$ varies linearly between $x - \frac{\Delta}{2}$ and $x + \frac{\Delta}{2}$. An important property of $\hat{f}(x)$ is that it agrees closely with $f(x)$ for small α but has a Fourier series which converges more rapidly than the series for $f(x)$. Furthermore, from its definition,

$$\hat{f}(x) \;=\; \sum_{j=-\infty}^{\infty} c_j \frac{1}{p\alpha} \int_{x-\frac{p\alpha}{2}}^{x+\frac{p\alpha}{2}} e^{2\pi i j x/p}\, dx$$

$$=\; \sum_{j=-\infty}^{\infty} \hat{c}_j e^{2\pi i j x/p}$$

where $\hat{c}_0 = c_0$ and $\hat{c}_j = c_j \sin(\pi j \alpha)/(\pi j \alpha)$ for $j \neq 0$. Evidently the Fourier coefficients of $\hat{f}(x)$ are easily obtainable from those of $f(x)$. When the series for $f(x)$ converges slowly, using the same number of terms in the series for $\hat{f}(x)$ often gives an approximation preferable to that provided by the series for $f(x)$. This process is called smoothing.

6.1.1 Trigonometric Interpolation and the FFT

Computing Fourier coefficients by numerical integration is very time consuming. Consequently, we are led to investigate alternative methods employing trigonometric polynomial interpolation through evenly spaced data. The resulting formulas are the basis of an important algorithm called the FFT (Fast Fourier Transform). Although the Fourier coefficients obtained by interpolation are approximate, these coefficients can be computed very rapidly when the number of sample points is an integer power of 2. We will discuss next the ideas behind trigonometric polynomial interpolation among evenly spaced data values.

Suppose we truncate the Fourier series and only use harmonics up to some order N. We assume $f(x)$ has period 2π so

$$f(x) = \sum_{j=-N}^{N} c_j e^{ijx}$$

This trigonometric polynomial satisfies $f(0) = f(2\pi)$ even though the original function might actually have a finite discontinuity at 0 and 2π. Consequently, we may choose to use, in place of $f(0)$, the limit as $\epsilon \to 0$ of $[f(\epsilon) + f(2\pi - \epsilon)]/2$.

It is well known that the functions e^{ijx} satisfy an orthogonality condition for integration over the interval 0 to 2π. They also satisfy an orthogonality condition regarding summation over equally spaced data. The latter condition is useful for deriving a discretized approximation of the integral formula for the exact Fourier coefficients. Let us choose data points

$$x_j = (\frac{2\pi}{2N})j = (\frac{\pi}{N})j \qquad 0 \le j \le (2N-1)$$

and write the simultaneous equations to make the trigonometric polynomial match the original function at the equally spaced data points. To shorten the notation we let

$$t = e^{i\pi/N} \qquad t^k = e^{ik\pi/N}$$

and write

$$f_k = \sum_{j=-N}^{N} c_j t^{kj}$$

Suppose we pick an arbitrary integer n in the range $-N < n < N$. Multiplying the last equation by t^{-kn} and summing from $k = 0$ to $2N - 1$ gives

$$\sum_{k=0}^{2N-1} f_k t^{-kn} = \sum_{k=0}^{2N-1} t^{-kn} \sum_{j=-N}^{N} c_j t^{kj}$$

Interchanging the summation order in the last equation yields

$$\sum_{k=0}^{2N-1} f_k t^{-kn} = \sum_{j=-N}^{N} c_j \sum_{k=0}^{2N-1} \zeta^k$$

where $\zeta = e^{i(j-n)\pi/N}$. Summing the inner geometric series gives

$$\sum_{k=0}^{2N-1} \zeta^k = \begin{cases} \frac{1-\zeta^{2N}}{1-\zeta} & \text{for } \zeta \neq 1 \\[2ex] 2N & \text{for } \zeta = 1 \end{cases}$$

We find, for all k and n in the stated range, that

$$\zeta^{2N} = e^{i2\pi(k-n)} = 1$$

Therefore we get

$$\sum_{k=0}^{2N-1} f_k t^{-kn} = 2N c_n \qquad \text{for } -N < n < N$$

In the cases where $n = \pm N$, the procedure just outlined only gives a relationship governing $c_N + c_{-N}$. Since the first and last terms cannot be computed uniquely, we customarily take N large enough to discard these last two terms and write simply

$$c_n = \frac{1}{2N} \sum_{k=0}^{2N-1} f_k t^{-kn} \qquad -N < n < N$$

This formula is the basis for fast algorithms (called FFT for Fast Fourier Transform) to compute approximate Fourier coefficients. The periodicity of the terms depending on various powers of $e^{i\pi/N}$ can be utilized to greatly reduce the number of trigonometric function evaluations. The case where N equals a power of 2 is especially attractive. The mathematical development is not provided here. However, the related theory was presented by Cooley and Tukey in 1965 [21] and has been expounded in many textbooks [52, 94]. The result is a remarkably concise algorithm which can be comprehended without studying the details of the mathematical derivation. For our present interests it is important to understand how to use MATLAB's intrinsic function for the FFT (**fft**).

Suppose a periodic function is evaluated at a number of equidistant points ranging over one period. It is preferable for computational speed that the number of sample points should equal an integer power of two ($n = 2^m$). Let the function values for argument vector

```
x=p/n*(0:n-1)
```

be an array f denoted by

$$f \iff [f_1, f_2, \cdots, f_n]$$

The function evaluation **fft**(f) produces an array of complex Fourier coefficients multiplied by n and arranged in a peculiar fashion. Let us illustrate this result for $n = 8$. If $f = [f_1, f_2, \cdots, f_8]$, then **fft**($f$)/8 produces $c = [c_0, c_1, c_2, c_3, c_*, c_{-3}, c_{-2}, c_{-1}]$. The term denoted by c_* actually turns out to equal $c_4 + c_{-4}$, so it would not be used in subsequent calculations. We generalize this procedure for arbitrary n as follows. Let $N = n/2 - 1$. In the transformed array, elements with indices of $1, \cdots, N + 1$ correspond to $c_0, \cdots, c_N$ and elements with indices of $n, n - 1, n - 2, \cdots, N + 3$ correspond to $c_{-1}, c_{-2}, c_{-3}, \cdots, c_{-N}$. It is also useful to remember that a real valued function has $c_{-n} = \mathbf{conj}(c_n)$. To fix our ideas about how to evaluate a Fourier series, suppose we want to sum an approximation involving harmonics from order zero to order ($nsum - 1$). We are dealing with a real valued function defined by **func** with a real argument vector x. The following code expands **func** and sums the series for argument x using $nsum$ terms.

```
function fouval=fftaprox(func,period,nfft,nsum,x)
fc=feval(func,period/nfft*(0:nfft-1));
fc=fft(fc)/nfft; fc(1)=fc(1)/2;
w=2*pi/period*(0:nsum-1);
fouval=2*real(exp(i*x(:)*w)*fc(:));
```

6.2   Some Applications

Applications of Fourier series arise in numerous practical situations such as structural dynamics, signal analysis, solution of boundary value problems, and image processing. Three examples are given below which illustrate use of the FFT. The first example calculates Bessel functions and the second problem studies forced dynamic response of a lumped mass system. The final example presents a program for constructing Fourier expansions and displaying graphical results for linearly interpolated or analytically defined functions.

6.2.1   Using the FFT to Compute Integer Order Bessel Functions

The FFT provides an efficient way to compute integer order Bessel functions $J_n(z)$ which are important in various physical applications [117]. Function $J_n(z)$ can be obtained as the complex Fourier coefficient of $e^{in\theta}$ in the generating function described by

$$e^{iz\sin(\theta)} = \sum_{n=-\infty}^{\infty} J_n(z)e^{in\theta}$$

Orthogonality conditions imply

$$J_n(z) = \frac{1}{2\pi} \int_0^{2\pi} e^{i(z\sin(\theta)-n\theta)} \, d\theta$$

The Fourier coefficients represented by $J_n(Z)$ can be computed approximately with the FFT. The infinite series converges very rapidly because the function it represents has continuous derivatives of all finite orders. Of course, $e^{iz\sin(\theta)}$ is highly oscillatory for large $|z|$, thereby requiring a large number of sample points in the FFT to obtain accurate results. For $n < 30$ and $|z| < 30$, a 128-point transform is adequate to give about ten digit accuracy for values of $J_n(z)$. The following code implements the above ideas and plots a surface showing how J_n changes in terms of n and z.

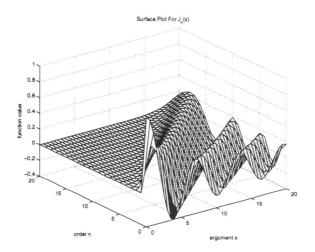

Figure 6.1. Surface Plot for $J_n(x)$

MATLAB Example

Script File plotjrun

```
 1: % Example: plotjrun
 2: % ~~~~~~~~~~~~~~~~~~~
 3: % This program computes integer order Bessel
 4: % functions of the first kind by use of the
 5: % FFT. The algorithm is often faster than
 6: % function BESSELN provided in MATLAB.
 7: %
 8: % User m functions required:
 9: %      jnft, genprint
10: %-------------------------------------------------
11:
12: x=0:.5:20; n=0:20; [J,tcp]=jnft(n,x);
13: surf(x,n,J');
14: title('Surface Plot For J_{n}(x)');
15: ylabel('order n'); xlabel('argument x');
16: zlabel('function value'); figure(gcf);
17: %genprint('plotjrun');
```

Function jnft

```
 1: function [J,tcp]=jnft(n,z,nft)
 2: %
 3: % [J,tcp]=jnft(n,z,nft)
 4: % ~~~~~~~~~~~~~~~~~~~~~
 5: % Integer order Bessel functions of the
 6: % first kind computed by use of the Fast
 7: % Fourier Transform (FFT).
 8: %
 9: % n   - integer vector defining the function
10: %       orders
11: % z   - a vector of values defining the
12: %       arguments
13: % nft - number of function evaluations used
14: %       in the FFT calculation. This value
15: %       should be an integer power of 2 and
16: %       should exceed twice the largest
17: %       component of n. When nft is omitted
18: %       from the argument list, then a value
19: %       equal to 128 is used. More accurate
```

```
20: %         values of J are computed as nft is
21: %         increased. For max(n) < 30 and
22: %         max(z) < 30, nft=128 gives about
23: %         ten digit accuracy.
24: % J    - a matrix of values for the integer
25: %         order Bessel function of the first
26: %         kind. Row position matches orders
27: %         defined by n, and column position
28: %         corresponds to arguments defined by
29: %         components of z.
30: % tcp - computer time required to make the
31: %         calculation
32: %
33: % User m functions called:  none.
34: %-----------------------------------------------
35:
36: tcp=cputime;
37: if nargin<3, nft=128; end;
38: J=exp(sin((0:nft-1)'* ...
39:    (2*pi/nft))*(i*z(:).'))/nft;
40: J=fft(J); J=J(1+n,:).';
41: if sum(abs(imag(z)))<max(abs(z))/1e10
42:    J=real(J);
43: end
44: tcp=cputime-tcp;
```

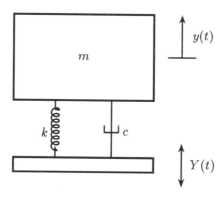

Figure 6.2. Mass System

6.2.2 Dynamic Response of a Mass on an Oscillating Foundation

Fourier series are often used to describe time dependent phenomena such as earthquake ground motion. Understanding the effects of foundation motions on an elastic structure is important in design. The model in Figure 6.2 embodies rudimentary aspects of this type of system and consists of a concentrated mass connected by a spring and viscous damper to a base which oscillates with known motion $Y(t)$. The system is assumed to have arbitrary initial conditions $y(0) = y_0$ and $\dot{y}(0) = v_0$ when the base starts moving. The resulting displacement and acceleration of the mass are to be computed.

We assume that $Y(t)$ can be represented well over some time interval p by a Fourier series of the form

$$Y(t) = \sum_{n=-\infty}^{\infty} c_n e^{i\omega_n t} \qquad \omega_n = \frac{2n\pi}{p}$$

where $c_{-n} = \mathbf{conj}(c_n)$ because Y is real valued. The differential equation governing this problem is

$$m\ddot{y} + c\dot{y} + ky = kY(t) + c\dot{Y}(t) = F(t)$$

where the forcing function can be expressed as

$$F(t) = \sum_{n=-\infty}^{\infty} c_n[k + ic\omega_n]e^{i\omega_n t}$$

$$= kc_0 + 2\,\textbf{real}\left(\sum_{n=1}^{\infty} f_n e^{i\omega_n t}\right)$$

and

$$f_n = c_n(k + ic\omega_n)$$

The corresponding steady-state solution of the differential equation is representable as

$$y_s(t) = \sum_{n=-\infty}^{\infty} y_n e^{i\omega_n t}$$

where $y_{-n} = \textbf{conj}(y_n)$ because $y_s(t)$ is real valued. Substituting the series solution into the differential equation and comparing coefficients of $e^{i\omega_n t}$ on both sides leads to

$$y_n = \frac{c_n(k + ic\omega_n)}{k - m\omega_n^2 + ic\omega_n}$$

These coefficients satisfy $y_{-n} = \textbf{conj}(y_n)$, so the displacement, velocity, and acceleration corresponding to the steady-state (also called particular) solution are

$$y_s(t) = c_0 + 2\,\textbf{real}\left(\sum_{n=1}^{\infty} y_n e^{i\omega_n t}\right)$$

$$\dot{y}_s(t) = 2\,\textbf{real}\left(\sum_{n=1}^{\infty} i\omega_n y_n e^{i\omega_n t}\right)$$

$$\ddot{y}_s(t) = -2\,\textbf{real}\left(\sum_{n=1}^{\infty} \omega_n^2 y_n e^{i\omega_n t}\right)$$

The initial conditions satisfied by y_s are

$$y_s(0) = c_0 + 2\,\textbf{real}\left(\sum_{n=1}^{\infty} y_n\right)$$

$$\dot{y}_s(0) = 2\,\textbf{real}\left(\sum_{n=1}^{\infty} i\omega_n y_n\right)$$

Because these values usually will not match the desired initial conditions, the total solution consists of $y_s(t)$ plus another function $y_h(t)$ satisfying the homogeneous differential equation

$$m\ddot{y}_h + c\dot{y}_h + ky_h = 0$$

The solution is

$$y_h = g_1 e^{s_1 t} + g_2 e^{s_2 t}$$

where s_1 and s_2 are roots satisfying

$$ms^2 + cs + k = 0$$

The roots are

$$s_1 = \frac{-c + \sqrt{c^2 - 4mk}}{2m} \qquad s_2 = \frac{-c - \sqrt{c^2 - 4mk}}{2m}$$

Since the total solution is

$$y(t) = y_s(t) + y_h(t)$$

the constants g_1 and g_2 are obtained by solving the two simultaneous equations

$$g_1 + g_2 = y(0) - y_s(0)$$

$$s_1 g_1 + s_2 g_2 = \dot{y}(0) - \dot{y}_s(0)$$

The roots s_1 and s_2 are equal when $c = 2\sqrt{mk}$. Then the homogeneous solution assumes an alternate form given by $(g_1 + g_2 t)e^{st}$ with $s = -c/(2m)$. In this special case we find that

$$g_1 = y(0) - y_s(0) \qquad g_2 = \dot{y}(0) - \dot{y}_s(0) - sg_1$$

It should be noted that even though roots s_1 and s_2 will often be complex numbers, this causes no difficulty since MATLAB handles the complex arithmetic automatically (just as it does when the FFT transforms real function values into complex Fourier coefficients).

The harmonic response solution works satisfactorily for a general forcing function as long as the damping coefficient c is nonzero. A special situation can occur when $c = 0$, because the forcing function may resonate with the natural frequency of the undamped system. If c is zero, and for some n we have $\sqrt{k/m} = 2\pi n/p$, a condition of harmonic resonance is produced and a value of zero in the denominator occurs when the corresponding y_n is computed. What actually happens in the undamped resonant case is the particular solution grows like $[te^{i\omega_n t}]$, quickly becoming large. Even when c is small and $\sqrt{k/m} \approx 2\pi n/p$, undesirably large values of y_n can result. Readers interested in the important phenomenon of resonance can find more detail in Meirovitch [67].

This example concludes by using a base motion resembling an actual earthquake excitation. Seismograph output employing about 2700 points recorded during the Imperial Valley, California, earthquake of

1940 provided the displacement history for Figure 6.3. The period used to describe the motion is 53.8 seconds. A program was written to analyze system response due to a simulated earthquake base excitation. The following program modules are used:

runimpv	sets data values and generates graphical results
fouaprox	generates Fourier series approximations for a general function
imptp	piecewise linear function approximating the Imperial Valley earthquake data
shkbftss	computes steady-state displacement and acceleration for a spring-mass-dashpot system subjected to base motion expandable in a Fourier series
hsmck	computes the homogeneous solution for the spring-mass-dashpot system subjected to general initial conditions

Numerical results were obtained for a system having a natural period close to one second ($2\pi/6 \approx 1.047$) and a damping factor of 5 percent. The function **imptp** was employed as an alternative to the actual seismograph data to provide a concisely expressible function which still embodies characteristics of a realistic base motion. Figure 6.4 shows a plot of function **imptp** along with its approximation by a twenty-term Fourier series. The series representation is surprisingly good considering the fact that such a small number of terms is used. The use of two-hundred terms gives an approximation which graphically does not deviate perceptibly from the actual function. Results showing how rapidly the Fourier coefficients diminish in magnitude with increasing order appear in Figure 6.5. The dynamical analysis produced displacement and acceleration values for the mass. Figure 6.6 shows both the total displacement as well as the displacement contributed from the homogeneous solution alone. Evidently, the steady-state harmonic response function captures well most of the motion and the homogeneous part could probably be neglected without serious error. Figure 6.7 also shows the total acceleration of the mass which is, of course, proportional to the resultant force on the mass due to the base motion.

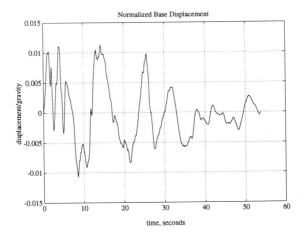

Figure 6.3. Normalized Base Displacement

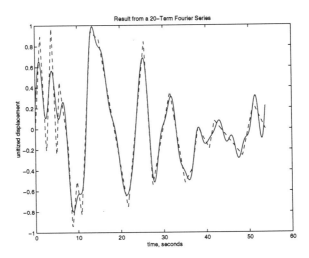

Figure 6.4. Result from a 20-Term Fourier Series

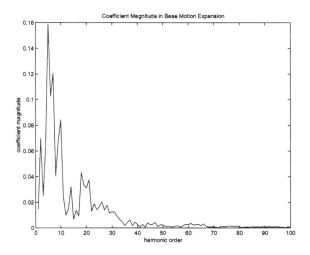

Figure 6.5. Coefficient Magnitude in Base Motion Expansion

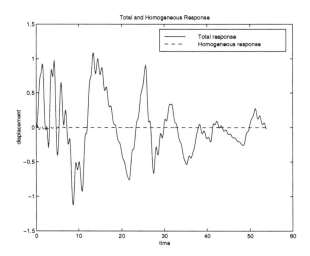

Figure 6.6. Total and Homogeneous Response

Before proceeding to the next example, the reader should be sure to appreciate the following important fact. Once a truncated Fourier series expansion of the forcing function using some appropriate number of terms is chosen, the truncated series defines an input function for which the response is computed exactly. If the user takes enough terms in the truncated series so that he/she is well satisfied with the function it approximates, then the computed response value for $y(t)$ will also be acceptable. This situation is distinctly different from the more complicated type of approximations occurring when finite difference or finite element methods produce discrete approximations for continuous field problems. Understanding the effects of grid size discretization error is more complex than understanding the effects of series truncation in the example given here.

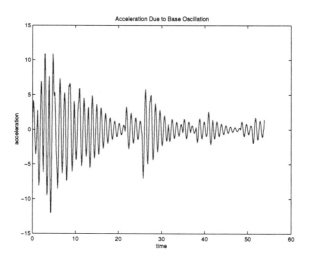

Figure 6.7. Acceleration Due to Base Oscillation

MATLAB Example

Script file runimpv

```
 1: % Example:  runimpv
 2: % ~~~~~~~~~~~~~~~~~~~
 3: % This is a driver program for the
 4: % earthquake example.
 5: %
 6: % User m functions required:
 7: %     fouaprox, imptp, hsmck, pauz
 8: %     shkbftss, lintrp, genprint
 9: %-------------------------------------------------
10:
11: % Make the undamped period about one
12: % second long
13: m=1; k=36;
14:
15: % Use damping equal to 5 percent of critical
16: c=.05*(2*sqrt(m*k));
17:
18: % Choose a period equal to length of
19: % Imperial Valley earthquake data
20: prd=53.8;
21:
22: nft=1024; tmin=0; tmax=prd;
23: ntimes=501; nsum=200;
24: tplt=linspace(0,prd,ntimes);
25: y20trm=fouaprox('imptp',prd,tplt,20);
26: plot(tplt,y20trm,'-',tplt,imptp(tplt),'--');
27: xlabel('time, seconds');
28: ylabel('unitized displacement');
29: title('Result from a 20-Term Fourier Series')
30: figure(gcf);
31: disp('Press [Enter] to continue'); pauz;
32: %genprint('20trmplt');
33:
34: % Show how magnitudes of Fourier coefficients
35: % decrease with increasing harmonic order
36:
37: fcof=fft(imptp((0:1023)/1024,1))/1024;
38: clf; plot(abs(fcof(1:100)));
39: xlabel('harmonic order');
40: ylabel('coefficient magnitude');
```

```
41: title(['Coefficient Magnitude in Base ' ...
42:         'Motion Expansion']); figure(gcf);
43: disp('Press [Enter] to continue'); pauz
44: %genprint('coefsize');
45:
46: % Compute forced response
47: [t,ys,ys0,vs0,as]= ...
48:    shkbftss(m,c,k,'imptp',prd,nft,nsum, ...
49:             tmin,tmax,ntimes);
50:
51: % Compute homogeneous solution
52: [t,yh,ah]= ...
53:    hsmck(m,c,k,-ys0,-vs0,tmin,tmax,ntimes);
54:
55: % Obtain the combined solution
56: y=ys(:)+yh(:); a=as(:)+ah(:);
57: clf; plot(t,y,'-',t,yh,'--');
58: xlabel('time'); ylabel('displacement');
59: title('Total and Homogeneous Response');
60: legend('Total response','Homogeneous response');
61: figure(gcf);
62: disp('Press [Enter] to continue'); pauz;
63: %genprint('displac');
64:
65: clf; plot(t,a,'-');
66: xlabel('time'); ylabel('acceleration')
67: title('Acceleration Due to Base Oscillation')
68: figure(gcf); %genprint('accel');
```

Function fouaprox

```
 1: function y=fouaprox(func,per,t,nsum,nft)
 2: %
 3: % y=fouaprox(func,per,t,nsum,nft)
 4: % ~~~~~~~~~~~~~~~~~~~~~~~~~~~~~~~~~
 5: % Approximation of a function by a Fourier
 6: % series.
 7: %
 8: % func    - function being expanded
 9: % per     - period of the function
10: % t       - vector of times at which the series
11: %             is to be evaluated
12: % nsum    - number of terms summed in the series
```

13: % nft   - number of function values used to
14: %   compute Fourier coefficients. This
15: %   should be an integer power of 2.
16: %   The default is 1024
17: %
18: % User m functions called: none.
19: %---
20:
21: if nargin<5, nft=1024; end;
22: nsum=min(nsum,fix(nft/2));
23: c=fft(feval(func,per/nft*(0:nft-1)))/nft;
24: c(1)=c(1)/2;
25: c=c(:); c=c(1:nsum);
26: w=2*pi/per*(0:nsum-1);
27: y=2*real(exp(i*t(:)*w)*c);

Function imptp

```
1: function ybase=imptp(t,period)
2: %
3: % ybase=imptp(t,period)
4: % ~~~~~~~~~~~~~~~~~~~~~
5: % This function defines a piecewise linear
6: % function resembling the ground motion of
7: % the earthquake which occurred in 1940 in
8: % the Imperial Valley of California. The
9: % maximum amplitude of base motion is
10: % normalized to equal unity.
11: %
12: % period - period of the motion
13: %           (optional argument)
14: % t      - vector of times between
15: %           tmin and tmax
16: % ybase  - piecewise linearly interpolated
17: %           base motion
18: %
19: % User m functions called:  lintrp
20: %----------------------------------------------
21:
22: tft=[ ...
23:    0.00    1.26    2.64    4.01    5.10 ...
24:    5.79    7.74;   8.65    9.74   10.77 ...
25:   13.06   15.07   21.60   25.49;  27.38 ...
```

```
26:    31.56     34.94     36.66     38.03     40.67 ...
27:    41.87;    48.40     51.04     53.80      0    ...
28:     0         0          0 ]';
29: yft=[ ...
30:     0         0.92     -0.25      1.00     -0.29 ...
31:     0.46     -0.16;    -0.97     -0.49     -0.83 ...
32:     0.95      0.86     -0.76      0.85;    -0.55 ...
33:     0.36     -0.52     -0.38      0.02     -0.19 ...
34:     0.08;    -0.26      0.24      0.00      0    ...
35:     0         0          0 ]';
36: tft=tft(:); yft=yft(:);
37: tft=tft(1:24); yft=yft(1:24);
38: if nargin == 2
39:    tft=tft*period/max(tft);
40: end
41: ybase=lintrp(tft,yft,t);
```

Function shkbftss

```
1: function [t,ys,ys0,vs0,as]=...
2:        shkbftss(m,c,k,ybase,prd,nft,nsum, ...
3:                 tmin,tmax,ntimes)
4: %
5: % [t,ys,ys0,vs0,as]=...
6: %    shkbftss(m,c,k,ybase,prd,nft,nsum, ...
7: %             tmin,tmax,ntimes)
8: % ~~~~~~~~~~~~~~~~~~~~~~~~~~~~~~~~~~~~~~~~~~
9: % This function determines the steady state
10: % solution of the scalar differential equation
11: %
12: %    m*y''(t) + c*y'(t) + k*y(t) =
13: %                    k*ybase(t) + c*ybase'(t)
14: %
15: % where ybase is a function of period prd
16: % which is expandable in a Fourier series
17: %
18: % m,c,k      - Mass, damping coefficient, and
19: %               spring stiffness
20: % ybase      - Function or vector of
21: %               displacements equally spaced in
22: %               time which describes the base
23: %               motion over a period
24: % prd        - Period used to expand xbase in a
```

```
25: %                     Fourier series
26: % nft        - The number of components used
27: %              in the FFT (should be a power
28: %              of two). If nft is input as
29: %              zero, then ybase must be a
30: %              vector and nft is set to
31: %              length(ybase)
32: % nsum       - The number of terms to be used
33: %              to sum the Fourier series
34: %              expansion of ybase. This should
35: %              not exceed nft/2.
36: % tmin,tmax  - The minimum and maximum times
37: %              for which the solution is to
38: %              be computed
39: % t          - A vector of times at which
40: %              the solution is computed
41: % ys         - Vector of steady state solution
42: %              values
43: % ys0,vs0    - Position and velocity at t=0
44: % as         - Acceleration ys''(t), if this
45: %              quantity is required
46: %
47: % User m functions called:  none.
48: %-------------------------------------------------
49:
50: if nft==0
51:     nft=length(ybase); ybft=ybase(:)
52: else
53:     tbft=prd/nft*(0:nft-1);
54:     ybft=fft(feval(ybase,tbft))/nft;
55:     ybft=ybft(:);
56: end
57: nsum=min(nsum,fix(nft/2)); ybft=ybft(1:nsum);
58: w=2*pi/prd*(0:nsum-1);
59: t=tmin+(tmax-tmin)/(ntimes-1)*(0:ntimes-1)';
60: etw=exp(i*t*w); w=w(:);
61: ysft=ybft.*(k+i*c*w)./(k+w.*(i*c-m*w));
62: ysft(1)=ysft(1)/2;
63: ys=2*real(etw*ysft); ys0=2*real(sum(ysft));
64: vs0=2*real(sum(i*w.*ysft));
65: if nargout > 4
66:     ysft=-ysft.*w.^2;
67:     as=2*real(etw*ysft);
68: end
```

Function hsmck

```
 1: function [t,yh,ah]= ...
 2:           hsmck(m,c,k,y0,v0,tmin,tmax,ntimes)
 3: %
 4: % [t,yh,ah]=hsmck(m,c,k,y0,v0,tmin,tmax,ntimes)
 5: % ~~~~~~~~~~~~~~~~~~~~~~~~~~~~~~~~~~~~~~~~~~~~~~~~
 6: % Solution of
 7: %       m*yh''(t) + c*yh'(t) + k*yh(t) = 0
 8: % subject to initial conditions of
 9: %       yh(0) = y0 and yh'(0) = v0
10: %
11: % m,c,k       -   mass, damping and spring
12: %                 constants
13: % y0,v0       -   initial position and velocity
14: % tmin,tmax   -   minimum and maximum times
15: % ntimes      -   number of times to evaluate
16: %                 solution
17: % t           -   vector of times
18: % yh          -   displacements for the
19: %                 homogeneous solution
20: % ah          -   accelerations for the
21: %                 homogeneous solution
22: %
23: % User m functions called:  none.
24: %-------------------------------------------------
25:
26: t=tmin+(tmax-tmin)/(ntimes-1)*(0:ntimes-1);
27: r=sqrt(c*c-4*m*k);
28: if r~=0
29:   s1=(-c+r)/(2*m); s2=(-c-r)/(2*m);
30:   g=[1,1;s1,s2]\[y0;v0];
31:   yh=real(g(1)*exp(s1*t)+g(2)*exp(s2*t));
32:   if nargout > 2
33:     ah=real(s1*s1*g(1)*exp(s1*t)+ ...
34:       s2*s2*g(2)*exp(s2*t));
35:   end
36: else
37:   s=-c/(2*m);
38:   g1=y0; g2=v0-s*g1; yh=(g1+g2*t).*exp(s*t);
39:   if nargout > 2
40:     ah=real(s*(2*g2+s*g1+s*g2*t).*exp(s*t));
41:   end
42: end
```

6.2.3 General Program to Construct Fourier Expansions

The final example in this chapter is a program to compute Fourier coefficients of general real valued functions and to display series with varying numbers of terms so that a user can see how rapidly such series converge. Since a truncated Fourier series is a continuous differentiable function, it cannot perfectly represent a discontinuous function such as a square wave. Near points where jump discontinuities occur, Fourier series approximations oscillate [18]. The same behavior occurs less seriously near points of slope discontinuity. Surprisingly, adding more terms does not cure the problem at jump discontinuities. The behavior, known as Gibbs phenomenon, produces approximations which overshoot the function on either side of the discontinuity. Illustrations of this behavior appear below.

A program was written to expand real functions of arbitrary period using Fourier series approximations defined by applying the FFT to data values equally spaced in time over a period. The function can either be piecewise linear or can be represented by a MATLAB M-file. For instance, a function varying like a sine curve with the bottom half cut off would be

```
function y=chopsine(x,period)
y=sin(pi*x/period).*(x<period)
```

The program consists of the following functions.

fouseris	main driver
sine	example for exact function input
lintrp	function for piecewise linear interpolation
fousum	sum a real valued Fourier series
read	reads several data items on one line

Comments within the program illustrate how to input data interactively. Details of different input options can be found by executing the program.

Let us see how well the FFT approximates a function of period 3 defined by piecewise linear interpolation through (x, y) values of $(0,1)$, $(1,1)$, $(1,-1)$, $(2,-1)$, $(3,1)$, and $(4,0)$. The function has jump discontinuities at $x = 0$, $x = 1$, and $x = 4$. A slope discontinuity also occurs at $x = 3$. Program results using a twenty-term approximation appear in Figure 6.8. Results produced by 100- and 250-term series plotted near $x = 1$ are shown in Figures 6.9 and 6.10. Clearly, adding more terms does not eliminate the oscillation. However, the oscillation at a jump discontinuity can be reduced with the Lanczos smoothing procedure. Results

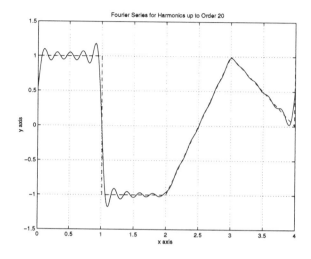

Figure 6.8. Fourier Series for Harmonics up to Order 20

for a series of 250 terms smoothed over an interval equal to the period times 0.01 appear in Figure 6.11. The oscillation is reduced at the cost of replacing the infinite slope at a discontinuity point by a steep slope of fifty-to-one for this case. Figure 6.12 shows a plot produced using an exact function definition as indicated in the second program execution. The reader may find it instructive to investigate how well Fourier series converge by running the program for other function choices.

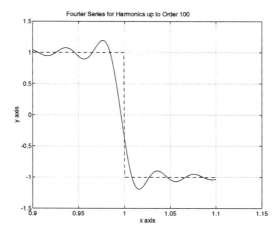

Figure 6.9. Fourier Series for Harmonics up to Order 100

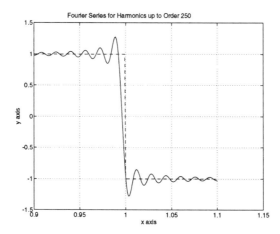

Figure 6.10. Fourier Series for Harmonics up to Order 250

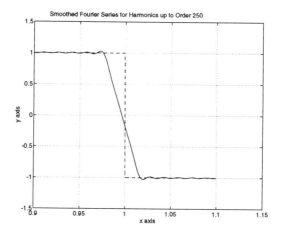

Figure 6.11. Smoothed Fourier Series for Harmonics to Order 250

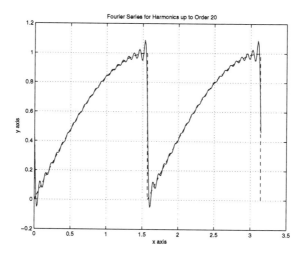

Figure 6.12. Exact Function Example for Harmonics up to Order 20

MATLAB Example

Output for Piecewise Linear Example

```
>> fouseris


FOURIER SERIES EXPANSION FOR A PIECEWISE LINEAR OR
          ANALYTICALLY DEFINED FUNCTION


Input the period of the function
 ? > 4


Input the number of data points to define the function
by piecewise linear interpolation (input a zero if the
function is defined analytically by the user).
 ? > 6


Input the x,y values one pair per line
 ? > 0,1
 ? > 1,1
 ? > 1,-1
 ? > 2,-1
 ? > 3,1
 ? > 4,0


To plot the series input xmin, xmax, and the highest
harmonic not exceeding 255 (input 0,0,0 to stop)
(Use a negative harmonic number to save your graph)
 ? > 0,4,20


To plot the series smoothed over a fraction of the
period, input the smoothing fraction
(give 0.0 for no smoothing).
 ? > 0


Press RETURN to continue


To plot the series input xmin, xmax, and the highest
harmonic not exceeding 255 (input 0,0,0 to stop)
(Use a negative harmonic number to save your graph)
 ? > 0,0,0
```

Output for Analytically Defined Example

```
>> fouseris
```

FOURIER SERIES EXPANSION FOR A PIECEWISE LINEAR OR
        ANALYTICALLY DEFINED FUNCTION

```
Input the period of the function
 ? > pi/2
```

```
Input the number of data points to define the function
by piecewise linear interpolation (input a zero if the
function is defined analytically by the user).
 ? > 0
```

Select the method used for exact function definition:

1 <=> Use an existing function with syntax defined by
the following example:

```
function y=sine(x,period)
%
% y=sine(x,period)
% ----------------
% This function specifies all or part of
% a sine wave.
%
%   x        - vector of argument values
%   period   - period of the function
%   y        - vector of function values
%
% User m functions called:  none
%--------------------------------------------
y=sin(rem(x,period));
```

or

2 <=> Use a one-line character string definition
involving argument x and period p. For example a sine
wave with the bottom cut off would be defined by:
sin(x*2*pi/p).*(x<p/2)

1 or 2 ? > 1

Enter the name of your function

```
 ? > sine
```

To plot the series input xmin, xmax, and the highest
harmonic not exceeding 255 (input 0,0,0 to stop)
(Use a negative harmonic number to save your graph)
 ? > 0,pi,-20

To plot the series smoothed over a fraction of the
period, input the smoothing fraction
(give 0.0 for no smoothing).
 ? > 0

Give a file name to save the current graph >
exactplt

Press RETURN to continue

To plot the series input xmin, xmax, and the highest
harmonic not exceeding 255 (input 0,0,0 to stop)
(Use a negative harmonic number to save your graph)
 ? > 0,0,0

Script File fouseris

```
 1: % Example: fouseris
 2: % ~~~~~~~~~~~~~~~~~
 3: % This program illustrates the convergence rate
 4: % of Fourier series approximations derived by
 5: % applying the FFT to a general function which
 6: % may be specified either by piecewise linear
 7: % interpolation in a data table or by
 8: % analytical definition in a function given by
 9: % the user. The linear interpolation model
10: % permits inclusion of jump discontinuities.
11: % Series having varying numbers of terms can
12: % be graphed to demonstrate Gibbs phenomenon
13: % and to show how well the truncated Fourier
14: % series represents the original function.
15: % Provision is made to plot the Fourier series
16: % of the original function or a smoothed
17: % function derived by averaging the original
18: % function over an arbitrary fraction of the
19: % total period.
20: %
21: % User m functions required:
22: %     fousum, lintrp, read, sine, genprint
23: %-------------------------------------------------
24:
25: % The following parameters control the number
26: % of fft points used and the number of points
27: % used for graphing.
28: nft=512; ngph=300; nmax=int2str(nft/2-1);
29:
30: fprintf('\nFOURIER SERIES EXPANSION FOR');
31: fprintf(' A PIECEWISE LINEAR OR');
32: fprintf('\n        ANALYTICALLY DEFINED ');
33: fprintf('FUNCTION\n');
34:
35: fprintf('\nInput the period of the function\n');
36: period=input('? > ');
37: xfc=(period/nft)*(0:nft-1)';
38: fprintf('\nInput the number of data ');
39: fprintf('points to define the function');
40: fprintf('\nby piecewise linear ');
41: fprintf('interpolation (input a zero if the');
42: fprintf('\nfunction is defined analytically');
```

```
43: fprintf(' by the user).\n');
44: nd=input('? > ');
45: if nd > 0, xd=zeros(nd,1); yd=xd;
46:    fprintf('\nInput the x,y values one ');
47:    fprintf('pair per line\n');
48:    for j=1:nd
49:       [xd(j),yd(j)]=read('? > ');
50:    end
51:
52: % Use nft interpolated data points to
53: % compute the fft
54:    yfc=lintrp(xd,yd,xfc); c=fft(yfc);
55: else
56:    fprintf('\nSelect the method used for ');
57:    fprintf('exact function definition:\n');
58:    fprintf('\n1 <=> Use an existing function ');
59:    fprintf('with syntax defined by');
60:    fprintf('\nthe following example:\n');
61:    type sine.m
62:    fprintf('or\n');
63:    fprintf('\n2 <=> Use a one-line character');
64:    fprintf(' string definition involving');
65:    fprintf('\nargument x and period p. For');
66:    fprintf(' example a sine wave with the');
67:    fprintf('\nbottom cut off would be ');
68:    fprintf('defined by: ');
69:    fprintf('sin(x*2*pi/p).*(x<p/2)\n');
70:    nopt=input('1 or 2 ? > ');
71:    if nopt == 1
72:       fprintf('\nEnter the name of your ');
73:       fprintf('function\n');
74:       fnam=input('? > ','s');
75:       yfc=feval(fnam,xfc,period); c=fft(yfc);
76:    else
77:       fprintf('\nInput the one-line definition');
78:       fprintf(' in terms of x and p\n');
79:       strng=input('? > ','s');
80:       x=xfc; p=period;
81:       yfc=eval(strng); c=fft(yfc);
82:    end
83: end
84:
85: while 1
86:    fprintf('\nTo plot the series input xmin,');
87:    fprintf(' xmax, and the highest');
```

```
88:     fprintf(['\nharmonic not exceeding ', ...
89:            nmax,' (press [Enter] to stop)']);
90:     fprintf('\n(Use a negative harmonic number');
91:     fprintf(' to save your graph)\n');
92:     [xl,xu,nh]=read('? > ');
93:     if norm([xl,xu,nh])==0, break; end
94:     pltsav=(nh < 0); nh=abs(nh);
95:     xtmp=xl+((xu-xl)/ngph)*(0:ngph);
96:     fprintf('\nTo plot the series smoothed ');
97:     fprintf('over a fraction of the');
98:     fprintf('\nperiod, input the smoothing ');
99:     fprintf('fraction');
100:    fprintf('\n(give 0.0 for no smoothing).\n');
101:    alpha=input('? > ');
102:    yfou=fousum(c,xtmp,period,nh,alpha);
103:    xxtmp=xtmp; idneg=find(xtmp<0);
104:    xng=abs(xtmp(idneg));
105:    xxtmp(idneg)=xxtmp(idneg)+ ...
106:               period*ceil(xng/period);
107:    if nd>0
108:      yexac=lintrp(xd,yd,rem(xxtmp,period));
109:    else
110:      if nopt == 1
111:        yexac=feval(fnam,xtmp,period);
112:      else
113:        x=xxtmp; yexac=eval(strng);
114:      end
115:    end
116:    in=int2str(nh);
117:    if alpha == 0
118:      titl=['Fourier Series for Harmonics ' ...
119:            'up to Order ',in];
120:    else
121:      titl=['Smoothed Fourier Series for ' ...
122:            'Harmonics up to Order ',in];
123:    end
124:    clf; plot(xtmp,yfou,'-',xtmp,yexac,'--');
125:    ylabel('y axis'); xlabel('x axis');
126:    title(titl); grid on; figure(gcf); disp(' ');
127:    input('Press [Enter] to continue ','s');
128:    if pltsav
129:      filnam=input(['Give a file name to ' ...
130:            'save the current graph > ? '],'s');
131:      if length(filnam) > 0
132:        genprint(filnam);
```

```
133:    end
134:  end
135: end
```

Function sine

```
1: function y=sine(x,period)
2: %
3: % y=sine(x,period)
4: % ~~~~~~~~~~~~~~~~~
5: % This function specifies all or part
6: % of a sine wave.
7: %
8: %   x        - vector of argument values
9: %   period  - period of the function
10: %   y        - vector of function values
11: %
12: % User m functions called:  none
13: %-----------------------------------------------
14:
15: y=sin(rem(x,period));
```

Function fousum

```
1: function yreal=fousum(c,x,period,k,alpha)
2: %
3: % yreal = fousum(c,x,period,k,alpha)
4: % ~~~~~~~~~~~~~~~~~~~~~~~~~~~~~~~~~~~~
5: % Sum the Fourier series of a real
6: % valued function.
7: %
8: %   x        - The vector of real values at
9: %              which the series is evaluated.
10: %   c        - A vector of length n containing
11: %              Fourier coefficients output by
12: %              the fft function
13: %   period - The period of the function
14: %   k        - The highest harmonic used in
15: %              the Fourier sum.  This must
16: %              not exceed n/2-1
17: %   alpha  - If this parameter is nonzero,
18: %              the Fourier coefficients are
```

```
19: %                    replaced by those of a function
20: %                    obtained by averaging the
21: %                    original function over alpha
22: %                    times the period
23: %    yreal  - The real valued Fourier sum
24: %                    for argument x
25: %
26: % The Fourier coefficients c must have been
27: % computed using the fft function which
28: % transforms the vector [y(1),...,y(n)] into
29: % an array of complex Fourier coefficients
30: % which have been multiplied by n and are
31: % arranged in the order:
32: %
33: %    [c(0),c(1),...,c(n/2-1),c(n/2),
34: %                    c(-n/2+1),...,c(-1)].
35: %
36: % The coefficient c(n/2) cannot be used
37: % since it is actually the sum of c(n/2) and
38: % c(-n/2). For a particular value of n, the
39: % highest usable harmonic is n/2-1.
40: %
41: % User m functions called:  none
42: %--------------------------------------------------
43:
44: x=x(:); n=length(c);
45: if nargin <4, k=n/2-1; alpha=0; end
46: if nargin <5, alpha=0; end
47: if nargin <3, period=2*pi; end
48: L=period/2; k=min(k,n/2-1); th=(pi/L)*x;
49: i=sqrt(-1); z=exp(i*th);
50: y=c(k+1)*ones(size(th)); pa=pi*alpha;
51: if alpha > 0
52:    jj=(1:k)';
53:    c(jj+1)=c(jj+1).*sin(jj*pa)./(jj*pa);
54: end
55: for j=k:-1:2
56:    y=c(j)+y.*z;
57: end
58: yreal=real(c(1)+2*y.*z)/n;
```

Chapter 7

Dynamic Response of Linear Second Order Systems

7.1 Solving the Structural Dynamics Equations for Periodic Applied Forces

The dynamics of a linear structure subjected to periodic forces obeys the matrix differential equation

$$M\ddot{X} + C\dot{X} + KX = F(t)$$

with initial conditions

$$X(0) = D_0 \qquad \dot{x}(0) = V_0$$

The solution vector $X(t)$ has dimension n and M, C, and K are real square matrices of order n. The mass matrix, M, the damping matrix, C, and the stiffness matrix, K, are all real. The forcing function $F(t)$, assumed to be real and having period L, can be approximated by a finite trigonometric series as

$$F(t) = \sum_{k=-N}^{N} c_k e^{i\omega_k t} \qquad i = \sqrt{-1} \qquad \omega_k = 2\pi k/L$$

The Fourier coefficients c_k are vectors which can be computed using the FFT. The fact that $F(t)$ is real also implies that $c_{-k} = \mathbf{conj}(c_k)$ and, therefore,

$$F(t) = c_0 + 2\,\mathbf{real}\left(\sum_{k=1}^{n} c_k e^{i\omega_k t}\right)$$

The solution of the differential equation is naturally resolvable into two distinct parts. The first is the so called particular or forced response which is periodic and has the same general mathematical form as the forcing function. Hence, we write

$$X_p = \sum_{k=-n}^{n} X_k e^{\imath \omega_k t} = X_0 + 2 \ \textbf{real} \left(\sum_{k=1}^{n} X_k e^{\imath \omega_k t} \right)$$

Substituting this series into the differential equation and matching coefficients of $e^{\imath \omega_k t}$ on both sides yields

$$X_k = (K - \omega_k^2 M + \imath \omega_k C)^{-1} c_k$$

The particular solution satisfies initial conditions given by

$$X_p(0) = X_0 + 2 \ \textbf{real} \left(\sum_{k=1}^{n} c_k \right)$$

and

$$\dot{X}_p(0) = 2 \ \textbf{real} \left(\sum_{k=1}^{n} \imath \omega_k c_k \right)$$

Since these conditions usually will not equal the desired values, the particular solution must be combined with what is called the homogeneous or transient solution X_k

$$M\ddot{X}_h + c\dot{X}_h + KX_h = 0$$

$$X_h(0) = D_0 - X_p(0) \qquad \dot{X}_h(0) = V_0 - \dot{X}_p(0)$$

The homogeneous solution can be constructed by reducing the original differential equation to first order form. Let Z be the vector of dimension $2n$ which is the concatenation of X and $\dot{X} = V$. Hence, $Z = [X; V]$ and the original equation of motion is

$$\frac{dZ}{dt} = AZ + P(t)$$

where

$$A = \begin{bmatrix} 0 & I \\ -M^{-1}K & -M^{-1}C \end{bmatrix} \qquad P = \begin{bmatrix} 0 \\ m^{-1}F \end{bmatrix}$$

The homogeneous differential equation resulting when $P = 0$ can be solved in terms of the eigenvalues and eigenvectors of matrix A. If we know eigenvalues λ_j and eigenvectors U_j satisfying

$$AU_j = \lambda_j U_j \qquad 1 \leq j \leq 2n$$

then the homogeneous solution can be written as

$$Z = \sum_{j=1}^{2n} z_j U_j e^{\imath \omega_j t}$$

The weighting coefficients z_j are computed to satisfy the desired initial conditions which require

$$\begin{bmatrix} U_1, & U_2, & \cdots, & U_{2n} \end{bmatrix} \begin{bmatrix} z_1 \\ \vdots \\ z_{2n} \end{bmatrix} = \begin{bmatrix} X_0 - X_p(0) \\ V_0 - \dot{X}_p(0) \end{bmatrix}$$

We solve this system of equations for $z_1, \cdots, z_{2n}$ and replace each U_j by $z_j U_j$. Then the homogeneous solution is

$$X_h = \sum_{j=1}^{n} U_j(1 : n) e^{\lambda_j t}$$

where $U_j(1 : n)$ means we take only the first n elements of column j.

In most practical situations, matrix C is nonzero and the eigenvalues $\lambda_1, \cdots, \lambda_{2n}$ have negative real parts. Then the exponential terms $e^{\lambda_j t}$ all decay rapidly, which explains the name, the transient solution, some authors give to X_h. In other cases, where the damping matrix C is zero, the eigenvalues λ_j are typically purely imaginary, and the homogeneous solution does not die out. In either instance, it is often customary in practical situations to ignore the homogeneous solution because it is usually small when compared to the contribution of the particular solution.

7.1.1 Application to Oscillations of a Vertically Suspended Cable

Let us solve the problem of small transverse vibrations of a vertically suspended cable. This system illustrates nicely how the natural frequencies and mode shapes of a linear system can be combined to satisfy general initial conditions on position and velocity.

The cable in Figure 7.1 is idealized as a series of n rigid links connected at frictionless joints. Two vectors, consisting of link lengths $[\ell_1, \ell_2, \cdots, \ell_n]$ and masses $[m_1, m_2, \cdots, m_n]$ lumped at the joints, characterize the system properties. The accelerations in the vertical direction will be negligibly small compared to transverse accelerations, because the transverse displacements are small. Consequently, the tension in the chain will remain close to the static equilibrium value. This means the tension in link $\imath$ is

$$T_\imath = g b_\imath \qquad b_\imath = \sum_{j=\imath}^{n} m_j$$

We assume that the transverse displacement y_i for mass m_i is small compared to the total length of the cable. A free body diagram for mass i is shown in Figure 7.2. The small deflection angles are related to the transverse deflections by $\theta_{i+1} = (y_{i+1} - y_i)\,\ell_{i+1}$ and $\theta_i = (y_i - y_{i-1})\,/\ell_i$. Summation of forces shows that the horizontal acceleration is governed by

$$m_i\ddot{y}_i \;\; = \;\; g(b_{i+1}/\ell_{i+1})\,(y_{i+1} - y_i) - g(b_i/\ell_i)\,(y_i - y_{i-1})$$

$$= \;\; g(b_i/\ell_i)y_{i-1} - g(b_i/\ell_i + b_{i+1}/\ell_{i+1})y_i + g(b_{i+1}/\ell_{i+1})y_{i+1}$$

In matrix form this equation is

$$M\ddot{Y} + KY = 0$$

where M is a diagonal matrix of mass coefficients and K is a symmetric tridiagonal matrix. The natural modes of free vibration are dynamical states where each element of the system simultaneously moves with harmonic motion of the same frequency. This means we seek motions of the form $Y = U\cos(\omega t)$, or equivalently $Y = U\sin(\omega t)$, which implies

$$KU_j = \lambda_j MU_j \qquad \lambda_j = \omega_j^2 \qquad 1 \le j \le n$$

Solving the eigenvalue problem $(M^{-1}K)U = \lambda U$ gives the natural frequencies $\omega_1, \cdots, \omega_n$ and the modal vectors $U_1, \cdots, U_n$. The response to general initial conditions is then obtained by superposition of the component modes. We write

$$Y = \sum_{j=1}^{n} \cos(\omega_j t)U_j c_j + \sin(\omega_j t)U_j d_j/\omega_j$$

where coefficients $c_1, \cdots, c_n$ and $d_1, \cdots, d_n$ (not to be confused with Fourier coefficients) are determined from the initial conditions as

$$\begin{bmatrix} U_1, & \cdots, & U_n \end{bmatrix} \begin{bmatrix} c_1 \\ \vdots \\ c_n \end{bmatrix} = Y(0) \qquad c = U^{-1}Y(0)$$

$$\begin{bmatrix} U_1, & \cdots, & U_n \end{bmatrix} \begin{bmatrix} d_1 \\ \vdots \\ d_n \end{bmatrix} = \dot{Y}(0) \qquad d = U^{-1}\dot{Y}(0)$$

and the system response is complete.

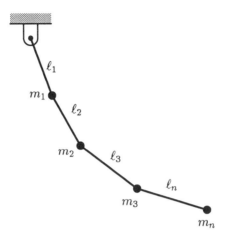

Figure 7.1. Transverse Cable Vibration

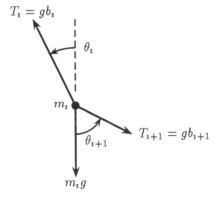

Figure 7.2. Forces on i'th Mass

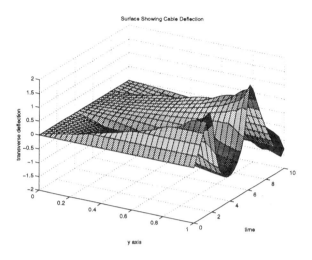

Figure 7.3. Surface Showing Cable Deflection

The following program determines the cable response for general initial conditions. The natural frequencies and mode shapes are computed along with an animation of the motion.

The cable motion produced when an initially vertical system is given the same initial transverse velocity for all masses was studied. Graphical results of the analysis appear in Figures 7.3 through 7.6. The surface plot in Figure 7.3 shows the cable deflection pattern in terms of longitudinal position and time. Figure 7.4 shows the deflection pattern at two times. Figure 7.5 traces the motion of the middle and the free end. At $t = 1$, the wave propagating downward from the support point is about halfway down the cable. By $t = 2$, the wave has reached the free end and the cable is about to swing back. Finally, traces of cable positions during successive stages of motion appear in Figure 7.6.

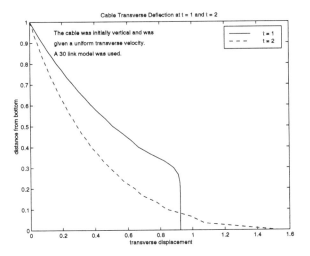

Figure 7.4. Cable Transverse Deflection at $t = 1$ and $t = 2$

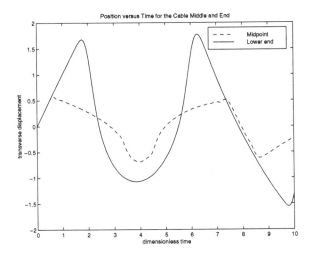

Figure 7.5. Position Versus Time for the Cable Middle and End

Trace of Cable Motion

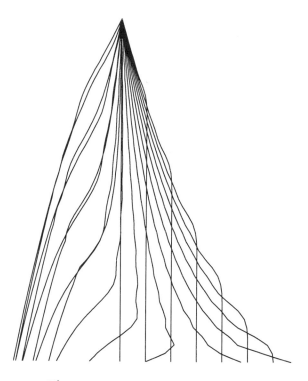

Figure 7.6. Trace of Cable Motion

MATLAB Example

Script File cablinea

```
 1: % Example: cablinea
 2: % ~~~~~~~~~~~~~~~~~
 3: % This program uses modal superposition to
 4: % compute the dynamic response of a cable
 5: % suspended at one end and free at the other.
 6: % The cable is given a uniform initial
 7: % velocity. Time history plots and animation
 8: % of the motion are provided.
 9: %
10: % User m functions required:
11: %    cablemk, udfrevib, canimate, genprint
12: %-----------------------------------------------
13:
14: % Initialize graphics
15: hold off; axis('normal'); clf; close;
16:
17: % Set physical parameters
18: n=30; gravty=1.; masses=ones(n,1)/n;
19: lengths=ones(n,1)/n;
20:
21: % Obtain mass and stiffness matrices
22: [m,k]=cablemk(masses,lengths,gravty);
23:
24: % Assign initial conditions & time limit
25: % for solution
26: dsp=zeros(n,1); vel=ones(n,1);
27: tmin=0; tmax=10; ntim=30;
28:
29: % Compute the solution by modal superposition
30: [t,u,modvc,natfrq]=...
31:    udfrevib(m,k,dsp,vel,tmin,tmax,ntim);
32:
33: % Interpret results graphically
34: nt1=sum(t<=tmin); nt2=sum(t<=tmax);
35: u=[zeros(ntim,1),u];
36: y=cumsum(lengths); y=[0;y(:)];
37:
38: % Plot deflection surface
39: clf; surf(y,t,u);
40: xlabel('y axis'); ylabel('time');
```

```
41: zlabel('transverse deflection');
42: title('Surface Showing Cable Deflection');
43: view([30,30]); figure(gcf);
44: %genprint('surface');
45: dum=input('Press [Enter] to continue','s');
46:
47: % Show deflection configuration at two times
48: % Use closer time increment than was used
49: % for the surface plots.
50: mtim=4*ntim;
51: [tt,uu,modvc,natfrq]=...
52:    udrevib(m,k,dsp,vel,tmin,tmax,mtim);
53: uu=[zeros(mtim,1),uu];
54: tp1=.1*tmax; tp2=.2*tmax;
55: s1=num2str(tp1); s2=num2str(tp2);
56: np1=sum(tt<=tp1); np2=sum(tt<=tp2);
57: u1=uu(np1,:); u2=uu(np2,:);
58: yp=flipud(y(:)); ym=max(yp);
59: plot(u1,yp,'-',u2,yp,'--');
60: ylabel('distance from bottom');
61: xlabel('transverse displacement');
62: title(['Cable Transverse Deflection ' ...
63:        'at t = ',s1,' and t = ',s2]);
64: legend('t = 1', 't = 2');
65: xm=.1*max([u1(:);u2(:)]);
66: text(xm,.95*ym, ...
67:    'The cable was initially vertical and was');
68: text(xm,.9*ym, ...
69:    'given a uniform transverse velocity.');
70: ntxt=int2str(n); n2=1+fix(n/2);
71: text(xm,.85*ym,...
72: ['A ',ntxt,' link model was used.']); figure(gcf);
73: dum=input('Press [Enter] to continue','s');
74: %genprint('twoposn');
75:
76: % Plot time history for the middle and the end
77: clf; plot(tt,uu(:,n2),'--',tt,uu(:,n+1),'-');
78: xlabel('dimensionless time');
79: ylabel('transverse displacement');
80: title(['Position versus Time for the ' ...
81:        'Cable Middle and End'])
82: legend('Midpoint','Lower end');
83: figure(gcf);
84: dum=input('Press [Enter] to continue','s');
85: %genprint('2timhist');
```

```
86:
87: % Plot animation of motion history
88: clf; canimate(y,u,t,0,.5*max(t),1);
89: %genprint('motntrac');
90:
```

Function cablemk

```
1: function [m,k]=cablemk(masses,lngths,gravty)
2: %
3: % [m,k]=cablemk(masses,lngths,gravty)
4: % ~~~~~~~~~~~~~~~~~~~~~~~~~~~~~~~~~~~~
5: % Form the mass and stiffness matrices for
6: % the cable.
7: %
8: % masses      - vector of masses
9: % lngths      - vector of link lengths
10: % gravty      - gravity constant
11: % m,k         - mass and stiffness matrices
12: %
13: % User m functions called:  none.
14: %-----------------------------------------------
15:
16: m=diag(masses);
17: b=flipud(cumsum(flipud(masses(:))))* ...
18:    gravty./lngths;
19: n=length(masses); k=zeros(n,n); k(n,n)=b(n);
20: for i=1:n-1
21:    k(i,i)=b(i)+b(i+1); k(i,i+1)=-b(i+1);
22:    k(i+1,i)=k(i,i+1);
23: end
```

Function udfrevib

```
1: function [t,u,mdvc,natfrq]=...
2:                 udfrevib(m,k,u0,v0,tmin,tmax,nt)
3: %
4: % [t,u,mdvc,natfrq]= ...
5: %                 udfrevib(m,k,u0,v0,tmin,tmax,nt)
6: % ~~~~~~~~~~~~~~~~~~~~~~~~~~~~~~~~~~~~~~~~~~~~~~~~
7: % This function computes undamped natural
8: % frequencies, modal vectors, and time response
```

```
 9: % by modal superposition.  The matrix
10: % differential equation and initial conditions
11: % are
12: %
13: %    m u'' + k u = 0,  u(0) = u0, u'(0) = v0
14: %
15: % m,k        - mass and stiffness matrices
16: % u0,v0      - initial position and velocity
17: %              vectors
18: % tmin,tmax - time limits for solution
19: %              evaluation
20: % nt         - number of times for solution
21: % t          - vector of solution times
22: % u          - matrix with row j giving the
23: %              system response at time t(j)
24: % mdvc       - matrix with columns which are
25: %              modal vectors
26: % natfrq     - vector of natural frequencies
27: %
28: % User m functions called:  none.
29: %------------------------------------------------
30:
31: % Call function eig to compute modal vectors
32: % and frequencies
33: [mdvc,w]=eig(m\k);
34: [w,id]=sort(diag(w)); w=sqrt(w);
35:
36: % Arrange frequencies in ascending order
37: mdvc=mdvc(:,id); z=mdvc\[u0(:),v0(:)];
38:
39: % Generate vector of equidistant times
40: t=linspace(tmin,tmax,nt);
41:
42: % Evaluate the displacement as a
43: % function of time
44: u=(mdvc*diag(z(:,1)))*cos(w*t)+...
45:   (mdvc*diag(z(:,2)./w))*sin(w*t);
46: t=t(:); u=u'; natfrq=w;
```

Function canimate

```
1: function canimate(y,u,t,tmin,tmax,norub)
2: %
```

```
 3: % canimate(y,u,t,tmin,tmax,norub)
 4: % ~~~~~~~~~~~~~~~~~~~~~~~~~~~~~~~~~~
 5: % This function draws an animated plot of
 6: % data values stored in array u. The
 7: % different columns of u correspond to position
 8: % values in vector y. The successive rows of u
 9: % correspond to different times. Parameter
10: % tpause controls the speed of the animation.
11: %
12: % u          - matrix of values for which
13: %                animated plots of u versus y
14: %                are required
15: % y          - spatial positions for different
16: %                columns of u
17: % t          - time vector at which positions
18: %                are known
19: % tmin,tmax - time limits for graphing of the
20: %                solution
21: % norub      - parameter which makes all
22: %                position images remain on the
23: %                screen. Only one image at a
24: %                time shows if norub is left out.
25: %                A new cable position appears each
26: %                time the user presses any key
27: %
28: % User m functions called:  none.
29: %-----------------------------------------------
30:
31: % If norub is input,
32: %    all images are left on the screen
33: if nargin < 6
34:    rubout = 1;
35: else
36:    rubout = 0;
37: end
38:
39: % Determine window limits
40: umin=min(u(:)); umax=max(u(:)); udif=umax-umin;
41: uavg=.5*(umin+umax);
42: ymin=min(y); ymax=max(y); ydif=ymax-ymin;
43: yavg=.5*(ymin+ymax);
44: ywmin=yavg-.55*ydif; ywmax=yavg+.55*ydif;
45: uwmin=uavg-.55*udif; uwmax=uavg+.55*udif;
46: n1=sum(t<=tmin); n2=sum(t<=tmax);
47: t=t(n1:n2); u=u(n1:n2,:);
```

```
48: u=fliplr (u); [ntime,nxpts]=size(u);
49:
50: hold off; cla; ey=0; eu=0; axis('square');
51: axis([uwmin,uwmax,ywmin,ywmax]);
52: axis off; hold on;
53: title('Trace of Cable Motion');
54:
55: % Plot successive positions
56: for j=1:ntime
57:   ut=u(j,:); plot(ut,y,'-');
58:   figure(gcf); pause(1);
59:
60:   % Erase image before next one appears
61:   if rubout & j < ntime, cla, end
62: end
```

7.2   Direct Integration Methods

Using stepwise integration methods to solve the structural dynamics equation provides an alternative to frequency analysis methods. If we invert the mass matrix and save the result for later use, the n degree-of-freedom system can be expressed concisely as a first order system in $2n$ unknowns for a vector by $z = [x; v]$, where v is the time derivative of x. The system can be solved by applying the variable step-size differential equation integrator **ode45** to the following function:

```
function zdot = sdeq(t,z)
global invmas_ damp_ stif_ forcname_
n=length(z)/2; x=z(1:n); v=z(n+1:2*n); fnam=forcname_;
zdot=[v;invmas_*(feval(fnam,t)-stif_*x-damp_*v)];
```

In this function, the inverted mass matrix has been stored in a global variable **invmas_**, the damping and stiffness matrices are in **damp_** and **stiff_**, and the forcing function name is stored in a character string called **forcname_** Although this approach is easy to implement, the resulting analysis can be very time consuming for large systems. Variable step integrators make adjustments to control stability and accuracy which often lead to very small integration steps. Consequently, alternative formulations employing fixed step-size are usually chosen. We will investigate two such algorithms derived from trapezoidal integration rules [7, 111]. The two fundamental integration formulas [25] needed are:

$$\int_a^b f(t)dt \;=\; \frac{h}{2}[f(a)+f(b)] - \frac{h^3}{12}f''(\epsilon_1) \qquad \left\{ \begin{array}{l} a \le \epsilon_1 \le b \\[4pt] h = (b-a) \end{array} \right.$$

and

$$\int_a^b f(t)dt \;=\; \frac{h}{2}[f(a)+f(b)] + \frac{h^2}{12}[f'(a)-f'(b)] \;+$$

$$\frac{h^5}{720}f^{(4)}(\epsilon_2) \qquad\qquad\qquad a \le \epsilon_2 \le b$$

The first formula, called the trapezoidal rule, gives a zero truncation error term when applied to a linear function. Similiarly, the second formula, called the trapezoidal rule with end correction, has a zero final term for a cubic integrand.

The idea is to multiply the differential equation by dt, integrate from t to $(t+h)$, and employ numerical integration formulas while observing that M, C, and K are constant matrices, or

$$M \int_t^{t+h} \dot{V} \, dt + C \int_t^{t+h} \dot{X} \, dt + K \int_t^{t+h} X \, dt = \int_t^{t+h} P(t) \, dt$$

and

$$\int_t^{t+h} \dot{X} \, dt = \int_t^{t+h} V \, dt$$

For brevity we utilize a notation characterized by $X(t) = X_0$, $X(t+h) = X_1$, $\tilde{X} = X_1 - X_0$. The trapezoidal rule immediately leads to

$$\left[M + \frac{h}{2}C + \frac{h^2}{4}K \right] \tilde{V} = \int P(t)dt -$$

$$h \left[CV_0 + K(X_0 + \frac{h}{2}V_0) \right] + O(h^3)$$

The last equation is a balance of impulse and momentum change involving the effective mass matrix

$$M_e = \left[M + \frac{h}{2}C + \frac{h^2}{4}K \right]$$

which can be inverted once and used repeatedly if the step-size is not changed.

To integrate the forcing function we can use the midpoint rule [25] which states

$$\int P(t) \, dt = hP \left(\frac{a+b}{2} \right) + O(h^3)$$

Solving for $\tilde{V}$ yields

$$\tilde{V} = \left[M + \frac{h}{2}C + \frac{h^2}{4}K \right]^{-1} \left[P \left(t + \frac{h}{2} \right) - CV_0 - \right.$$

$$\left. K \left(X_0 + \frac{h}{2}V_0 \right) \right] h + O(h^3)$$

The velocity and position at $(t + h)$ are then computed as

$$V_1 = V_0 + \tilde{V} \qquad X_1 = X_0 + \frac{h}{2}[V_0 + V_1] + O(h^3)$$

A more accurate formula with truncation error of order h^5 can be developed from the extended trapezoidal rule. This leads to

$$M\tilde{V} + C\tilde{X} + K \left[\frac{h}{2}(\tilde{X} + 2X_0) - \frac{h^2}{12}\tilde{V} \right] = \int P(t)dt + O(h^5)$$

and

$$\tilde{X} = \frac{h}{2}[\tilde{V} + 2V_0] + \frac{h^2}{12}[\dot{V_0} - \dot{V_1}] + O(h^5)$$

Multiplying the last equation by M and employing the differential equation to reduce the $\dot{V_0} - \dot{V_1}$ terms gives

$$M\tilde{X} = \frac{h}{2}M[\tilde{V} + 2V_0] + \frac{h^2}{12}[-\tilde{P} + C\tilde{V} + K\tilde{X}] + O(h^5)$$

These results can be arranged into a single matrix equation to be solved for $\tilde{X}$ and $\tilde{V}$, or

$$\begin{bmatrix} -(\frac{h}{2}M + \frac{h^2}{12}C) & (M - \frac{h^2}{12}K) \\ \\ (M - \frac{h^2}{12}K) & (C + \frac{h}{2}K) \end{bmatrix} \begin{bmatrix} \tilde{V} \\ \\ \tilde{X} \end{bmatrix} =$$

$$\begin{bmatrix} hMV_0 + \frac{h^2}{12}(P_0 - P_1) \\ \\ \int P dt - hKX_0 \end{bmatrix} + O(h^5)$$

A Gauss two-point formula [25] evaluates the force integral consistent with the desired error order so that

$$\int_t^{t+h} P(t)dt = \frac{h}{2}\left[P(t + \alpha h) + P(t + \beta h)\right] + O(h^5)$$

where $\alpha = \frac{3-\sqrt{3}}{6}$ and $\beta = \frac{3+\sqrt{3}}{6}$.

7.2.1   Example on Cable Response by Direct Integration

Functions implementing the last two algorithms appear in the following program which solves the previously considered cable dynamics example by direct integration. Questions of computational efficiency and numerical accuracy are examined for two different step-sizes. Figures 7.7 and 7.8 present solution times as multiples of the times needed for a modal response solution. The accuracy measures employed are described next. Note that the displacement response matrix has rows describing system positions at successive times. Consequently, a measure of the difference between approximate and exact solutions is given by the vector

```
error_vector = sqrt(sum(((x_aprox-x_exact).^2)'));
```

Typically this vector has small initial components (near $t = 0$) and larger components (near the final time). The error measure is compared

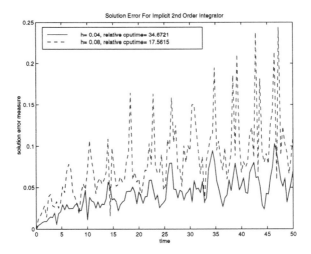

Figure 7.7. Solution Error for Implicit 2nd Order Integrator

for different integrators and time steps in the figures. Note that the fourth order integrator is more efficient than the second order integrator because a larger integration step can be taken without excessive accuracy loss. Using $h = 0.4$ for **mckde4i** achieved nearly the same accuracy as that given by **mckde2i** with $h = 0.067$. However, the computation time for **mckde2i** was several times as large as that for **mckde4i**.

In the past it has been traditional to use only second order methods for solving the structural dynamics equation. This may have been dictated by considerations on computer memory. Since workstations widely available today have relatively large memories and can invert a matrix of order two hundred in about a second, it appears that use of high order integrators may gain in popularity.

The following computer program concludes our chapter on solution of linear, constant-coefficient matrix differential equations. The next chapter deals with integration of nonlinear problems.

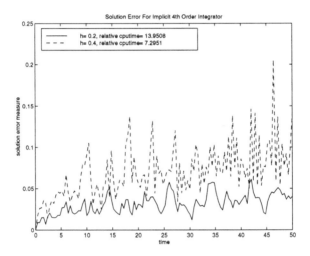

Figure 7.8. Solution Error for Implicit 4th Order Integrator

MATLAB Example

Script File deislner

```
1: % Example:  deislner
2: % ~~~~~~~~~~~~~~~~~~
3: % Solution error for simulation of cable
4: % motion using a second or a fourth order
5: % implicit integrator.
6: %
7: % This program uses implicit second or fourth
8: % order integrators to compute the dynamical
9: % response of a cable which is suspended at
10: % one end and is free at the other end. The
11: % cable is given a uniform initial velocity.
12: % A plot of the solution error is given for
13: % two cases where approximate solutions are
14: % generated using numerical integration rather
15: % than modal response which is exact.
16: %
17: % User m functions required:
18: %    mckde2i, mckde4i, cablemk, udfrevib,
19: %    genprint, pauz, plterror
20: %-------------------------------------------------
21:
22: % Choose a model having twenty links of
23: % equal length
24:
25: fprintf(...
26: '\nPlease wait: solution takes a while\n')
27: clear all
28: n=20; gravty=1.; n2=1+fix(n/2);
29: masses=ones(n,1)/n; lengths=ones(n,1)/n;
30:
31: % First generate the exact solution by
32: % modal superposition
33: [m,k]=cablemk(masses,lengths,gravty);
34: c=zeros(size(m));
35: dsp=zeros(n,1); vel=ones(n,1);
36: t0=0; tfin=50; ntim=126; h=(tfin-t0)/(ntim-1);
37:
38: % Numbers of repetitions each solution is
39: % performed to get accurate cpu times for
40: % the chosen step sizes are shown below.
```

```
41: % Parameter jmr may need to be increased to
42: % give reliable cpu times on fast computers
43:
44: jmr=500;
45: j2=fix(jmr/50); J2=fix(jmr/25);
46: j4=fix(jmr/20); J4=fix(jmr/10);
47:
48: % Loop through all solutions repeatedly to
49: % obtain more reliable timing values on fast
50: % computers
51: tcpmr=clock;
52: for j=1:jmr;
53:     [tmr,xmr]=udfrevib(m,k,dsp,vel,t0,tfin,ntim);
54: end
55: tcpmr=etime(clock,tcpmr); tcpmr=tcpmr/jmr;
56:
57: % Second order implicit results
58: i2=10; h2=h/i2; tsav=clock;
59: for j=1:j2
60:     [t2,x2]=mckde2i(m,c,k,t0,dsp,vel,tfin,h2,i2);
61: end
62: tcp2=etime(clock,tsav)/j2; tr2=tcp2/tcpmr;
63:
64: I2=5; H2=h/I2; tsav=clock;
65: for j=1:J2
66:     [T2,X2]=mckde2i(m,c,k,t0,dsp,vel,tfin,H2,I2);
67: end
68: Tcp2=etime(clock,tsav)/J2; Tr2=Tcp2/tcpmr;
69:
70: % Fourth order implicit results
71: i4=2; h4=h/i4; tsav=clock;
72: for j=1:j4
73:     [t4,x4]=mckde4i(m,c,k,t0,dsp,vel,tfin,h4,i4);
74: end
75: tcp4=etime(clock,tsav)/j4; tr4=tcp4/tcpmr;
76:
77: I4=1; H4=h/I4; tsav=clock;
78: for j=1:J4
79:     [T4,X4]=mckde4i(m,c,k,t0,dsp,vel,tfin,H4,I4);
80: end
81: Tcp4=etime(clock,tsav)/J4; Tr4=Tcp4/tcpmr;
82:
83: % Plot error measures for each solution
84: plterror
```

Script File plterror

```
 1: % Script file: plterror
 2: % ~~~~~~~~~~~~~~~~~~~~~~~
 3: % Plot error measures showing how different
 4: % integrators and time steps compare with
 5: % the exact solution using modal response.
 6: %
 7: % User m functions called:  none
 8: %-------------------------------------------------
 9:
10: % Compare the maximum error in any component
11: % at each time with the largest deflection
12: % occurring during the complete time history
13: maxd=max(abs(xmr(:)));
14: er2=max(abs(x2-xmr)')/maxd;
15: Er2=max(abs(X2-xmr)')/maxd;
16: er4=max(abs(x4-xmr)')/maxd;
17: Er4=max(abs(X4-xmr)')/maxd;
18:
19: plot(t2,er2,'-',T2,Er2,'--');
20: title(['Solution Error For Implicit ',...
21:         '2nd Order Integrator']);
22: xlabel('time');
23: ylabel('solution error measure');
24: lg1=['h= ', num2str(h2),    ...
25:         ', relative cputime= ', num2str(tr2)];
26: lg2=['h= ', num2str(H2),    ...
27:         ', relative cputime= ', num2str(Tr2)];
28: legend(lg1,lg2,2); figure(gcf);
29: disp('Press [Enter] to continue'); pauz;
30: %genprint('deislne2');
31:
32: plot(t4,er4,'-',T4,Er4,'--');
33: title(['Solution Error For Implicit ',...
34:         '4th Order Integrator']);
35: xlabel('time');
36: ylabel('solution error measure');
37: lg1=['h= ', num2str(h4),    ...
38:         ', relative cputime= ', num2str(tr4)];
39: lg2=['h= ', num2str(H4),    ...
40:         ', relative cputime= ', num2str(Tr4)];
41: legend(lg1,lg2,2); figure(gcf);
42: %genprint('deislne4');
```

Function mckde2i

```
 1: function [t,x,tcp] = ...
 2:       mckde2i(m,c,k,t0,x0,v0,tmax,h,incout,forc)
 3: %
 4: % [t,x,tcp]= ...
 5: %     mckde2i(m,c,k,t0,x0,v0,tmax,h,incout,forc)
 6: % ~~~~~~~~~~~~~~~~~~~~~~~~~~~~~~~~~~~~~~~~~~~~~~~~~~
 7: % This function uses a second order implicit
 8: % integrator % to solve the matrix differential
 9: % equation
10: %             m x'' + c x' + k x = forc(t)
11: % where m,c, and k are constant matrices and
12: % forc is an externally defined function.
13: %
14: % Input:
15: % ------
16: % m,c,k    mass, damping and stiffness matrices
17: % t0       starting time
18: % x0,v0    initial displacement and velocity
19: % tmax     maximum time for solution evaluation
20: % h        integration stepsize
21: % incout   number of integration steps between
22: %          successive values of output
23: % forc     externally defined time dependent
24: %          forcing function. This parameter
25: %          should be omitted if no forcing
26: %          function is used.
27: %
28: % Output:
29: % -------
30: % t        time vector going from t0 to tmax
31: %          in steps of
32: % x        h*incout to yield a matrix of
33: %          solution values such that row j
34: %          is the solution vector at time t(j)
35: % tcp      computer time for the computation
36: %
37: % User m functions called:  none.
38: %------------------------------------------------
39:
40: if (nargin > 9); force=1; else, force=0; end
41: if nargout ==3, tcp=clock; end
42: hbig=h*incout;
```

```
43: t=(t0:hbig:tmax)'; n=length(t);
44: ns=(n-1)*incout; ts=t0+h*(0:ns)';
45: xnow=x0(:); vnow=v0(:);
46: nvar=length(x0);
47: jrow=1; jstep=0; h2=h/2;
48:
49: % Form the inverse of the effective
50: % stiffness matrix
51: mnv=h*inv(m+h2*(c+h2*k));
52:
53: % Initialize the output matrix for x
54: x=zeros(n,nvar); x(1,:)=xnow';
55: zroforc=zeros(length(x0),1);
56:
57: % Main integration loop
58: for j=1:ns
59:    tj=ts(j);tjh=tj+h2;
60:    if force
61:       dv=feval(forc,tjh);
62:    else
63:       dv=zroforc;
64:    end
65:    dv=mnv*(dv-c*vnow-k*(xnow+h2*vnow));
66:    vnext=vnow+dv;xnext=xnow+h2*(vnow+vnext);
67:    jstep=jstep+1;
68:    if jstep == incout
69:       jstep=0; jrow=jrow+1; x(jrow,:)=xnext';
70:    end
71:    xnow=xnext; vnow=vnext;
72: end
73: if nargout ==3
74:    tcp=etime(clock,tcp);
75: else
76:    tcp=[];
77: end
```

Function mckde4i

```
1: function [t,x,tcp] = ...
2:       mckde4i(m,c,k,t0,x0,v0,tmax,h,incout,forc)
3: %
4: % [t,x,tcp]= ...
5: %       mckde4i(m,c,k,t0,x0,v0,tmax,h,incout,forc)
```

```
 6: %  ~~~~~~~~~~~~~~~~~~~~~~~~~~~~~~~~~~~~~~~~~~~~~~
 7: % This function uses a fourth order implicit
 8: % integrator with fixed stepsize to solve the
 9: % matrix differential equation
10: %           m x'' + c x' + k x = forc(t)
11: % where m,c, and k are constant matrices and
12: % forc is an externally defined function.
13: %
14: % Input:
15: % ------
16: % m,c,k    mass, damping and stiffness matrices
17: % t0       starting time
18: % x0,v0    initial displacement and velocity
19: % tmax     maximum time for solution evaluation
20: % h        integration stepsize
21: % incout   number of integration steps between
22: %          successive values of output
23: % forc     externally defined time dependent
24: %          forcing function. This parameter
25: %          should be omitted if no forcing
26: %          function is used.
27: %
28: % Output:
29: % -------
30: % t        time vector going from t0 to tmax
31: %          in steps of h*incout
32: % x        matrix of solution values such
33: %          that row j is the solution vector
34: %          at time t(j)
35: % tcp      computer time for the computation
36: %
37: % User m functions called:  none.
38: %-------------------------------------------------
39:
40: if nargin > 9, force=1; else, force=0; end
41: if nargout ==3, tcp=clock; end
42: hbig=h*incout; t=(t0:hbig:tmax)';
43: n=length(t); ns=(n-1)*incout; nvar=length(x0);
44: jrow=1; jstep=0; h2=h/2; h12=h*h/12;
45:
46: % Form the inverse of the effective stiffness
47: % matrix for later use.
48:
49: m12=m-h12*k;
50: mnv=inv([[(-h2*m-h12*c),m12];
```

```
51:        [m12,(c+h2*k)]]);
52:
53: % The forcing function is integrated using a
54: % 2 point Gauss rule
55: r3=sqrt(3); b1=h*(3-r3)/6; b2=h*(3+r3)/6;
56:
57: % Initialize output matrix for x and other
58: % variables
59: xnow=x0(:); vnow=v0(:);
60: tnow=t0; zroforc=zeros(length(x0),1);
61:
62: if force
63:    fnow=feval(forc,tnow);
64: else
65:    fnow=zroforc;
66: end
67: x=zeros(n,nvar); x(1,:)=xnow'; fnext=fnow;
68:
69: % Main integration loop
70: for j=1:ns
71:    tnow=t0+(j-1)*h; tnext=tnow+h;
72:    if force
73:      fnext=feval(forc,tnext);
74:      di1=h12*(fnow-fnext);
75:      di2=h2*(feval(forc,tnow+b1)+ ...
76:              feval(forc,tnow+b2));
77:      z=mnv*[(di1+m*(h*vnow)); (di2-k*(h*xnow))];
78:      fnow=fnext;
79:    else
80:      z=mnv*[m*(h*vnow); -k*(h*xnow)];
81:    end
82:    vnext=vnow + z(1:nvar);
83:    xnext=xnow + z((nvar+1):2*nvar);
84:    jstep=jstep+1;
85:
86:    % Save results every incout steps
87:    if jstep == incout
88:      jstep=0; jrow=jrow+1; x(jrow,:)=xnext';
89:    end
90:
91:    % Update quantities for next step
92:    xnow=xnext; vnow=vnext; fnow=fnext;
93: end
94: if nargout==3
95:    tcp=etime(clock,tcp);
```

```
96: else
97:    tcp=[];
98: end
```

Chapter 8

Integration of Nonlinear Initial Value Problems

8.1 General Concepts on Numerical Integration of Nonlinear Matrix Differential Equations

Methods for solving differential equations numerically are one of the most valuable analysis tools now available. Less expensive computer power and more friendly software are stimulating wider use of digital simulation methods. At the same time, intelligent use of numerically integrated solutions requires appreciation of inherent limitations of the techniques employed. The present chapter discusses the widely used Runge-Kutta method and applies it to some specific examples.

When physical systems are described by mathematical models, it is common that various system parameters are only known approximately. For example, to predict the response of a building undergoing earthquake excitation, simplified formulations may be necessary to handle the elastic and frictional characteristics of the soil and the building. Our observation that simple models are used often to investigate behavior of complex systems does not necessarily amount to a rejection of such procedures. In fact, good engineering analysis depends critically on development of reliable models which can capture salient features of a process without employing unnecessary complexity. At the same time, analysts need to maintain proper caution regarding trustworthiness of answers produced with computer models. Many nonlinear systems respond strongly to small changes in physical parameters. Scientists today realize that, in dealing with highly nonlinear phenomena such as weather prediction, it is simply impossible to make reliable long term forecasts [44] because of various unalterable factors. Among these are a) uncertainty about initial conditions, b) uncertainty about the adequacy of mathematical models describing relevant physical processes, c) uncertainty about error contributions arising from use of spatial and time discretizations in construction of approximate numerical solutions, and d) uncertainty about effects of arithmetic roundoff error. In light of the criticism and cautions being stated about the dangers of using numerical solutions, the thrust of the discussion is that idealized models must not be regarded as infallible, and no numerical solution should be accepted as credible without adequately investigating effects of parameter perturbation within uncertainty limits of the parameters. To illustrate how sensitive a system can be to initial conditions, we might consider a very simple model concerning motion of a pendulum of length ℓ given an initial velocity v_0 starting from a vertically downward position. If v_0 exceeds $2\sqrt{g\ell}$, the pendulum will reach a vertically upward position and will go over the top. If v_0 is less than $2\sqrt{g\ell}$, the vertically upward position is never reached. Instead, the pendulum oscillates about the bottom position. Consequently, initial velocities of $1.999\sqrt{g\ell}$ and $2.001\sqrt{g\ell}$ produce quite different system behavior with only a tiny change in initial velocity. Other examples illustrating the difficulties of computing the response of nonlinear systems are cited below. These examples are not chosen to discourage use of the powerful tools now available for numerical integration of differential equations. Instead, the intent is to encourage users of these methods to exercise proper caution so that confidence in the reliability of results is fully justified.

Many important physical processes are governed by differential equations. Typical cases include dynamics of rigid and flexible bodies, heat conduction, and electrical current flow. Solving a system of differential equations subject to known initial conditions allows us to predict the future behavior of the related physical system. Since very few important differential equations can be solved in closed form, approximations which are directly or indirectly founded on series expansion methods have been developed. The basic problem addressed is that of accurately computing $Y(t + h)$ when $Y(t)$ is known, along with a differential equation governing system behavior from time t to $(t + h)$. Recursive application of a satisfactory numerical approximation procedure, with possible adjustment of step-size to maintain accuracy and stability, allows approximate prediction of system response subsequent to the starting time.

Numerical methods for solving differential equations are extremely important tools for analyzing engineering systems. Although valuable algorithms have been developed which facilitate construction of approximate solutions, all available methods are vulnerable to limitations inherent in the underlying approximation processes. The essence of the difficulty lies in the fact that, as long as a finite integration step-size is used, integration error occurs at each time step. In many instances, these errors have an accumulative effect which grows exponentially and eventually destroys solution validity. To some extent, accuracy problems can be limited by regulating step-size to keep local error within a desired tolerance. Typically, decreasing an integration tolerance increases the time span over which a numerical solution is valid. However, high costs for computer time to analyze large and complex systems sometimes preclude generation of long time histories which may be more expensive than is practically justifiable.

8.2 Runge-Kutta Methods and the ODE45 Integrator Provided in MATLAB

Formulation of one method to solve differential equations is discussed in this section. Suppose a function $y(x)$ satisfies a differential equation of the form $y'(x) = f(x, y)$, subject to $y(x_0) = y_0$, where f is a known differentiable function. We would like to compute an approximation of $y(x_0 + h)$ which agrees with a Taylor's series expansion up to a certain order of error. Hence,

$$y(x_0 + h) = \tilde{y}(x_0, h) + O(h^{n+1})$$

where $O(h^{n+1})$ denotes a quantity which decreases at least as fast as h^{n+1} for small h. Taylor's theorem allows us to write

$$y(x_0 + h) = y(x_0) + y'(x_0)h + \frac{1}{2}y''(x_0)h^2 + O(h^3)$$

$$= y_0 + f(x_0, y_0)h + \frac{1}{2}[f_x(x_0, y_0) +$$

$$f_y(x_0, y_0)f_0]h^2 + O(h^3)$$

where $f_0 = f(x_0, y_0)$. The last formula can be used to compute a second order approximation $\hat{y}(x_0+h)$, provided the partial derivatives f_x and f_y can be evaluated. However, this may be quite difficult since the function $f(x, y)$ may not even be known explicitly.

The idea leading to Runge-Kutta integration is to compute $y(x_0 + h)$ by making several evaluations of function f instead of having to differentiate that function. Let us seek an approximation in the form

$$\tilde{y}(x_0 + h) = y_0 + h[k_0 f_0 + k_1 f(x_0 + \alpha h, y_0 + \beta h f_0)]$$

We choose k_0, k_1, α, and β to make $\tilde{y}(x_0+h)$ match the series expansion of $y(x)$ as well as possible. Evidently,

$$f(x_0 + \alpha h, y_0 + \beta h f_0) = f_0 + [f_x(x_0, y_0)\alpha + f_y(x_0, y_0)f_0\beta]h + O(h^2)$$

and therefore,

$$\tilde{y}(x_0 + h) = y_0 + h[[(k_0 + k_1)f_0 + k_1(f_x(x_0, y_0)\alpha +$$

$$f_y(x_0, y_0)\beta f_0)]h + O(h^2)$$

$$= y_0 + (k_0 + k_1)f_0 h + [f_x(x_0, y_0)\alpha k_1 +$$

$$f_y(x_0, y_0)f_0 \beta k_1]h^2 + O(h^3)$$

The last relation shows that

$$y(x_0 + h) = \tilde{y}(x_0 + h) + O(h^3)$$

provided

$$k_0 + k_1 = 1 \qquad \alpha k_1 = \frac{1}{2} \qquad \beta k_1 = \frac{1}{2}$$

This system of three equations in four unknowns has an infinite number of solutions; one of these is $k_0 = k_1 = \frac{1}{2}$, $\alpha = \beta = 1$. This implies that

$$y(x_0 + h) = y(x_0) + \frac{1}{2}[f_0 + f(x_0 + h, y_0 + hf_0)]h + O(h^3)$$

Neglecting the truncation error $O(h^3)$ gives a difference approximation known as Heun's method [60], which is classified as a second order Runge-Kutta method. Reducing the step-size by h reduces the truncation error by about a factor of $(\frac{1}{2})^3 = \frac{1}{8}$. Of course, the formula can be used recursively to compute approximations to $y(x_0 + h)$, $y(x_0 + 2h)$, $y(x_0 + 3h)$, ... In most instances, the solution accuracy decreases as the number of integration steps is increased and results eventually become unreliable. Decreasing h and taking more steps within a fixed time span helps, but this also has practical limits governed by computational time and arithmetic roundoff error.

The idea leading to Heun's method can be extended further to develop higher order formulas. One of the best known is the fourth order Runge-Kutta method described as follows

$$y(x_0 + h) = y(x_0) + h[k_1 + 2k_2 + 2k_3 + k_4]/6$$

where

$$k_1 = f(x_0, y_0) \qquad k_2 = f(x_0 + \frac{h}{2}, y_0 + k_1\frac{h}{2})$$

$$k_3 = f(x_0 + \frac{h}{2}, y_0 + k_2\frac{h}{2}) \qquad k_4 = f(x_0 + h, y_0 + k_3 h)$$

The truncation error for this formula is order h^5, so the error is reduced by about a factor of $\frac{1}{32}$ when the step-size is halved. The development of the fourth order Runge-Kutta method is algebraically quite complicated [42]. We note that accuracy of order four is achieved with four evaluations of f for each integration step. This situation does not extend to higher orders. For instance, an eighth order formula may require twelve evaluations per step. This price of more function evaluations may be worthwhile provided the resulting truncation error is small enough to permit much larger integration steps than could be achieved with formulas of lower order. MATLAB provides function **ode45** which uses variable step-size and employs formulas of order four and five. (Note: In MATLAB 5.x the integrators can output results for an arbitrary time vector using, for instance, even time increments. We have chosen not to employ this feature to simplify conversion of this code to MATLAB 4.x.)

8.3   Step-size Limits Necessary to Maintain Numerical Stability

It can be shown that, for many numerical integration methods, taking too large a step-size produces absurdly large results which increase exponentially with successive time steps. This phenomenon, known as numerical instability, can be illustrated with the simple differential equation

$$y'(t) = f(t, y) = \lambda y$$

which has the solution $y = ce^{\lambda t}$. If the real part of λ is positive, the solution becomes unbounded with increasing time. However, a pure imaginary λ produces a bounded oscillatory solution, whereas the solution decays exponentially for $\mathbf{real}(\lambda) < 0$. Applying Heun's method [42] gives

$$y(t + h) = y(t)[1 + (\lambda h) + \frac{(\lambda h)^2}{2}]$$

This shows that at each integration step the next value of y is obtained by multiplying the previous value by a factor

$$p = 1 + (\lambda h) + \frac{(\lambda h)^2}{2}$$

which agrees with the first three Taylor series terms of $e^{\lambda h}$. Clearly, the difference relation leads to

$$y_n = y_0 p^n$$

As n increases, y_n will approach infinity unless $|p| \le 1$. This stability condition can be interpreted geometrically by regarding λh as a complex variable z and solving for all values of z such that

$$1 + z + \frac{z^2}{2} = \zeta e^{i\theta} \qquad |\zeta| \le 1 \qquad 0 \le \theta \le 2\pi$$

Taking $\zeta = 1$ identifies the boundary of the stability region, which is normally a closed curve lying in the left half of the complex plane. Of course, h is assumed to be positive and the real part of λ is nonpositive. Otherwise, even the exact solution would grow exponentially. For a given λ, the step-size h must be taken small enough to make $|\lambda h|$ lie within the stability zone. The larger $|\lambda|$ is, the smaller h must be to prevent numerical instability.

The idea illustrated by Heun's method can be easily extended to a Runge-Kutta method of arbitrary order. A Runge-Kutta method of order n reproduces the exact solution through terms of order n in the Taylor series expansion. The differential equation $y' = \lambda y$ implies

$$y(t + h) = y(t)e^{\lambda h}$$

and

$$e^{\lambda h} = \sum_{k=0}^{n} \frac{(\lambda h)^k}{k!} + O(h^{n+1})$$

Consequently, points on the boundary of the stability region for a Runge-Kutta method of order n are found by solving the polynomial

$$1 - e^{i\theta} + \sum_{k=1}^{n} \frac{z^k}{k!} = 0$$

for a dense set of θ-values ranging from zero to 2π. Using MATLAB's intrinsic function **roots** allows easy calculation of the polynomial roots which may be plotted to show the stability boundary. The following short program accomplishes the task. Program output for integrators of order four and six are shown in Figures 8.1 and 8.2. Note that the region for order 4 resembles a semicircle with radius close to 2.8. Using $|\lambda h| > 2.8$, with Runge-Kutta of order 4, would give results which rapidly become unstable. The figures also show that the stability region for Runge-Kutta of order 6 extends farther out on the negative real axis than Runge-Kutta of order 4 does. The root finding process also introduces some meaningless stability zones in the right half plane which should be ignored.

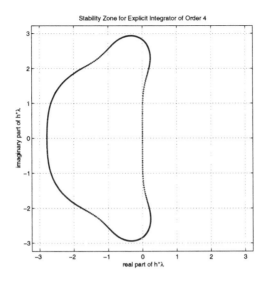

Figure 8.1. Stability Zone for Explicit Integrator of Order 4

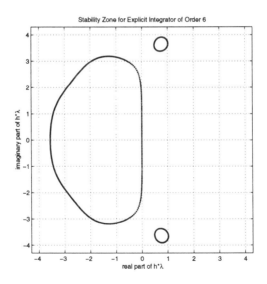

Figure 8.2. Stability Zone for Explicit Integrator of Order 6

MATLAB Example

Script File rkdestab

```
 1: % Example:  rkdestab
 2: % ~~~~~~~~~~~~~~~~~~~
 3: % This program plots the boundary of the region
 4: % of the complex plane governing the maximum
 5: % step size which may be used for stability of
 6: % a Runge-Kutta integrator of arbitrary order.
 7: %
 8: % npts  - a value determining the number of
 9: %          points computed on the stability
10: %          boundary of an explicit Runge-Kutta
11: %          integrator.
12: % xrang - controls the square window within
13: %          which the diagram is drawn.
14: %          [ -3, 3, -3, 3] is appropriate for
15: %          the fourth order integrator.
16: %
17: % User m functions required: genprint
18: %-------------------------------------------------
19:
20: clf; close;
21: fprintf('\nSTABILITY REGION FOR AN ');
22: fprintf('EXPLICIT RUNGE-KUTTA');
23: fprintf('\n    INTEGRATOR OF ARBITRARY ');
24: fprintf('ORDER\n\n');
25: nordr=input('Give the integrator order ? > ');
26: fprintf('\nInput the number of points ');
27: fprintf('used to define\n');
28: npts=input('the boundary (100 is typical) ? > ');
29: r=zeros(npts,nordr); v=1./gamma(nordr+1:-1:2);
30: d=2*pi/(npts-1); i=sqrt(-1);
31:
32: % Generate polynomial roots to define the
33: % stability boundary
34: for j=1:npts
35:    % polynomial coefficients
36:    v(nordr+1)=1-exp(i*(j-1)*d);
37:    % complex roots
38:    t=roots(v); r(j,:)=t(:).';
39: end
40:
```

```matlab
41: % Plot the boundary
42: rel=real(r(:)); img=imag(r(:));
43: w=1.1*max(abs([rel;img]));
44: plot(rel,img,'.');
45: axis([-w,w,-w,w]); axis('square');
46: xlabel('real part of h*\lambda');
47: ylabel('imaginary part of h*\lambda');
48: ns=int2str(nordr);
49: st=['Stability Zone for Explicit ' ...
50:     'Integrator of Order ',ns];
51: title(st); grid on; figure(gcf);
52: %genprint('rkdestab');
53: disp(' '); disp('All Done');
```

8.4   Discussion of Procedures to Maintain Accuracy by Varying Integration Step-size

When we solve a differential equation numerically, our first inclination is to seek output at even increments of the independent variable. However, this is not the most natural form of output appropriate to maintain integration accuracy. Whenever solution components are changing rapidly, a small time step may be needed, whereas using a small time step might be quite inefficient at times where the solution remains smooth. Most modern ODE programs employ variable step-size algorithms which decrease the integration step-size whenever some local error tolerance is violated and conversely increase the step-size when the increase can be performed without loss of accuracy. If results at even time increments are needed, these can be determined by interpolation of the non-equidistant values.

Although the derivation of algorithms to regulate step-size is an important topic, development of these methods is not presented here. Several references [42, 45, 50, 60] discuss this topic with adequate detail. The primary objective in regulating step-size is to gain computational efficiency by taking as large a step-size as possible while maintaining accuracy and minimizing the number of function evaluations.

Practical problems involving a single first order differential equation are rarely encountered. More commonly, a system of second order equations occurs which is then transformed into a system involving twice as many first order equations. Several hundred, or even several thousand dependent variables may be involved. Evaluating the necessary time derivatives at a single time step may require computationally intensive tasks such as matrix inversion. Furthermore, performing this fundamental calculation several thousand times may be necessary in order to construct time responses over time intervals of practical interest. Integrating large systems of nonlinear differential equations is one of the most important and most resource intensive aspects of scientific computing.

Instead of deriving the algorithms used for step-size control in **ode45**, we will outline briefly the ideas employed to integrate $y'(t) = f(t, y)$ from t to $(t + h)$. It is helpful to think of y as a vector. For a given time step and y value, the program makes six evaluations of f. These values allow evaluation of two Runge-Kutta formulas, each having different truncation errors. These formulas permit estimation of the actual truncation error and proper step-size adjustment to control accuracy. If the estimated error is too large, the step-size is decreased until the error tolerance is satisfied or an error condition occurs because the necessary step-size has fallen below a set limit. If the estimated error is found to be smaller than necessary, the integration result is accepted and the step-size is increased for the next pass. Even though this type of process may not be extremely interesting to discuss, it is nevertheless an essential part of any well designed program for integrating differential equations numerically. Readers should become familiar with the error control feature of ODE solvers they employ. Printing and studying the code for **ode45** is worthwhile. It should also be remembered that solutions generated by tools such as **ode45** always contain accumulated error effects from roundoff and arithmetic truncation. Such errors eventually make results sufficiently far from the starting time of the solution invalid.

This chapter concludes with the analysis of several realistic nonlinear problems having certain properties of their exact solutions known. These known properties are compared with numerical results to assess error growth. The first problem studies an inverted pendulum for which the loading function produces a simple exact displacement function. Examples concerning top dynamics, a projectile trajectory, and a falling chain are presented.

8.5 Example on Forced Oscillations of an Inverted Pendulum

The inverted pendulum in Figure 8.3 involves a weightless rigid rod of length l which has a mass m attached to the end. Attached to the mass is a spring with stiffness constant k and an unstretched length of γl. The spring has length l when the pendulum is in the vertical position. Externally applied loads consist of a driving moment $M(t)$, the particle weight, and a viscous damping moment $cl^2\dot\theta$.

The differential equation governing the motion of this system is found to be

$$\ddot\theta = -(c/m)\dot\theta + (g/l)\sin(\theta) + M(t)/(ml^2) - (2k/m)\sin(\theta)(1 - \alpha/\lambda)$$

where

$$\lambda = \sqrt{5 - 4\cos(\theta)}$$

This system can be changed to a more convenient form by introducing dimensionless variables. We let $t = (\sqrt{l/g})\tau$ where τ is dimensionless time. Then

$$\ddot\theta = -\alpha\dot\theta + \sin(\theta) + P(\tau) - \beta\sin(\theta)(1 - \gamma/\lambda)$$

where

$$
\begin{aligned}
\alpha &= (c/m)\sqrt{l/g} = \text{viscous damping factor} \\
\beta &= 2(k/m)/(g/l) \\
\lambda &= \sqrt{5 - 4\cos(\theta)} \\
\gamma &= \text{(unstretched spring length)}/l \\
P(\tau) &= M/(mgl) = \text{dimensionless driving moment}
\end{aligned}
$$

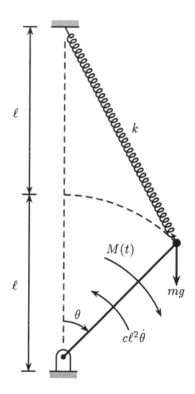

Figure 8.3. Forced Vibration of an Inverted Pendulum

It is interesting to test how well a numerical method can reconstruct a known exact solution for a nonlinear function. Let us assume that the driving moment $M(\tau)$ produces a motion having the equation

$$\theta_e(\tau) = \theta_0 \sin(\omega\tau)$$

for arbitrary θ_0 and ω. Then

$$\dot{\theta}_e(\tau) = \omega\theta_0 \cos(\omega\tau)$$

and

$$\ddot{\theta}_e(\tau) = -\omega^2\theta_e$$

Consequently, the necessary driving moment is

$$P(\tau) \quad = \quad -\omega^2\theta_e - \sin(\theta_e) + \gamma\omega\theta_0 \cos(\omega\tau) +$$

$$\beta\sin(\theta_e)\left[1 - \gamma/\sqrt{5 - 4\cos(\theta_e)}\right]$$

Applying this forcing function, along with the initial conditions

$$\theta(0) = 0 \qquad \dot{\theta}(0) = \theta_0\omega$$

should return the solution $\theta = \theta_e(\tau)$. For a specific numerical example we choose $\theta_0 = \pi/8$, $\omega = 0.5$, and four different combinations of β, γ, and *tol*. The second order differential equation has the form $\ddot{\theta} = f(\tau, \theta, \dot{\theta})$. This is expressed as a first order matrix system by letting $y_1 = \theta$, $y_2 = \dot{\theta}$, which gives

$$\dot{y}_1 = y_2 \qquad \dot{y}_2 = f(\tau, y_1, y_2)$$

A function describing the system for solution by **ode45** is provided at the end of this section. Parameters θ_0, ω_0, α, ζ, and β are passed as global variables.

We can examine how well the numerically integrated θ match θ_e by using the error measure

$$|\theta(\tau) - \theta_e(\tau)|$$

Furthermore, the exact solution satisfies

$$\theta_e^2 + (\dot{\theta}_e/\omega)^2 = \theta_0^2$$

so plotting $\dot{\theta}/(\theta_0\omega)$ on a horizontal axis and θ/θ_0 on a vertical axis should produce a unit circle. Violation of that condition signals loss of solution accuracy.

How certain physical parameters and numerical tolerances affect terms in this problem can be demonstrated by the following four data cases.

1. The spring is soft and initially unstretched. A liberal integration tolerance is used.

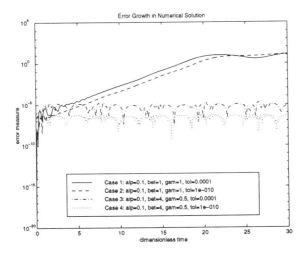

Figure 8.4. Error Growth in Numerical Solution

2. The spring is soft and initially unstretched. A stringent integration tolerance is used.

3. The spring is stiff and initially stretched. A liberal integration tolerance is used.

4. The spring is stiff and initially stretched. A stringent integration tolerance is used.

The curves in Figure 8.4 show the following facts:

1. When the spring is unstretched initially, the numerical solution goes unstable quickly.

2. Stretching the spring initially and increasing the spring constant improves numerical stability of the solution.

3. Decreasing the integration tolerance increases the time period over which the solution is valid.

An additional curve illustrating the numerical inaccuracy of results for Case 1 appears in Figure 8.4. A plot of $\theta(\tau)$ versus $\dot\theta(\tau)/\omega$ should produce a circle. However, solution points quickly depart from the desired locus.

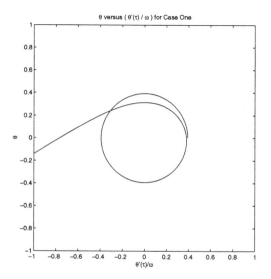

Figure 8.5. θ versus $(\theta'(\tau)/\omega)$ for Case One

MATLAB Example

Script File prun

```
 1: % Example: prun
 2: % ~~~~~~~~~~~~~~
 3: % Dynamics of an inverted pendulum integrated
 4: % by use of ode45.
 5: %
 6: % User m functions required:
 7: %    pinvert, mom, genprint
 8: %-----------------------------------------------
 9:
10: global th0_ w_ alp_ bet_ gam_ ncal_
11: th0_=pi/8; w_=.5; tmax=30; ncal_=0;
12:
13: fprintf('\nFORCED OSCILLATION OF AN ');
14: fprintf('INVERTED PENDULUM\n');
15: fprintf('\nNote: Several minutes may be ');
16: fprintf('required to generate');
17: fprintf('\nthe four sets of numerical ');
18: fprintf('results needed.\n');
19:
20: % loose spring with liberal tolerance
21: alp_=0.1; bet_=1.0; gam_=1.0; tol=1.e-4;
22: a1=num2str(alp_); b1=num2str(bet_);
23: g1=num2str(gam_); e1=num2str(tol);
24: options=odeset('RelTol',tol);
25: [t1,z1]= ...
26:   ode45('pinvert',[0,tmax],[0;w_*th0_],...
27:         options);
28: n1=ncal_; ncal_=0;
29:
30: % loose spring with stringent tolerance
31: alp_=0.1; bet_=1.0; gam_=1.0; tol=1.e-10;
32: a2=num2str(alp_); b2=num2str(bet_);
33: g2=num2str(gam_); e2=num2str(tol);
34: options=odeset('RelTol',tol);
35: [t2,z2]= ...
36:   ode45('pinvert',[0,tmax],[0;w_*th0_],...
37:         options);
38: n2=ncal_; ncal_=0;
39:
40: % tight spring with liberal tolerance
```

```
41: alp_=0.1; bet_=4.0; gam_=0.5; tol=1.e-4;
42: a3=num2str(alp_); b3=num2str(bet_);
43: g3=num2str(gam_); e3=num2str(tol);
44: options=odeset('RelTol',tol);
45: [t3,z3]= ...
46:   ode45('pinvert',[0,tmax],[0;w_*th0_],...
47:         options);
48: n3=ncal_; ncal_=0;
49:
50: % tight spring with stringent tolerance
51: alp_=0.1; bet_=4.0; gam_=0.5; tol=1.e-10;
52: a4=num2str(alp_); b4=num2str(bet_);
53: g4=num2str(gam_); e4=num2str(tol);
54: options=odeset('RelTol',tol);
55: [t4,z4]= ...
56:   ode45('pinvert',[0,tmax],[0;w_*th0_],...
57:         options);
58: n4=ncal_; ncal_=0; save pinvert.mat;
59:
60: % Plot results
61: clf; semilogy( ...
62:   t1,abs(z1(:,1)/th0_-sin(w_*t1)),'-r',...
63:   t2,abs(z2(:,1)/th0_-sin(w_*t2)),'--g',...
64:   t3,abs(z3(:,1)/th0_-sin(w_*t3)),'-.b',...
65:   t4,abs(z4(:,1)/th0_-sin(w_*t4)),':m');
66: title('Error Growth in Numerical Solution')
67: xlabel('dimensionless time');
68: ylabel('error measure');
69: c1=['Case 1: alp=',a1,', bet=',b1,', gam=', ...
70:     g1,', tol=',e1];
71: c2=['Case 2: alp=',a2,', bet=',b2,', gam=', ...
72:     g2,', tol=',e2];
73: c3=['Case 3: alp=',a3,', bet=',b3,', gam=', ...
74:     g3,', tol=',e3];
75: c4=['Case 4: alp=',a4,', bet=',b4,', gam=', ...
76:     g4,', tol=',e4];
77: legend(c1,c2,c3,c4);
78: dum=input('\nPress [Enter] to continue\n','s');
79: %genprint('pinvert');
80:
81: % plot a phase diagram for case 1
82: clf; plot(z1(:,2)/w_,z1(:,1));
83: axis('square'); axis([-1,1,-1,1]);
84: xlabel('\theta''(\tau)/\omega'); ylabel('\theta');
85: title(['\theta versus ( \theta''(\tau) / ' ...
```

```
86:          '\omega ) for Case One']); figure(gcf);
87: %genprint('crclplt');
88: disp(' '); disp('All Done');
```

Function pinvert

```
1: function zdot=pinvert(t,z)
2: %
3: % zdot=pinvert(t,z)
4: % ~~~~~~~~~~~~~~~~~
5: % Equation of motion for the pendulum
6: %
7: % t    - time value
8: % z    - vector [theta ; theta_dot]
9: % zdot - time derivative of z
10: %
11: % User m functions called:  mom
12: %-------------------------------------------------
13:
14: global alp_ bet_ gam_ ncal_
15: ncal_=ncal_+1; th=z(1); thd=z(2);
16: c=cos(th); s=sin(th); lam=sqrt(5-4*c);
17: zdot=[thd;
18:       mom(t)+s-alp_*thd-bet_*s*(1-gam_/lam)];
```

Function mom

```
1: function me=mom(t)
2: %
3: % me=mom(t)
4: % ~~~~~~~~~
5: % t - time
6: % me - driving moment needed to produce
7: %      exact solution
8: %
9: % User m functions called:  none.
10: %-------------------------------------------------
11:
12: global th0_ w_ alp_ bet_ gam_
13: th=th0_*sin(w_*t);
14: thd=w_*th0_*cos(w_*t); thdd=-th*w_^2;
15: s=sin(th); c=cos(th); lam=sqrt(5-4*c);
16: me=thdd-s+alp_*thd+bet_*s*(1-gam_/lam);
```

8.6 Dynamics of a Spinning Top

The dynamics of a symmetrical spinning top can be analyzed simply by computing the path followed by the gravity center in cartesian coordinates. Consider a top spinning with its apex (or tip) constrained to remain at the origin. The gravity center lies at position r along the axis of symmetry and the only applied forces are the weight $-mg\hat{k}$ through the gravity center and the support reaction at the tip of the top. The inertial properties involve a moment of inertia J_a about the symmetry axis and a transverse inertial moment J_t relative to an axis normal to the symmetry axis and passing through the apex of the top. The velocity of the gravity center and the angular velocity Ω are related by[1]

$$v = \dot{r} = \Omega \times r$$

This implies that Ω can be expressed in terms of radial and transverse components as

$$\Omega = \ell^{-2} r \times v + \ell^{-1} \omega_a r$$

where $\ell = |r|$ and ω_a is the magnitude of the angular velocity component in the radial direction. The angular momentum with respect to the origin is therefore

$$H = J_t \ell^{-2} r \times v + J_a \ell^{-1} \omega_a r$$

and the potential plus kinetic energy is given by

$$K = mgz + \frac{J_t \ell^{-2} v \cdot v + J_a \omega_a^2}{2}$$

where z is the height of the gravity center above the origin.

The equations of motion can be found using the principle that the moment of all applied forces about the origin must equal the time rate of change of the corresponding angular momentum. Hence

$$M = J_t \ell^{-2} r \times a + J_a \ell^{-1} \left[\omega_a v + \dot{\omega}_a r \right]$$

where $a = \dot{v} = \ddot{r}$ is the total acceleration of the gravity center. The radial component of the last equation is obtainable by a dot product with r to give

$$r \cdot M = J_a \ell \dot{\omega}_a$$

where simplifications result because $r \cdot (r \times a) = 0$ and $r \cdot v = 0$. The remaining components of M for the transverse direction result by taking $r \times M$ and noting that

$$r \times (r \times a) = (r \cdot a)r - \ell^2 a = -\ell^2 a_t$$

---

[1] In this section the quantities v, r, Ω, H, M, and a all represent vector quantities.

where a_t is the vector component of total acceleration normal to the direction of r. This leads to

$$r \times M = -J_t a_t + J_a \ell^{-1} \omega_a \, r \times v$$

Since the gravity center moves on a spherical surface of radius ℓ centered at the origin, the radial acceleration is given by

$$a_r = -v \cdot v \, \ell^{-2} r$$

and the total acceleration equation becomes

$$a = -\frac{r \times M}{J_t} + \frac{J_a \ell^{-1} \omega_a}{J_t} \, r \times v - v \cdot v \, \ell^{-2} r$$

In the case studied here, only the body weight $-mg\hat{k}$ causes a moment about the origin so

$$M = -mg \, r \times \hat{k} \qquad r \cdot M = 0$$

and

$$r \times M = -mg \left[zr - \ell^2 \hat{k} \right]$$

The radial component of the moment equation simply gives $\dot{\omega}_a = 0$, so the axial component of angular velocity retains its initial value throughout the motion.

Integrating the differential equations

$$\dot{v} = a \qquad \dot{r} = v$$

numerically subject to appropriate initial conditions produces a trajectory of the gravity center motion. The simple formulation presented here treats x, y, and z as if they were independent variables even though

$$x^2 + y^2 + z^2 = \ell^2 \qquad x v_x + y v_y + z v_z = 0$$

are implied. The type of analysis traditionally used in advanced dynamics books [47] would employ Euler angles, thereby assuring exact satisfaction of $|r| = \ell$. The accuracy of the solution method proposed here can be checked by finding a) whether the total energy of the system remains constant and b) whether the component of angular momentum in the z-direction remains constant. However, even when constraint conditions are satisfied exactly, reliability of digital simulations of nonlinear systems over long time periods becomes questionable due to accumulated inaccuracies caused by arithmetic roundoff and the approximate nature of integration formulas.

The program **toprun** integrates the equations of motion and interprets the results. This program reads data to specify properties of a conical top along with the initial position and the angular velocity. Intrinsic function **ode45** is employed to integrate the motion equation defined in function **topde**. The path followed by the gravity center is plotted and error measures regarding conservation of energy and angular momentum are computed. Figures 8.6 and 8.7 show results for a top having properties given by the test case suggested in the interactive data input. A top which has its symmetry axis initially horizontal along the y-axis is given an angular velocity of $[0, 10, 2]$. Integrating the equation of motion with an error tolerance of 10^{-8} leads to the response shown in the Figure 8.6. Error measures computed regarding the fluctuation in predicted values of total energy and angular momentum about the z-axis fluctuate about one part in 100,000. It appears that the analysis employing cartesian coordinates does produce good results.

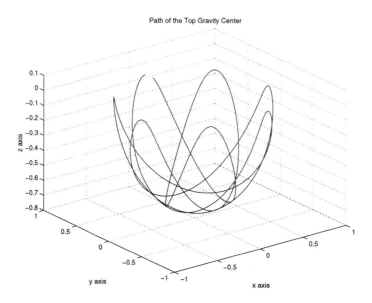

Figure 8.6. Path of the Top Gravity Center

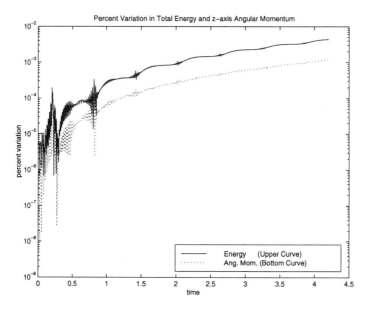

Figure 8.7. Variation in Total Energy and z-axis Angular Momentum

8.6.1   Program Output and Code

Script File toprun

```
 1: % Example: toprun
 2: % ~~~~~~~~~~~~~~~
 3: %
 4: % Example which analyzes the response of a
 5: % spinning conical top.
 6: %
 7: % User m functions required:
 8: %     topde, cubrange, read, genprint
 9: %------------------------------------------------
10:
11: global uz c1 c2
12: disp(' ');
13: disp(['*** Dynamics of a Homogeneous ', ...
14:       'Conical Top ***']); disp(' ');
15: disp(['Input the gravity constant and the ', ...
16:       'body weight (try 32.2,5)']);
17: [grav,wt]=read('? ');
18: mass=wt/grav; tmp=zeros(3,1);
19: disp(' ');
20: disp(['Input the height and base radius ', ...
21:       '(try 1,.5)']);
22: [ht,rb]=read('? '); len=.75*ht;
23: jtrans=3*mass/20*(rb*rb+4*ht*ht);
24: jaxial=3*mass*rb*rb/10;
25: disp(' ');
26: disp(['Input a vector along the initial ', ...
27:       'axis direction (try 0,1,0)']);
28: [tmp(1),tmp(2),tmp(3)]=read('? ');
29: e3=tmp(:)/norm(tmp); r0=len*e3;
30: disp(' ');
31: disp(['Input the initial angular velocity ', ...
32:       '(try 0,10,2)']);
33: [tmp(1),tmp(2),tmp(3)]=read('? '); omega0=tmp;
34: omegax=e3'*omega0(:); rdot0=cross(omega0,r0);
35: z0=[r0(:);rdot0(:)]; uz=[0;0;1];
36: c1=wt*len^2/jtrans; c2=omegax*jaxial/jtrans;
37: disp(' ');
38: disp(['Input tfinal,and the integration ', ...
39:       'tolerance (try 4.2, 1e-8)']);
40: [tfinl,tol]=read('? '); disp(' ');
```

```
41: fprintf( ...
42:   'Please wait for solution of equations.\n');
43:
44: % Integrate the equations of motion
45: odeoptn=odeset('RelTol',tol);
46: [tout,zout]=ode45('topde',[0,tfinl],z0,odeoptn);
47: t=tout; x=zout(:,1); y=zout(:,2); z=zout(:,3);
48: vx=zout(:,4); vy=zout(:,5); vz=zout(:,6);
49:
50: % Compute total energy and angular momentum
51: c3=jtrans/(len*len); taxial=jaxial/2*omegax^2;
52: r=zout(:,1:3)'; v=zout(:,4:6)';
53: etotal=(wt*r(3,:)+taxial+c3/2*sum(v.*v))';
54: h=(jaxial*omegax/len*r+c3*cross(r,v))';
55:
56: % Plot the path of the gravity center
57: clf; axis('equal');
58: axis(cubrange([x(:),y(:),z(:)])); plot3(x,y,z);
59: title('Path of the Top Gravity Center');
60: xlabel('x axis'); ylabel('y axis');
61: zlabel('z axis'); grid on; figure(gcf);
62: disp(' ');
63: dumy=input('Press [Enter] to continue','s');
64: %genprint('toppath');
65: n=2:length(t);
66:
67: % Compute energy and angular momentum error
68: % quantities and plot results
69: et=etotal(1); enrger=abs(100*(etotal(n)-et)/et);
70: hzs=abs(h(1,3));
71: angmzer=abs(100*(h(n,3)-hzs)/hzs);
72: vec=[enrger(:);angmzer(:)];
73: minv=min(vec); maxv=max(vec);
74:
75: clf;
76: semilogy(t(n),enrger,'-r',t(n),angmzer,':m');
77: axis('normal'); xlabel('time');
78: ylabel('percent variation');
79: title(['Percent Variation in Total Energy ', ...
80:        'and z-axis Angular Momentum']);
81: legend(' Energy      (Upper Curve)', ...
82:        ' Ang. Mom. (Bottom Curve)',4);
83: figure(gcf); %genprint('topvar');
```

Function topde

```
1: function zdot=topde(t,z)
2: %
3: % zdot=topde(t,z)
4: % ~~~~~~~~~~~~~~~~
5: %
6: % This function defines the equation of motion
7: % for a symmetrical top. The vector z equals
8: % [r(:);v(:)] which contains the Cartesian
9: % components of the gravity center radius and
10: % its velocity.
11: %
12: % t    - the time variable
13: % z    - the vector [x; y; z; vx; vy; vz]
14: %
15: % zdot - the time derivative of z
16: %
17: % User m functions called:  none
18: %-------------------------------------------------
19:
20: global uz c1 c2
21: % The global variables are defined as follows:
22: %    uz=[0;0;1];
23: %    c1=wt*len^2/jtrans;
24: %    c2=omegax*jaxial/jtrans;
25:
26: z=z(:); r=z(1:3); len=norm(r); ur=r/len;
27:
28: % Make certain the input velocity is
29: % perpendicular to r
30: v=z(4:6); v=v-(ur'*v)*ur;
31: vdot=-c1*(uz-ur*ur(3))+c2*cross(ur,v)- ...
32:      ((v'*v)/len)*ur;
33: zdot=[v;vdot];
```

8.7   Motion of a Projectile

The problem of aiming a projectile to strike a distant target involves integrating a system of differential equations governing the motion and adjusting the initial inclination angle to achieve the desired hit [99]. A reasonable model for the projectile motion assumes atmospheric drag proportional to the square of the velocity. Consequently, the equations of motion are

$$\dot{v}_x = -cvv_x \qquad \dot{v}_y = -g - cvv_y \qquad \dot{x} = v_x \qquad \dot{y} = v_y$$

where g is the gravity constant and c is a ballistic coefficient depending on such physical properties as the projectile shape and air density.

The natural independent variable in the equations of motion is time. However, horizontal position x is a more desirable independent variable, since the target will be located at some distant point (x_f, y_f) relative to the initial position $(0,0)$ where the projectile is launched. We can formulate the differential equations in terms of x by using the relationship

$$dx = v_x \, dt \qquad \frac{dt}{dx} = \frac{1}{v_x}$$

Then

$$\frac{dy}{dx} = \frac{v_y}{v_x} \qquad \frac{dv_y}{dt} = v_x \frac{dv_y}{dx} \qquad \frac{dv_x}{dt} = v_x \frac{dv_x}{dx}$$

and the equations of motion become

$$\frac{dy}{dx} = \frac{v_y}{v_x} \qquad \frac{dt}{dx} = \frac{1}{v_x}$$

$$\frac{dv_x}{dx} = -cv \qquad \frac{dv_y}{dx} = \frac{-(g + cvv_y)}{v_x}$$

Taking a vector z defined by

$$z = [v_x; \; v_y; \; y; \; t]$$

leads to a first order matrix differential equation

$$\frac{dz}{dx} = \frac{[-cvv_x; \; -(g + cvv_y); \; v_y; \; 1]}{v_x}$$

where

$$v = \sqrt{v_x^2 + v_y^2}$$

The reader should note that an ill-posed problem can occur if the initial velocity of the projectile is not large enough so that the maximum desired value of x is reached before v_x is reduced to zero from atmospheric drag. Consequently, error checking is needed to handle such

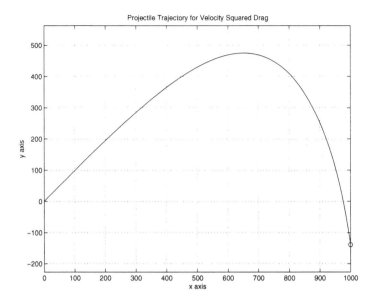

Figure 8.8. Projectile Trajectory for v^2 Drag Condition

a circumstance. The functions **traject** and **projcteq** employ intrinsic function **ode45** to compute the projectile trajectory. Graphical results produced by the default data case appear in Figure 8.8. The function **traject** will be employed again in Chapter 13 for an optimization problem where a search procedure is used to compute the initial inclination angle needed to hit a target at some specified distant position. In this section we simply provide the functions to integrate the equations of motion.

8.7.1   Program Output and Code

Function traject

```
 1: function [y,x,t]=traject ...
 2:         (angle,vinit,gravty,cdrag,xfinl,noplot)
 3: %
 4: % [y,x,t]=traject ...
 5: %         (angle,vinit,gravty,cdrag,xfinl,noplot)
 6: % ~~~~~~~~~~~~~~~~~~~~~~~~~~~~~~~~~~~~~~~~~~~~~~~
 7: %
 8: % This function integrates the dynamical
 9: % equations for a projectile subjected to
10: % gravity loading and atmospheric drag
11: % proportional to the square of the velocity.
12: %
13: % angle  - initial inclination of the
14: %          projectile in degrees
15: % vinit  - initial velocity of the projectile
16: %          (muzzle velocity)
17: % gravty - the gravitational constant
18: % cdrag  - drag coefficient specifying the drag
19: %          force per unit mass which equals
20: %          cdrag*velocity^2.
21: % xfinl  - the projectile is fired toward the
22: %          right from x=0.  xfinl is the
23: %          largest x value for which the
24: %          solution is computed. The initial
25: %          velocity must be large enough that
26: %          atmospheric damping does not reduce
27: %          the horizontal velocity to zero
28: %          before xfinl is reached.  Otherwise
29: %          an error termination will occur.
30: % noplot - plotting of the trajectory is
31: %          omitted when this parameter is
32: %          given an input value
33: %
34: % y,x,t  - the y, x and time vectors produced
35: %          by integrating the equations of
36: %          motion
37: %
38: % Global variables:
39: %
40: % grav,  - two constants replicating gravty and
```

```
41: % dragc    cdrag, for use in function projcteq
42: % vtol    - equal to vinit/1e6, used in projcteq
43: %            to check whether the horizontal
44: %            velocity has been reduced to zero
45: %
46: % User m functions called: projcteq
47: %-----------------------------------------------
48:
49: global grav dragc vtol
50:
51: % Default data case generated when input is null
52: if nargin ==0
53:   angle=45; vinit=600; gravty=32.2;
54:   cdrag=0.002; xfinl=1000;
55: end;
56:
57: % Assign global variables and evaluate
58: % initial velocity
59: grav=gravty; dragc=cdrag; ang=pi/180*angle;
60: vtol=vinit/1e6;
61: z0=[vinit*cos(ang); vinit*sin(ang); 0; 0];
62:
63: % Integrate the equations of motion defined
64: % in function projcteq
65: deoptn=odeset('RelTol',1e-6);
66: [x,z]=ode45('projcteq',[0,xfinl],z0,deoptn);
67:
68: y=z(:,3); t=z(:,4); n=length(x);
69: xf=x(n); yf=y(n);
70:
71: % Plot the trajectory curve
72: if nargin < 6
73:   plot(x,y,'-',xf,yf,'o');
74:   xlabel('x axis'); ylabel('y axis');
75:   title(['Projectile Trajectory for ', ...
76:         'Velocity Squared Drag']);
77:   axis('equal'); grid on; figure(gcf);
78:   %genprint('trajplot');
79: end
```

Function projcteq

```
 1: function zp=projcteq(x,z)
 2: %
 3: % zp=projcteq(x,z)
 4: % ~~~~~~~~~~~~~~~~~
 5: %
 6: % This function defines the equation of motion
 7: % for a projectile loaded by gravity and
 8: % atmospheric drag proportional to the square
 9: % of the velocity.
10: %
11: % x     -  the horizontal spatial variable
12: % z     -  a vector containing [vx; vy; y; t];
13: %
14: % zp    -  the derivative dz/dx which equals
15: %          [vx'(x); vy'(x); y'(x); t'(x)];
16: %
17: % Global variables:
18: %
19: % grav  -  the gravity constant
20: % dragc -  the drag coefficient divided by
21: %          gravity
22: % vtol  -  a global variable used to check
23: %          whether vx is zero
24: %
25: % User m functions called:  none
26: %-------------------------------------------------
27:
28: global grav dragc vtol
29: vx=z(1); vy=z(2); v=sqrt(vx^2+vy^2);
30:
31: % Check to see whether drag reduced the
32: % horizontal velocity to zero before the
33: % xfinl was reached.
34: if abs(vx) < vtol
35:     disp(' ');
36:     disp('***********************************');
37:     disp('ERROR in function projcteq. The ');
38:     disp('  initial velocity of the projectile');
39:     disp('  was not large enough for xfinal to');
40:     disp('  be reached.');
41:     disp('EXECUTION IS TERMINATED.');
42:     disp('***********************************');
```

```
43:    disp(' '),error(' ');
44: end
45: zp=[-dragc*v; -(grav+dragc*v*vy)/vx; ...
46:     vy/vx; 1/vx];
```

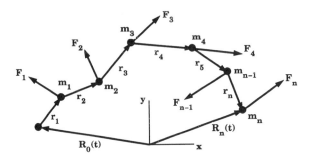

Figure 8.9. Chain with Specified End Motion

8.8   Example on Dynamics of a Chain with Specified End Motion

The dynamics of flexible cables is often modeled using a chain of rigid links connected by frictionless joints. A chain having specified end motions illustrates the behavior of a system governed by nonlinear equations of motion and auxiliary algebraic constraints. In particular, we will study a gravity loaded cable fixed at both ends. The total cable length exceeds the distance between supports, so that the static deflection configuration would resemble a catenary.

A simple derivation of the equations of motion employing principles of rigid body dynamics is given next. Readers not versed in principles of rigid body dynamics [47] may nevertheless understand the subsequent programs by analyzing the equations of motion which have a concise mathematical form. The numerical solutions vividly illustrate some numerical difficulties typically encountered in multibody dynamical studies. Such problems are both computationally intensive, as well as highly sensitive to accumulated effects of numerical error.

The mathematical model of interest is the two-dimensional motion of a cable (or chain) having n rigid links connected by frictionless joints. A typical link $\imath$ has its mass $m_\imath$ concentrated at one end. The geometry is depicted in Figure 8.9. The chain ends undergo specified motions $R_0(t) = [X_0(t) \; ; \; Y_0(t)]$ for the first link and $R_n(t) = [X_n(t) \; ; \; Y_n(t)]$ for the last link. The direction vector along link $\imath$ is described by $r_\imath = [x_\imath \; ; \; y_\imath] = \ell_\imath[\cos(\theta_\imath) \; ; \; \sin(\theta_\imath)]$. We assume that each joint $\imath$ is subjected to a force $F_\imath = [f_{x\imath} \; ; \; f_{y\imath}]$ where $0 \le \imath \le n$. Index values $\imath = 0$ and $\imath = n$ denote unknown constraint forces which must act at the outer ends of the first and last links to achieve the required end displacements. The forces applied at the interior joints are arbitrary. It is convenient to characterize the dynamics of each link in terms of its direction angle.

Thus
$$\dot{r}_i = r_i'\dot{\theta}_i \qquad \ddot{r}_i = r_i'\ddot{\theta}_i + r_i''\dot{\theta}_i^2 = r_i'\ddot{\theta}_i - r_i\dot{\theta}_i^2$$

where primes and dots denote differentiation with respect to θ_i and t, respectively. Therefore

$$\dot{r}_i = [-y_i \; ; \; x_i]\dot{\theta}_i \qquad \ddot{r}_i = [-\ddot{y}_i \; ; \; x_i]\ddot{\theta}_i - [x_i \; ; \; y_i]\dot{\theta}_i^2$$

The global position vector of joint i is

$$R_i = R_0 + \sum_{j=1}^{i} r_j = R_0 + \sum_{j=1}^{n} <i-j> r_j$$

where the symbol $<k>$ means one for $k \geq 0$, and zero for negative k. Consequently, the velocity and acceleration of joint i are

$$\dot{R}_i = \dot{R}_0 + \sum_{j=1}^{n} <i-j> r_j'\dot{\theta}_j$$

$$\ddot{R}_i = \ddot{R}_0 + \sum_{j=1}^{n} <i-j> r_j'\ddot{\theta}_j - \sum_{j=1}^{n} <i-j> r_j\dot{\theta}_j^2$$

The ends of the chain each have specified motions so, not all of the inclination angles are independent and consequently

$$\sum_{j=1}^{n} r_j = R_n - R_0$$

$$\sum_{j=1}^{n} r_j'\dot{\theta}_j = \dot{R}_n - \dot{R}_0$$

$$\sum_{j=1}^{n} r_j'\ddot{\theta}_j - \sum_{j=1}^{n} r_j\dot{\theta}_j^2 = \ddot{R}_n - \ddot{R}_0$$

Combining the last constraint equation with equations of motion written for masses $m_1, \cdots, m_n$ yields a complete system of $(n+2)$ equations determining $\theta_1, \cdots, \theta_n$ and the components of F_n. The fact that all masses are concentrated at frictionless joints shows that link i is a two-force member carrying an internal load directed along r_i. Consequently, the D'Alembert principle [47] implies that the sum of all external and inertial loads from joints $i, i+1, \cdots, n$ must give a resultant passing through joint i in the direction of r_i. Since r_i' and r_i are perpendicular, requiring a vector to be in the direction of r_i is equivalent to making it normal to r_i'. Therefore

$$r_i' \cdot \left[\sum_{j=1}^{n} <j-i> \left\{ F_j - m_j \ddot{R}_j \right\} \right] = 0 \qquad 1 \leq i \leq n$$

The last n equations involve $\ddot{\theta}_i$ and two end force components f_{xn} and f_{yn}. Some algebraic rearrangement results in a matrix differential equation of concise form containing several auxiliary coefficients defined as follows:

$$b_i = \sum_{k=i}^{n} m_k \qquad m_{ij} = m_{ji} = b_i \qquad 1 \le i \le n \qquad 1 \le j \le i$$

$$a_{ij} = m_{ij}(x_i x_j + y_i y_j) \qquad 1 \le i \le n \qquad 1 \le j \le n$$

$$b_{ij} = m_{ij}(x_i y_j - x_j y_i) \qquad 1 \le i \le n \qquad 1 \le j \le n$$

$$p_{xi} = \sum_{j=i}^{n-1} f_{xi} \qquad p_{yi} = \sum_{j=i}^{n-1} f_{yi} \qquad 1 \le i \le n$$

For $i = n$, the last two sums mean $p_{xn} = p_{yn} = 0$. Furthermore, we denote the acceleration components of the chain ends as $\ddot{R}_0 = [a_{x0} \; ; \; a_{y0}]$ and $\ddot{R}_n = [a_{xn} \; ; \; a_{yn}]$. Using the various quantities just defined, the equations of motion become

$$\sum_{j=1}^{n} a_{ij}\ddot{\theta}_j + y_i f_{xn} - x_i f_{yn} = \sum_{j=1}^{n} b_{ij}\dot{\theta}_j^2 + x_i(p_{yi} - b_i a_{y0}) -$$

$$y_i(p_{xi} - b_i a_{x0})$$

$$= e_i \qquad 1 \le i \le n$$

The remaining two components of the constraint equations completing the system are

$$\sum_{j=1}^{n} y_j \ddot{\theta}_j = -\sum_{j=1}^{n} x_j \dot{\theta}_j^2 - a_{xn} + a_{x0} = e_{n+1}$$

$$\sum_{j=1}^{n} x_j \ddot{\theta}_j = \sum_{j=1}^{n} y_j \dot{\theta}_j^2 + a_{yn} - a_{y0} = e_{n+2}$$

Consequently, we get the following symmetric matrix equation to solve for $\ddot{\theta}_1, \cdots, \ddot{\theta}_n, f_{xn}$ and f_{yn}.

$$\begin{bmatrix} A & X & Y \\ X^T & 0 & 0 \\ Y^T & 0 & 0 \end{bmatrix} \begin{bmatrix} \ddot{\theta} \\ f_{xn} \\ -f_{yn} \end{bmatrix} = \begin{bmatrix} E \end{bmatrix}$$

where X, Y, E and θ are column matrices, and the matrix $A = [a_{ij}]$ is symmetric. Because most numerical integrators for differential equations solve first order systems, it is convenient to employ the vector $Z = [\theta \; ; \; \dot{\theta}]$ having $2n$ components. Then the differential equation $\dot{Z} = H(t, Z)$ is completely defined when $\ddot{\theta}$ has been computed for known Z. The system is integrated numerically to give θ and $\dot{\theta}$ as functions of time. These quantities can then be used to compute the global cartesian coordinates of the link configurations, thereby completely describing the time history of the chain.

The general equations of motion simplify somewhat when the chain ends are fixed and the external forces only involve gravity loads. Then $p_{xi} = 0$ and $p_{yi} = -g(b_i - b_n)$ which gives

$$\sum_{j=1}^{n} m_{ij}(x_i x_j + y_i y_j)\ddot{\theta}_j - x_i f_{yn} + y_i f_{xn} =$$

$$g(b_i - b_n) + \sum_{j=1}^{n} m_{ij}(x_i y_j - x_j y_i)\dot{\theta}_j^2 \qquad 1 \leq i \leq n$$

The last two equations to complete the set are:

$$\sum_{j=1}^{n} x_j \ddot{\theta}_j = \sum_{j=1}^{n} y_j \dot{\theta}_j^2 \qquad \sum_{j=1}^{n} y_j \ddot{\theta}_j = -\sum_{j=1}^{n} x_j \dot{\theta}_j^2$$

A program was written to simulate motion of a cable fixed at both ends and released from rest. The cable falls under the influence of gravity from an initially elevated position. Function **ode45** is used to perform the numerical integration. The program involves the following modules.

cablenl	driver program to set initial physical constants and numerical tolerances
equamo	function which defines the equations of motion for use by **ode45**
pltxmidl	function to plot the horizontal position of the middle
unsymerr	function to compute and plot a measure of how much the deflection platform loses symmetry with passing time
plotmotn	function producing an animated plot of the cable position for a sequence of times
eventime	function which linearly interpolates **ode45** output to produce position values corresponding to equidistant time intervals
lintrp	function performing piecewise linear interpolation

A configuration with eight identical links was specified. For simplicity, the total mass, total cable length, and gravity constant were all normalized to equal unity. Results of the simulation are shown below. Figure 8.10 shows cable positions during the early stages of motion when results of the numerical integration are reliable. However, the numerical solution eventually becomes worthless due to accumulated numerical inaccuracies yielding the motion predictions indicated in Figure 8.11. The nature of the error growth can be seen clearly in Figure 8.12 which plots the x-coordinate of the chain midpoint as a function of time. Since the chosen mass distribution and initial deflection is symmetrical about the middle, the subsequent motion will remain symmetrical unless the numerical solution becomes invalid. Although the midpoint coordinate should remain at a constant value of $0.5\sqrt{2}$, it appears to abruptly go unstable near $t = 17$. More careful examination indicates that this numerical instability does not actually occur suddenly. Instead, it grows exponentially from the outset of the simulation. The error is caused by the accumulation of truncation errors intrinsic to the numerical integration process allowing errors at each step which are regulated within a small but finite tolerance. A global measure of symmetry loss of the deflection pattern is plotted on a semilog scale in Figure 8.13. Note that the error curve has a nearly linear slope until the solution degenerates completely near $t = 18$. The reader can verify that choosing a less stringent error tolerance such as tol = 1E-3 will produce a solution which degenerates much sooner than $t = 18$. It should also be observed that this dynamical model exhibits another important characteristic of highly nonlinear systems, namely, extreme sensitivity to physical properties. Note that shortening the last link by only one part in ten thousand makes the system deflection quickly lose all appearance of symmetry by $t = 6$. Hence, two systems having nearly identical physical parameters and initial conditions may behave very differently a short time after motion is initiated. The conclusion implied is that analysts should thoroughly explore how parameter variations affect response predictions in nonlinear models.

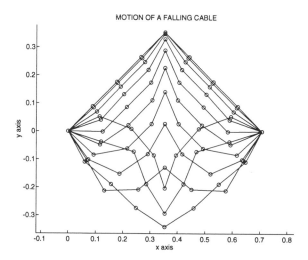

Figure 8.10. Motion During Initial Phase

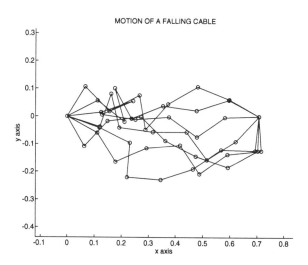

Figure 8.11. Motion After Solution Degenerates

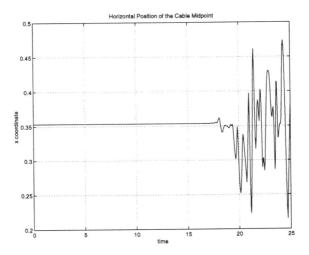

Figure 8.12. Horizontal Position of the Cable Midpoint

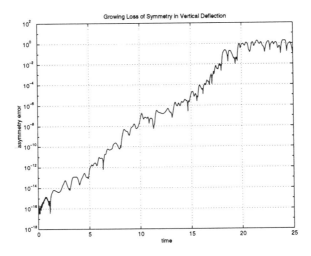

Figure 8.13. Growing Loss of Symmetry in Vertical Deflection

MATLAB Example
Script File cablenl

```
 1: % Example: cablenl
 2: % ~~~~~~~~~~~~~~~~~
 3: % Numerical integration of the matrix
 4: % differential equations for the nonlinear
 5: % dynamics of a cable fixed at both ends.
 6: %
 7: % User m functions required:
 8: %    unsymerr, plotmotn, eventime, equamo,
 9: %    lintrp, pltxmidl, genprint
10: %-------------------------------------------
11:
12: clear; clf;
13:
14: % Make variables global for use
15: % by other routines
16: global first_ n_ m_ len_ grav_ b_ mas_ py_
17:
18: fprintf('\nNONLINEAR DYNAMICS OF A ');
19: fprintf('FALLING CABLE\n');
20: fprintf('\nPlease wait: calculations will ');
21: fprintf('take several minutes\n');
22:
23: % Set up data for a cable of n_ links,
24: % initially arranged in a triangular
25: % deflection configuration.
26:
27: % parameter controlling initialization of
28: % auxiliary parameters used in function
29: % equamo
30: first_=1;
31: % number of links in the cable
32: n_=8; n=n_; nh=n_/2;
33: % vector of lengths and gravity constant
34: len_=1/n*ones(n,1); grav_=1;
35: % vector of mass constants
36: m_=ones(1,n_)/n_;
37:
38: % initial position angles
39: th0=pi/4*[ones(nh,1);-ones(nh,1)];
40: td0=zeros(size(th0)); z0=[th0;td0];
```

```
41:
42: % time limits, integration tolerances, and
43: % number of plot points
44: tfinl=25; tmin=0; tmax=25;
45: tol=1e-6; nplot=500;
46: trac=0; len=len_;
47:
48: % Perform the numerical integration using a
49: % variable stepsize Runge-Kutta integrator
50:
51: quadtime=cputime; flps=flops;
52: odeopts=odeset('RelTol',tol);
53: [tu,zu]=ode45('equamo',[0,tfinl],z0,odeopts);
54: thu=zu(:,1:n);
55: quadtime=cputime-quadtime; flps=flops-flps;
56:
57: % Interpolate for results evenly spaced in time
58: [tevn,xevn,yevn]= ...
59:    eventime(tu,thu,len,nplot,tmin,tmax);
60: save cablenl.mat
61:
62: % Plot the horizontal position of the
63: % cable midpoint
64: pltxmidl(tevn,xevn(:,1+n_ /2));
65: dumy=input('Press [Enter] to continue','s');
66: close;
67:
68: % Show error growth indicated by symmetry
69: % loss of the vertical deflection pattern
70: [terr,yerr]=unsymerr(tu,thu,len,tmin,tmax);
71: close;
72:
73: % Plot successive deflection positions of
74: % the cable
75: plotmotn(xevn,yevn,1);
76:
77: close; fprintf('\nAll Done\n');
```

Function equamo

```
1: function zdot=equamo(t,z)
2: %
3: % zdot=equamo(t,z)
```

```
 4: %  ~~~~~~~~~~~~~~~~~
 5: % Equation of motion for a cable fixed at
 6: % both ends and loaded by gravity forces only
 7: %
 8: % t          current time value
 9: % z          column vector defined by
10: %            [thet(t);theta'(t)]
11: % zdot       column vector defined by
12: %            the concatenation
13: %            z'(t) = [theta'(t);theta''(t)]
14: %
15: % User m functions called:  none.
16: %------------------------------------------------
17:
18: % Values accessed as global variables
19: global first_ n_ m_ len_ grav_ b_ mas_ py_
20:
21: % Initialize parameters first time
22: % function is called
23: if first_==1, first_=0;
24: % mass parameters
25:   m_=m_(:); b_=flipud(cumsum(flipud(m_)));
26:   mas_=b_(:,ones(n_,1));
27:   mas_=tril(mas_)+tril(mas_,-1)';
28: % load effects from gravity forces
29:   py_=-grav_*(b_-b_(n_));
30: end
31:
32: % Solve for zdot = [theta'(t); theta''(t)];
33: n=n_; len=len_;
34: th=z(1:n); td=z(n+1:2*n); td2=td.*td;
35: x=len.*cos(th); y=len.*sin(th);
36:
37: % Matrix of mass coefficients and
38: % constraint conditions
39: amat=[[mas_.*(x*x'+y*y'),x,y];
40:       [x,y;zeros(2,2)]'];
41:
42: % Right side vector involves applied forces
43: % and inertial terms
44: bmat=x*y'; bmat=mas_.*(bmat-bmat');
45:
46: % Solve for angular acceleration.
47: % Most computation occurs here.
48: soln=amat\[x.*py_+bmat*td2; y'*td2; -x'*td2];
```

```
49:
50: % Final result needed for use by the numerical
51: % integrator
52: zdot=[td; soln(1:n)];
```

Function unsymerr

```
 1: function [terr,yerr]= ...
 2:                   unsymerr(t,thta,len,tmin,tmax)
 3: %
 4: % [terr,yerr]=unsymerr(t,thta,len,tmin,tmax)
 5: % ~~~~~~~~~~~~~~~~~~~~~~~~~~~~~~~~~~~~~~~~~~~
 6: % This function computes an error measure which
 7: % shows how the initally symmetric deflection
 8: % configuration progressively loses symmetry
 9: % due to growing error in the numerical
10: % solution.
11: %
12: % t            - vector of times for solution
13: %                evaluation
14: % thta         - matrix with successive rows
15: %                defining sets of theta values
16: %                which specify a configuration of
17: %                the cable
18: % len          - vector of lengths of the
19: %                cable links
20: % tmin,tmax    - time limits over which solution
21: %                error is evaluated
22: % terr         - subset of t values between tmin
23: %                and tmax at which the solution
24: %                error is computed
25: % yerr         - vector specifying the solution
26: %                error defined in the following
27: %                manner. Let Y be a column vector
28: %                of length n-1 representing the
29: %                vertical deflection of the cable
30: %                (excluding the fixed ends). A
31: %                corresponding element of yerr
32: %                would be defined as
33: %                norm(Y-flipud(Y))/norm(Y).
34: %
35: % User m functions called:  genprint
36: %-------------------------------------------------
```

```
37:
38: [nt,n]=size(thta);
39: if nargin < 5, tmin=min(t); tmax=max(t); end
40:
41: % Compute time within specified limits
42: n1=sum(t<=tmin); n2=sum(t<=tmax);
43: terr=t(n1:n2); thta=thta(n1:n2,:);
44:
45: % Compute values of the vertical deflection.
46: nte=length(terr); len=len(:)';
47: y=cumsum((len(ones(nte,1),:).*sin(thta))')';
48:
49: % Evaluate growth in unsymmetrical character
50: % of the deflection pattern
51: yy=y(:,1:n-1); ydif=yy-yy(:,n-1:-1:1);
52: yerr=sqrt(sum((ydif.*ydif)'))./ ...
53:        sqrt(sum((yy.*yy)'));
54:
55: % Graph the solution error. An approximately
56: % linear trend on a semilog plot would
57: % indicate exponential growth of error with
58: % passing time.
59: hold off; axis('normal'); clf;
60: semilogy(terr,yerr);
61: xlabel('time'); ylabel('asymmetry error');
62: title(['Growing Loss of Symmetry in ' ...
63:        'Vertical Deflection']);
64: grid on; figure(gcf);
65: dumy=input('Press [Enter] to continue','s');
66: %genprint('unsymerr');
```

Function plotmotn

```
 1: function plotmotn(x,y,isave)
 2: %
 3: % plotmotn(x,y,ifsave)
 4: % ~~~~~~~~~~~~~~~~~~~~
 5: % This function plots the cable time
 6: % history described by coordinate values
 7: % stored in the rows of matrices x and y.
 8: %
 9: % x,y    - matrices having successive rows
10: %          which describe position
```

```
11: %              configurations for the cable
12: % isave - parameter controlling the form
13: %              of output. When isave is not input,
14: %              successive positions are plotted.
15: %              The next position is plotted when
16: %              the user presses any key. If isave
17: %              is given a value, then successive
18: %              positions are not erased so that
19: %              the sequence of positions are all
20: %              left showing.
21: %
22: % User m functions called: genprint
23: %------------------------------------------------
24:
25: % Set a square window to contain all
26: % possible positions
27: [nt,n]=size(x);
28: xmin=min(x(:)); xmax=max(x(:));
29: ymin=min(y(:)); ymax=max(y(:));
30: w=max(xmax-xmin,ymax-ymin)/2;
31: xmd=(xmin+xmax)/2; ymd=(ymin+ymax)/2;
32: hold off; clf; axis('normal'); axis('equal');
33: axis([xmd-w,xmd+w,ymd-w,ymd+w]);
34: title('Motion of a Falling Cable');
35: xlabel('x axis'); ylabel('y axis');
36:
37: % Plot successive positions describing
38: % time history
39: istr=['Press [Enter] for next position, ' ...
40:       'Q [Enter] to exit'];
41: axis off; hold on;
42: for j=1:nt
43:    fprintf('\nPosition %6.0f of %6.0f\n', j, nt);
44:    xj=x(j,:); yj=y(j,:);
45:    if nargin==2
46:      % Plot and then erase
47:      plot(xj,yj,'-b',xj,yj,'ob'); figure(gcf);
48:      pause(1); more=input(istr,'s');
49:      if isempty(more), more=' '; end
50:      if more == 'Q' | more == 'q', break, end
51:      cla;
52:    else
53:      % Plot and leave trace
54:      plot(xj,yj,'-b',xj,yj,'ob');
55:      figure(gcf); pause(1); more=input(istr,'s');
```

```
56:     if isempty(more), more=' '; end
57:       if more == 'Q' | more == 'q', break, end
58:   end
59: end
60:
61: % Save plot history for subsequent printing
62: %genprint('plotmotn');
63: hold off; close;
```

Function eventime

```
1: function [te,xe,ye]= ...
2:                 eventime(t,th,len,nte,tmin,tmax)
3: %
4: % [te,xe,ye]=eventime(t,th,len,nte,tmin,tmax)
5: % ~~~~~~~~~~~~~~~~~~~~~~~~~~~~~~~~~~~~~~~~~~~~~
6: % This function computes cable position
7: % coordinates for a series of evenly spaced
8: % time values
9: %
10: % t           unevenly spaced time values
11: %             produced by the differential
12: %             equation integrator
13: % th          theta values output by the
14: %             differential equation integrator
15: % len         vector of lengths for the
16: %             cable links
17: % nte         number of evenly spaced time
18: %             values to be used
19: % tmin,tmax   maximum and minimum times
20: %             for output
21: % te          vector of output times
22: % xe,ye       matrices containing interpolated
23: %             coordinate values corresponding
24: %             to even time increments
25: %
26: % User m functions called:  lintrp
27: %-----------------------------------------------
28:
29: % Compute position vectors xe, ye corresponding
30: % to evenly spaced times from tmin to tmax
31: if nargin < 6, tmin=min(t); tmax=max(t); end
32: [nt,n]=size(th); len=len(:)';
```

```
33:
34: % Generate vector of equally spaced times
35: te=tmin+(tmax-tmin)/(nte-1)*(0:nte-1)';
36: the=zeros(nte,n);
37:
38: % Compute theta values at desired times
39: for j=1:n, the(:,j)=lintrp(t,th(:,j),te); end
40:
41: % Generate global position coordinates
42: % for the desired times
43: xe=cumsum((len(ones(nte,1),:).*cos(the))')')';
44: ye=cumsum((len(ones(nte,1),:).*sin(the))')')';
45: xe=[zeros(nte,1),xe]; ye=[zeros(nte,1),ye];
```

Function pltxmidl

```
1: function pltxmidl(t,x)
2: %
3: % pltxmidl(t,x)
4: % ~~~~~~~~~~~~~~
5: % t - time vector
6: % x - horizontal position of the cable midpoint
7: %
8: % User m functions called:  genprint
9: %------------------------------------------------
10:
11: clf; plot(t,x);
12: ylabel('x coordinate'); xlabel('time')
13: title(['Horizontal Position of the ' ...
14:        'Cable Midpoint'])
15: grid on; figure(gcf); %genprint('xmidl');
```

Chapter 9

Boundary Value Problems for Linear Partial Differential Equations

9.1 Several Important Partial Differential Equations

Many physical phenomena are characterized by linear partial differential equations. Such equations are attractive to study because (a) principles of superposition apply in the sense that linear combinations of component solutions can often be used to build more general solutions and (b) finite difference or finite element approximations lead to systems of linear equations amenable to solution by matrix methods. The table below lists several frequently encountered equations and some applications. We only show one- or two-dimensional forms, although some of these equations have relevant applications in three dimensions.

In most practical applications the differential equations must be solved within a finite region of space while simultaneously prescribing boundary conditions on the function and its derivatives. Furthermore, initial conditions may exist. In dealing with the initial value problem, we are trying to predict future system behavior when initial conditions, boundary conditions, and a governing physical process are known. Solutions to such problems are seldom obtainable in a closed finite form.

Equation	Equation Name	Applications
$u_{xx} + u_{yy} = \alpha u_t$	Heat	Transient heat conduction
$u_{xx} + u_{yy} = \alpha u_{tt}$	Wave	Transverse vibrations of membranes and other wave type phenomena
$u_{xx} + u_{yy} = 0$	Laplace	Steady-state heat conduction and electrostatics
$u_{xx} + u_{yy} = f(x, y)$	Poisson	Stress analysis of linearly elastic bodies
$u_{xx} + u_{yy} + \omega^2 u = 0$	Helmholtz	Steady-state harmonic vibration problems
$EI y_{xxxx} = -A\rho y_{tt} + f(x, t)$	Beam	Transverse flexural vibrations of elastic beams

Even when series solutions are developed, an infinite number of terms may be needed to provide generality. For example, the problem of transient heat conduction in a circular cylinder leads to an infinite series of Bessel functions employing characteristic values which can only be computed approximately. Hence, the notion of an "exact" solution expressed as an infinite series of transcendental functions is deceiving. At best, we can hope to produce results containing insignificantly small computation errors.

The present chapter studies seven problems. Six of these are solved by series methods. The last problem is distinctive from the others since exact natural frequencies of an elastic beam are compared with approximations produced by finite difference and finite element methods.

9.2 Solving the Laplace Equation Inside a Rectangular Region

Functions which satisfy Laplace's equation are encountered often in practice. Such functions are called harmonic; and the problem of determining a harmonic function subject to given boundary values is known as the Dirichlet problem [117]. In a few cases with simple geometries, the Dirichlet problem can be solved explicitly. One instance is a rectangular region with the boundary values of the function being expandable in a Fourier sine series. The following program employs the FFT to construct a solution for boundary values represented by piecewise linear interpolation. Surface and contour plots of the resulting field values are also presented.

The problem of interest satisfies the differential equation

$$\frac{\partial^2 u}{\partial x^2} + \frac{\partial^2 u}{\partial y^2} = 0 \qquad 0 < x < a \qquad 0 < y < b$$

with the boundary conditions of the form

$$
\begin{aligned}
u(x,0) &= F(x) & 0 < x < a \\
u(x,b) &= G(x) & 0 < x < a \\
u(0,y) &= P(y) & 0 < y < b \\
u(a,y) &= Q(y) & 0 < y < b
\end{aligned}
$$

The series solution can be represented as

$$u(x,y) = \sum_{n=1}^{\infty} f_n a_n(x,y) + g_n a_n(x,b-y) +$$

$$p_n b_n(x,y) + q_n b_n(a-x,y)$$

where

$$a_n(x, y) = \sin\left[\frac{n\pi x}{a}\right] \sinh\left[\frac{n\pi(b - y)}{a}\right] / \sinh\left[\frac{n\pi b}{a}\right]$$

$$b_n(x, y) = \sinh\left[\frac{n\pi(a - x)}{b}\right] \sin\left[\frac{n\pi y}{b}\right] / \sinh\left[\frac{n\pi a}{b}\right]$$

and the constants f_m, g_m, p_n, and q_n are coefficients in the Fourier sine expansions of the boundary value functions. This implies that

$$F(x) = \sum_{n=1}^{\infty} f_n \sin\left[\frac{n\pi x}{a}\right] \qquad G(x) = \sum_{n=1}^{\infty} g_n \sin\left[\frac{n\pi x}{a}\right]$$

$$P(y) = \sum_{n=1}^{\infty} p_n \sin\left[\frac{n\pi y}{b}\right] \qquad Q(y) = \sum_{n=1}^{\infty} q_n \sin\left[\frac{n\pi y}{b}\right]$$

The coefficients in the series can be computed by integration as

$$f_n = \frac{2}{a} \int_0^a F(x) \sin\left[\frac{n\pi x}{a}\right] dx \qquad g_n = \frac{2}{a} \int_0^a G(x) \sin\left[\frac{n\pi x}{a}\right] dx$$

$$p_n = \frac{2}{a} \int_0^b P(y) \sin\left[\frac{n\pi y}{b}\right] dy \qquad q_n = \frac{2}{a} \int_0^b Q(y) \sin\left[\frac{n\pi y}{b}\right] dy$$

or approximate coefficients can be obtained using the FFT. The latter approach is chosen here and the solution is evaluated for an arbitrary number of terms in the series.

The chosen problem solution has the disadvantage of employing eigenfunctions which vanish at the ends of the expansion intervals. Consequently, it is desirable to combine the series with an additional term allowing exact satisfaction of the corner conditions for cases where the boundary value functions for adjacent corners agree. This implies requirements such as $F(a) = Q(0)$ and three other similar conditions. It is evident that the function

$$u_p(x, y) = c_1 + c_2 x + c_3 y + c_4 xy$$

is harmonic and varies linearly along each side of the rectangle. Constants $c_1, \cdots, c_4$ can be computed to satisfy the corner values and the total solution is represented as u_p plus a series solution involving modified boundary conditions.

The following program solves the boundary value problem and plots results. The functions used in this program are described below.

laplarec	sets data parameters, establishes global variables, calls other functions, and generates output
setup	generates coefficients defining a particular solution to satisfy corner values
ulbc	the harmonic function pertaining to corner conditions
sinfft	generates coefficients in a Fourier sine series
laprec	sums the series solution of Laplace's equation
lintrp	piecewise linear interpolation function
fbot,gtop,plft,qrht	boundary value functions for bottom, top, left side, and right side of the rectangle. These are specified as piecewise linear.
read	utility function to read data
genprint	utility function to save graph to file

The example data set defined in the driver program has no particular significance other than to produce interesting surface and contour plots. Different boundary conditions can be handled by slight modifications of the input data.

In this example only 30 series terms were used to illustrate how Fourier series based solutions behave near discontinuities in boundary values. Oscillations in solution values occur at corners and other points where jump discontinuities occur. This behavior is typical of Fourier series discussed in Chapter 6.

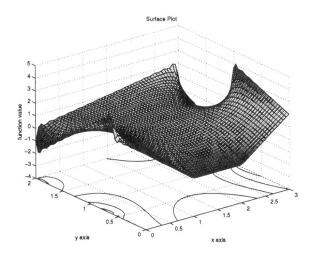

Figure 9.1. Surface Plot

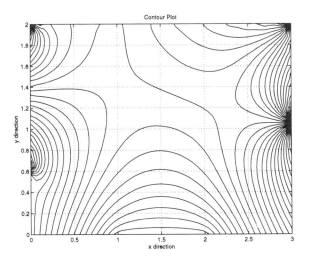

Figure 9.2. Contour Plot

MATLAB Example

Script file laplarec

```
 1: % Example: laplarec
 2: % ~~~~~~~~~~~~~~~~~~
 3: % This program uses Fourier series methods
 4: % to solve Laplace's equation in a rectangle.
 5: % Boundary conditions are defined by piecewise
 6: % linear interpolation of boundary data. The
 7: % program can easily be changed to deal with
 8: % problems where the boundary conditions are
 9: % expressed by analytically defined functions.
10: % Surface and contour plots of the solution
11: % are also provided.
12: %
13: % User m functions required:
14: %    setup, ulbc, sinfft, laprec, fbot, gtop,
15: %    plft, qrht, lintrp, genprint
16: %--------------------------------------------------

17:
18: clear
19: global a_ b_ xbot_ ubot_ xtop_ utop_
20: global ylft_ ulft_ yrht_ urht_
21: global u1_ u2_
22:
23: a_=3; b_=2; % Rectangle side lengths
24:
25: % Data for piecewise linear
26: % boundary conditions
27: xtop_=[0;    3];    utop_= [0;    2];
28: ntop=length(xtop_);
29: xbot_=[0;1;2;3];    ubot_=[2;-1;-1;2];
30: nbot=length(xbot_);
31: ylft_=[0;0.6;0.6;2]; ulft_=[2;2;4;-2];
32: nlft=length(ylft_);
33: yrht_=[0;1;1.1;2];   urht_=[2;4;-2; 0];
34: nrht=length(ylft_);
35:
36: % Create the constants needed in
37: % function ulbc
38: cc=setup;
39:
40: % Adjust boundary data to give zero
```

```
41: % end conditions
42: for k=1:nbot
43:   ubot_(k)=ubot_(k)-ulbc(cc,xbot_(k),0);end
44: for k=1:ntop
45:   utop_(k)=utop_(k)-ulbc(cc,xtop_(k),b_);end
46: for k=1:nlft
47:   ulft_(k)=ulft_(k)-ulbc(cc,0,ylft_(k));end
48: for k=1:nrht
49:   urht_(k)=urht_(k)-ulbc(cc,a_,yrht_(k));end
50:
51: % Generate Fourier coefficients for the
52: % modified boundary conditions
53: sbot=sinfft('fbot',a_,9);
54: stop=sinfft('gtop',a_,9);
55: slft=sinfft('plft',b_,9);
56: srht=sinfft('qrht',b_,9);
57:
58: % Generate a grid of interior points
59: % for solution evaluation
60: nin=51; ntrms=30;
61: xin=linspace(0,a_,nin); yin=linspace(0,b_,nin);
62:
63: % Evaluate the solution having zero
64: % corner values
65: uin=laprec(...
66:     sbot,stop,slft,srht,a_,b_,xin,yin,ntrms);
67:
68: % Correct results for nonzero corner values
69:
70: uin=uin+ulbc(cc,xin,yin); uin=flipud(uin');
71:
72: % Display surfaces showing function
73: % values on the grid
74:
75: clf; surfc(xin,yin,flipud(uin));
76: xlabel('x axis'); ylabel('y axis');
77: zlabel('function value');
78: title('Surface Plot'); figure(gcf);
79: dumy=input('Press [Enter] to continue','s');
80: %genprint('lapsrfac');
81:
82: clf; contour(xin,yin,flipud(uin),30);
83: title('Contour Plot');
84: xlabel('x direction'); ylabel('y direction');
85: figure(gcf); %genprint('contur');
```

```
86: disp('All Done');
```

Function setup

```
1: function cc=setup
2: %
3: % cc=setup
4: % ~~~~~~~~
5: % cc - coefficients used to define the
6: %       particular solution to satisfy
7: %       corner conditions
8: %
9: % User m functions called:
10: %     fbot, qrht, plft, gtop
11: %-----------------------------------------------
12:
13: global   a_ b_
14:
15: s=(a_+b_)/1e10;
16: c1=(fbot(s)+plft(s))/2;
17: c2=(fbot(a_-s)+qrht(s))/2;
18: c3=(plft(b_-s)+gtop(s))/2;
19: c4=(gtop(a_-s)+qrht(b_-s))/2;
20: mat=[[1,0,0,0];[1,a_,0,0];
21:      [1,0,b_,0];[1,a_,b_,a_*b_]];
22: vec=[c1;c2;c3;c4]; cc=mat\vec;
```

Function ulbc

```
1: function u=ulbc(c,x,y)
2: %
3: % u=ulbc(c,x,y)
4: % ~~~~~~~~~~~~~
5: % This function determines a harmonic function
6: % satisfying boundary conditions which vary
7: % linearly on the sides of a rectangle.
8: %
9: % User m functions called:  none
10: %-----------------------------------------------
11:
12: x=x(:); y=y(:)'; nx=length(x); ny=length(y);
13: x=x*ones(1,ny); y=ones(nx,1)*y;
```

```
14: u=c(1)*ones(nx,ny)+c(2)*x+c(3)*y+c(4)*x.*y;
```

Function sinfft

```
1: function sincof = sinfft(fun,hafper,powr2)
2: %
3: % sincof = sinfft(fun,hafper,powr2)
4: % ~~~~~~~~~~~~~~~~~~~~~~~~~~~~~~~~~~~~~
5: % This function determines coefficients in
6: % the Fourier sine series of a general real
7: % valued function.
8: %
9: %  hafper - the half  period over which the
10: %            expansion applies
11: %  fun    - the real valued function being
12: %            expanded. This function must be
13: %            defined for vector arguments with
14: %            components between zero
15: %            and hafper.
16: %  powr2  - the power of 2 determining the
17: %            number of function values used
18: %            in the FFT (number of
19: %            values = 2^powr2).  When powr2
20: %            is 9, then 512 function values
21: %            are used and 255 Fourier
22: %            coefficients are computed.
23: %
24: % User m functions called:  none
25: %-------------------------------------------------
26:
27: n=2^powr2; period=2*hafper; n2=n/2;
28: x=(period/n)*(0:n2);
29: fval=feval(fun,x);
30: fval=fval(:); fval=[fval;-fval(n2:-1:2)];
31: foucof=fft(fval);
32: sincof=-(2/n)*imag(foucof(2:n2));
```

Function laprec

```
1: function u=laprec(f,g,p,q,a,b,x,y,ntrms)
2: %
3: % u=laprec(f,g,p,q,x,y,ntrms)
```

```
 4: % ~~~~~~~~~~~~~~~~~~~~~~~~~~~~~
 5: % This function sums the series which solves
 6: % Laplace's equation in a rectangle.
 7: %
 8: %  f,g,p,q - Fourier sine series coefficients
 9: %             for the boundary conditions on
10: %             the bottom, top, left, and
11: %             right sides
12: %   a,b    - the horizontal and vertical
13: %             side lengths
14: %   x,y    - vectors containing coordinates
15: %             of points defining a rectangular
16: %             grid on which the solution is
17: %             to be evaluated.
18: %   ntrms  - number of series terms used (not
19: %             exceeding length(f));
20: %
21: % User m functions called:  none
22: %-------------------------------------------------
23:
24: nt=length(f);
25: if nargin==8, ntrms=nt; end;
26: ntrms=min(nt,ntrms);
27: x=x(:); y=y(:)';
28: n=1:ntrms; nx=length(x); ny=length(y);
29: if nt>ntrms
30:   f=f(n); g=g(n); p=p(n); q=q(n); end
31: a2=2*a; b2=2*b; na=(pi/a)*n; nb=(pi/b)*n;
32: denomx=1-exp(-b2*na(:));
33: f=f(:)./denomx; g=g(:)./denomx;
34: denomy=1-exp(-a2*nb(:)');
35: p=p(:)'./denomy; q=q(:)'./denomy;
36:
37: u1_=(f*ones(1,ny)).* ...
38:     (exp(-na'*y)-exp(-na'*(b2-y)));
39: u2_=(g*ones(1,ny)).* ...
40:     (exp(-na'*(b-y))-exp(-na'*(b+y)));
41: u=sin(x*na)*( u1_+u2_ );
42: u3_=(exp(-x*nb)-exp(-(a2-x)*nb)).* ...
43:     (ones(nx,1)*p);
44: u4_=(exp(-(a-x)*nb)-exp(-(a+x)*nb)).* ...
45:     (ones(nx,1)*q);
46: u=u+(u3_+u4_)*sin(nb'*y);
```

Function fbot

```
1: function ubot=fbot(x)
2: %
3: % ubot=fbot(x)
4: % ~~~~~~~~~~~~~
5: % x     - vector argument
6: % ubot - function value on bottom side
7: %
8: % User m functions called:  lintrp
9: %--------------------------------------------
10:
11: global xbot_ ubot_
12:
13: ubot=lintrp(xbot_,ubot_,x);
```

Function gtop

```
1: function utop=gtop(x)
2: %
3: % utop=gbot(x)
4: % ~~~~~~~~~~~~~
5: % x     - vector argument
6: % gtop - function value on top side
7: %
8: % User m functions called:  lintrp
9: %--------------------------------------------
10:
11: global xtop_ utop_
12:
13: utop=lintrp(xtop_,utop_,x);
```

Function plft

```
1: function ulft=plft(y)
2: %
3: % ulft=plft(y)
4: % ~~~~~~~~~~~~~
5: % y     - vector argument
6: % ulft - function value on left side
7: %
```

```
8: % User m functions called:  lintrp
9: %-----------------------------------------------
10:
11: global ylft_ ulft_
12:
13: ulft=lintrp(ylft_,ulft_,y);
```

Function qrht

```
1: function urht=qrht(y)
2: %
3: % urht=qrht(y)
4: % ~~~~~~~~~~~~
5: % y    - vector argument
6: % urht - function value on right side
7: %
8: % User m functions called:  lintrp
9: %-----------------------------------------------
10:
11: global yrht_ urht_
12:
13: urht=lintrp(yrht_,urht_,y);
```

9.3 The Vibrating String

Transverse motion of a tightly stretched string illustrates one of the simplest occurrences of one-dimensional wave propagation. The transverse deflection satisfies the wave equation

$$a^2 \frac{\partial^2 u}{\partial X^2} = \frac{\partial^2 u}{\partial T^2}$$

where $u(X, T)$ satisfies initial conditions

$$u(X, 0) = F(X) \qquad \frac{\partial u(X, 0)}{\partial T} = G(X)$$

with boundary conditions

$$u(0, T) = 0 \qquad u(\ell, T) = 0$$

where ℓ is the string length. If we introduce the dimensionless variables $x = X/\ell$ and $t = T/(\ell/a)$ the differential equation becomes

$$u_{xx} = u_{tt}$$

where subscripts denote partial differentiation. The boundary conditions become

$$u(0, t) = u(1, t) = 0$$

and the initial conditions become

$$u(x, 0) = f(x) \qquad u_t(x, 0) = g(x)$$

Let us restrict attention to the case where the string is released from rest initially so

$$u(x, 0) = f(x) \qquad u_t(x, 0) = 0$$

The solution can be found in series form as

$$u(x, t) = \sum_{n=1}^{\infty} a_n p_n(x) \cos(\omega_n t)$$

where ω_n are natural frequencies and satisfaction of the differential equation of motion requires

$$p_n''(x) + \omega_n^2 p_n(x) = 0$$

so

$$p_n = A_n \sin(\omega_n x) + B_n \cos(\omega_n x)$$

The boundary condition $p_n(0) = B_n = 0$ and $p_n(1) = A_n \sin(\omega_n)$ requires $A_n \neq 0$ and $\omega_n = n\pi$, where n is an integer. This leads to a solution in the form

$$u(x, t) = \sum_{n=1}^{\infty} a_n \sin(n\pi x) \cos(n\pi t)$$

The remaining condition on initial conditions requires

$$\sum_{n=1}^{\infty} a_n \sin(n\pi x) = f(x) \qquad 0 < x < 1$$

Therefore, the coefficients a_n are obtainable from the Fourier series expansion of $f(x)$ where $f(-x) = -f(x)$ and $f(x + 2) = f(x)$. It follows that a_n are obtainable by integration as

$$a_n = 2 \int_0^1 f(x) \sin(n\pi x) \, dx$$

However, an easier way to compute the coefficients is to use the FFT. A solution will be given for an arbritrary piecewise linear initial condition.

Before developing the Fourier series solution, let us examine the case of an infinite string governed by

$$a^2 U_{XX} = U_{TT} \qquad -\infty < X < \infty$$

and initial conditions

$$U(X, 0) = F(X) \qquad U_T(X, 0) = G(X)$$

The reader can verify directly that the solution of this problem is given by

$$U(X, T) = \frac{1}{2}[F(X - aT) + F(X + aT)] + \frac{1}{2a} \int_{X-aT}^{X+aT} G(x) \, dx$$

When the string is released from rest, $G(X)$ is zero and the solution reduces to

$$\frac{F(X - aT) + F(X + aT)}{2}$$

which shows that the initial deflection splits into two parts with one half translating to the left at speed a and the other half moving to the right at speed a. This solution can also be adapted to solve the problem for a string of length ℓ fixed at each end. The condition $U(0, T) = 0$ implies

$$F(-aT) = -F(aT)$$

which shows that $F(X)$ must be odd valued. Similarly, $U(\ell, T) = 0$ requires

$$F(\ell - aT) + F(\ell + aT) = 0$$

Combining this condition with $F(X) = -F(X)$ shows that

$$F(X + 2\ell) = F(X)$$

so $F(X)$ must have a period of 2ℓ. In the string of length ℓ, $F(X)$ is only known for $0 < X < \ell$, and we must take

$$F(X) = -F(2\ell - X) \qquad \ell < X < 2\ell$$

Furthermore the solution has the form

$$u(X, T) = \frac{F(x_p) + F(x_m)}{2}$$

where $x_p = X + aT$ and $x_m = X - aT$. The quantity x_p will always be positive and x_m can be both positive and negative. The necessary sign change and periodicity can be achieved by evaluating $F(X)$ as

$$\mathbf{sign}(X).*F(\mathbf{rem}(\mathbf{abs}(X)), 2 * \ell)$$

where **rem** is the intrinsic remainder function used in the exact solution implemented in function **stwav**.

A computer program based on the ideas outlined was written for an initial deflection which is piecewise linear. The Fourier series solution allows the user to select varying numbers of terms in the series in order to illustrate how well the initial deflection configuration is represented by a truncated sine series. A function animating the time response shows clearly how the initial deflection splits in two parts translating in opposite directions. In the Fourier solution, dimensionless variables are employed to make the string length and the wave speed both equal one. Consequently, the time required for the motion to exactly return to the starting position equals two, representing how long it takes for a disturbance to propagate from one end to the other and back. When the motion is observed for $0 < x < 1$, it is evident that waves reflected from a wall are inverted. It is interesting to note, however, that the Fourier series defines a solution valid for $-\infty < x < \infty$ having the periodicity properties discussed above. If the solution is evaluated taking $-1 < x < 2$, for example, we can visualize the solution as two wave forms of infinite length which combine to make the deflection remain zero at $x = 0$ and $x = 1$. The program consists of the following functions.

rundfl	script to input initial deflection data
sincof	uses **fft** to generate coefficients in a sine series
initdefl	defines the initial deflection by piecewise linear interpolation
strvib	evaluates the series solution for general x and t
smotion	animates the string motion
read	facilitates interactive data input
lintrp	performs interpolation to evaluate a piecewise linear function
stwav	a function to evaluate $u(x,t)$ exactly from the translating wave form of solution. This function serves the same purpose as **strvib**. It is a stand alone unit not employed by **rundfl**.

Results are shown below for a string which was deflected initially in a square wave. This example was selected to illustrate the approximation produced when a small number of Fourier coefficients, in this case 40 terms, are used. Ripples are clearly evident in the surface plot of $u(x,t)$ in Figure 9.3. The deflection configuration of the string at $t = 1$ when the initial deflection form has passed through half a period of motion appears in Figure 9.4. One other example given in Figure 9.5 shows output which function **stwav** produces when it is executed with no input data. The surface describes $u(x,t)$ through one period of motion starting with a triangular deflection pattern.

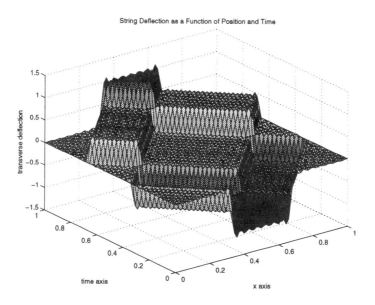

Figure 9.3. String Deflection as a Function of Position and Time

Figure 9.4. Wave Propagation in a String

Deflection Surface for a Vibrating String

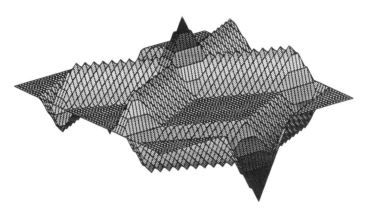

Figure 9.5. Deflection Surface for a Vibrating String

9.3.1 Program Output and Code

Output from Example rundfl

```
          Fourier Series Solution for
Vibration of a String with Linearly Interpolated
        Initial Deflection and Fixed Ends


Enter the number of interior data points (the fixed
end point coordinates are added automatically)
? 4


The string stretches between fixed endpoints at
x=zero and x=one. Enter 4 sets of x,y to specify
interior initial deflections (one pair per line)
? 0.33,0
? 0.33,-1
? 0.67,-1
? 0.67,0


To examine the resulting motion, enter values of
xmin, xmax and the number of x values (typical
values are 0, 1, 51). Enter 0 to stop.
? 0,1,71


Also give values of tmin, tmax, and the number of
of times (typical values are 0, 2, 51)
? 0,1,71


How many series terms are to be used? 40

Press [Enter] for a surface plot describing the motion

Animation of the motion is shown next - Press [Enter]

To examine the resulting motion, enter values of,
xmin, xmax and the number of x values (typical
values are 0, 1, 51). Enter 0 to stop.
? 0,0,0
```

Script File rundfl

```
 1: %
 2: % Example: rundfl
 3: % ~~~~~~~~~~~~~~~
 4: % This program analyzes wave motion of a string
 5: % having arbitrary initial deflection.
 6: %
 7: % User m functions required:
 8: %     sincof, initdefl, strvib, smotion, read,
 9: %     lintrp, genprint
10: %------------------------------------------------
11:
12: global xdat ydat
13:
14: fprintf ...
15:    ('\n          Fourier Series Solution for');
16: fprintf ...
17:    ('\nVibration of a String with Linearly ');
18: fprintf('Interpolated');
19: fprintf('\n        Initial Deflection and ');
20: fprintf('Fixed Ends\n');
21: fprintf('\nEnter the number of interior ');
22: fprintf('data points (the fixed');
23: fprintf('\nend point coordinates are ');
24: fprintf('added automatically)\n');
25:
26: n=input('? ');
27: xdat=zeros(n+2,1); ydat=xdat; xdat(n+2)=1;
28:
29: fprintf('\nThe string stretches between ');
30: fprintf('fixed endpoints at');
31: fprintf('\nx=zero and x=one. ');
32: fprintf('Enter %g sets of x,y to specify',n);
33: fprintf('\ninterior initial deflections ');
34: fprintf('(one pair per line)\n');
35: for j=2:n+1, [xdat(j),ydat(j)]=read; end; nx=1;
36: a=sincof('initdefl',1,1024); % sine coefficients
37:
38: while 1
39:    fprintf('\nTo examine the resulting ');
40:    fprintf('motion, enter values of');
41:    fprintf('\nxmin, xmax and the number of x ');
42:    fprintf('values (typical');
```

```
43:    fprintf('\nvalues are 0, 1, 51). ');
44:    fprintf('Enter 0 to stop.\n');
45:    [xmin,xmax,nx]=read; if nx==0, break, end
46:    x=linspace(xmin,xmax,nx);
47:    xx=linspace(xmin,xmax,151);
48:    fprintf('\nAlso give values of tmin, ');
49:    fprintf('tmax, and the number of');
50:    fprintf('\nof times (typical values ');
51:    fprintf('are 0, 2, 51)\n');
52:    [tmin,tmax,nt]=read;
53:    t=linspace(tmin,tmax,nt); disp(' ')
54:    ntrms=input(...
55:    'How many series terms are to be used ? ');
56:    y=strvib(a,t,x,1,ntrms); % time history
57:    yy=strvib(a,t,xx,1,ntrms);
58:    fprintf('\nPress [Enter] for a surface' );
59:    fprintf(' plot describing the motion\n');
60:    pause; hold off; surf(x,t,y);
61:    grid on; colormap('default');
62:    xlabel('x axis'); ylabel('time axis');
63:    zlabel('transverse deflection');
64:    title(['String Deflection as a Function ', ...
65:       'of Position and Time']); figure(gcf);
66:    fprintf('\nAnimation of the motion is ');
67:    dumy=input('shown next - Press [Enter]','s');
68:    %genprint('strdefl');
69:    smotion(xx,yy,'Wave Propagation in a String');
70:    disp(''); pause(1);
71: end
72: %genprint('strwave');
```

Function initdefl

```
 1: function y=initdefl(x)
 2: %
 3: % y=initdefl(x)
 4: % ~~~~~~~~~~~~~~
 5: % This function defines the linearly
 6: % interpolated initial deflection
 7: % configuration.
 8: %
 9: % x - a vector of points at which the initial
10: %       deflection is to be computed
11: %
12: % y - transverse initial deflection value for
13: %       argument x
14: %
15: % xdat, ydat - global data vectors used for
16: %                   linear interpolation
17: %
18: % User m functions required:  lintrp
19: %-------------------------------------------------
20:
21: global xdat ydat
22: xd=xdat; yd=ydat; y=lintrp(xd,yd,x);
```

Function strvib

```
 1: function y=strvib(a,t,x,hp,n)
 2: %
 3: % y=strvib(a,t,x,hp,n)
 4: % ~~~~~~~~~~~~~~~~~~~~~
 5: % Sum the Fourier series for the string motion.
 6: %
 7: % a   - Fourier coefficients of initial
 8: %         deflection
 9: % t,x - vectors of time and position values
10: % hp  - the half period for the series
11: %         expansion
12: % n   - the number of series terms used
13: %
14: % y   - matrix with y(i,j) equal to the
15: %         deflection at position x(i) and
```

```
16: %          time t(j)
17: %
18: % User m functions required: none
19: %-----------------------------------------------
20:
21: w=pi/hp*(1:n); a=a(1:n); a=a(:)';
22: x=x(:); t=t(:)';
23: y=((a(ones(length(x),1),:).* ...
24:    sin(x*w))*cos(w(:)*t))';
```

Function sincof

```
 1: function a=sincof(func,hafper,nft)
 2: %
 3: % a=sincof(func,hafper,nft)
 4: % ~~~~~~~~~~~~~~~~~~~~~~~~~~
 5: % This function calculates the sine
 6: % coefficients.
 7: %
 8: % func   - the name of a function  defined over
 9: %          a half period
10: % hafper - the length of the half period of the
11: %          function
12: % nft    - the number of function values used
13: %          in the Fourier series
14: %
15: % a      - the vector of Fourier sine series
16: %          coefficients
17: %
18: % User m functions required:  none
19: %-----------------------------------------------
20:
21: n2=nft/2; x=hafper/n2*(0:n2);
22: y=feval(func,x); y=y(:);
23: a=fft([y;-y(n2:-1:2)]); a=-imag(a(2:n2))/n2;
```

Function smotion

```
 1: function smotion(x,y,titl)
 2: %
 3: % smotion(x,y,titl)
 4: % ~~~~~~~~~~~~~~~~~
```

```
 5: % This function animates the string motion.
 6: %
 7: % x    - a vector of position values along the
 8: %        string
 9: % y    - a matrix of transverse deflection
10: %        values where successive rows give
11: %        deflections at successive times
12: % titl - a title shown on the plot (optional)
13: %
14: % User m functions required: none
15: %-------------------------------------------------
16:
17: if nargin < 3, titl=' '; end
18: xmin=min(x); xmax=max(x);
19: ymin=min(y(:)); ymax=max(y(:));
20: [nt,nx]=size(y); clf reset;
21: for k=1:2
22:   for j=1:nt
23:     plot(x,y(j,:),'b');
24:     axis([xmin,xmax,2*ymin,2*ymax]);
25:     axis('off'); title(titl);
26:     drawnow; figure(gcf);
27:   end
28: end
```

Function stwav

```
 1: function [Y,T,X]=stwav(t,x,xd,yd,len,a)
 2: %
 3: % [Y,T,X]=stwav(t,x,xd,yd,len,a)
 4: % ~~~~~~~~~~~~~~~~~~~~~~~~~~~~~~~~~~~
 5: % This function computes the dynamic response
 6: % of a vibrating string released from rest with
 7: % a given initial deflection defined by
 8: % piecewise linear interpolation. The analysis
 9: % employs the translating wave solution for
10: % arbitrary initial deflection.
11: %
12: % xd,yd - data values defining the initial
13: %         deflection.  Values in xd must be
14: %         increasing and must lie between 0
15: %         and len.
16: % t     - a vector of time values at which the
```

```
17: %          solution is to be computed
18: % x      - a vector of x values at which the
19: %          solution is to be computed
20: % len    - the length of the string
21: % a      - the velocity of wave propagation in
22: %          the string
23: %
24: % Y      - matrix of transverse deflection
25: %          values such that Y(i,j) is the
26: %          deflection at time T(i,1) and
27: %          position X(1,j)
28: % T      - matrix of time values at which the
29: %          solution is computed. This matrix is
30: %          output from function meshgrid and has
31: %          all rows identical.
32: % X      - matrix of position values at which
33: %          the solution is computed. This matrix
34: %          is output from function meshgrid and
35: %          has all columns identical.
36: %
37: % User m functions required:
38: %     lintrp, smotion, genprint
39: %-----------------------------------------------
40:
41: if nargin < 6, a=1; end;
42: if nargin < 5, len=1; end;
43:
44: % Generate a triangle wave when no input
45: if nargin==0
46:    xd=[0;.33; .5; .67; 1];
47:    yd=[0;  0; -1;   0; 0];
48:    x=linspace(0,len,61);
49:    t=linspace(0,2*len/a,61);
50: end
51:
52: % Compute the solution as
53: %     Y=[F(X+a*T)+F(X-a*T)]/2
54: % where
55: %     F(-X)=F(X) and F(X+2*len)=F(X).
56: % Thus, F(X) is an odd valued function of 2*len.
57: % The initial condition Y(X,0)=F(X) is initially
58: % defined for 0 <= X < =len, so values of ABS(X)
59: % must be reduced modulo 2*len and the values
60: % ABS(X)>len can then be evaluated as
61: %     -sign(X)*F(2*len-ABS(X))
```

```
62:
63:  xd=[xd(:);flipud(2*len-xd(:))];
64:  yd=[yd(:);-flipud(yd(:))]; [T,X]=meshgrid(t,x);
65:  [n1,n2]=size(X); xp=X(:)+a*T(:); xm=X(:)-a*T(:);
66:
67:  Y=(lintrp(xd,yd,rem(xp,2*len))+...
68:      lintrp(xd,yd,rem(abs(xm),2*len)).*sign(xm))/2;
69:  Y=reshape(Y,n1,n2);
70:
71:  if nargin ==0   % Plot surface for default data
72:     surf(X,T,Y);
73:     title(['Deflection Surface for a ', ...
74:            'Vibrating String']);
75:     colormap('jet'); axis('off'); figure(gcf);
76:     disp(' '); %genprint('deflsurf');
77:     dumy=input(...
78:            'Press [Enter] for animated motion','s');
79:     smotion(X(:,1),Y');
80:  end
```

9.4   Forced Vibration of a Pile Embedded in an Elastic Medium

Structures are often supported by piles embedded in soil foundations. The response of these systems when the foundation experiences oscillation, such as that occurring in a earthquake, has considerable practical interest. Let us examine a simple model approximating a single pile connected to an overlying structure. The pile is treated as a beam of uniform cross section which is buried in an elastic medium. An attached mass at the top causes inertial resistance to translation and rotation. The beam, shown in Figure 9.6 in a deflected position, has length ℓ with $x = 0$ denoting the lower end and $x = \ell$ denoting the top. Rotating the member $90°$ from the vertical is done to agree with the coordinate referencing traditionally used in beam analysis. The transverse bending response is to be computed when the surrounding elastic medium experiences an oscillatory motion of the form

$$y_f = y_o e^{\imath \omega t}$$

The differential equation governing transverse oscillations of the beam is

$$EI\frac{\partial^4 y(x,t)}{\partial x^4} = -A\rho\frac{\partial^2 y(x,t)}{\partial t^2} + k\left(y_o e^{\imath \omega t} - y\right)$$

where EI is the product of the elastic modulus and inertial moment of the beam; $A\rho$ is the product of cross section area and mass per unit volume; and k describes the foundation stiffness in terms of force per unit length per unit of transverse deflection. The shear V and moment M in the beam are related to the deflection $y(x,t)$ by

$$V = EI\frac{\partial^3 y(x,t)}{\partial x^3} \qquad M = EI\frac{\partial^2 y(x,t)}{\partial x^2}$$

In the current analysis we consider forced response of frequency ω described in the form

$$y(x,t) = f(x)e^{\imath \omega t}$$

so

$$V = EIf'''(x)e^{\imath \omega t} \qquad M = EIf''(x)e^{\imath \omega t}$$

The boundary conditions at $x = 0$ require vanishing moment and shear so

$$f''(0) = 0 \qquad f'''(0) = 0$$

The boundary conditions at $x = \ell$ are more involved because inertial resistance of the end mass must be handled. We assume that the gravity center of the end mass is located along the axis of the beam at a distance h above the top end. Furthermore, the attached body has a mass m_o and

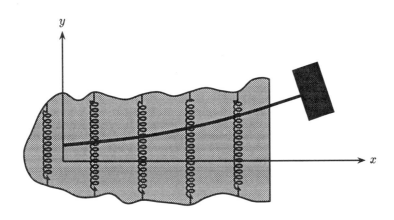

Figure 9.6. Forced Vibration of a Pile in an Elastic Medium

inertial moement J_o about its gravity center. The angular acceleration $\ddot{\theta}$ and the transverse acceleration a_m are expressible as

$$\ddot{\theta} = \frac{\partial^3 y(\ell, t)}{\partial x \partial t^2} = -\omega^2 f'(\ell) e^{\iota \omega t}$$

and

$$a_m = \frac{\partial^2 y(\ell, t)}{\partial t^2} + h\ddot{\theta} = -\omega^2 e^{\iota \omega t} [f(\ell) + h f'(\ell)]$$

Writing equations of motion for the end mass gives

$$m_o a_m = V(\ell, t)$$

and

$$J_o \ddot{\theta}_m = -h V(\ell, t) - M(\ell, t)$$

Representing these conditions in terms of $f(x)$ yields

$$-\omega^2 m_o [f(\ell) + h f'(\ell)] = EI f'''(\ell)$$

and

$$\omega^2 J_o f'(\ell) = EI[f''(\ell) + h f'''(\ell)]$$

Furthermore, the fraction $e^{\iota \omega t}$ cancels out of the differential equation

$$EI f''''(x) = (A\rho\omega^2 - k) f(x) + y_o k$$

The general solution of this fourth order linear differential equation is expressed as

$$f(x) = \frac{y_o k}{k - A\rho\omega^2} \left[1 + \sum_{j=1}^{4} c_j e^{s_j x} \right]$$

where s_j are complex roots given by

$$s_j = \left(\frac{A\rho\omega^2 - k}{EI} \right)^{1/4} e^{\iota(j-1)\pi/2} \qquad j = 1, 2, 3, 4$$

The conditions of zero moment and shear at $x = 0$ lead to

$$\sum_{j=1}^{4} s_j^2 c_j = 0 \qquad \sum_{j=0}^{4} s_j^3 c_j = 0$$

The shear and moment conditions at $x = \ell$ require

$$\sum_{j=1}^{4} s_j^3 e^{s_j \ell} c_j = -m_o \omega^2 \left[1 + \sum_{j=1}^{4} (1 + h s_j) e^{s_j \ell} c_j \right]$$

and

$$\sum_{\jmath=1}^{4}(s_\jmath^2 + hs_\jmath^3)e^{s_\jmath\ell}c_\jmath = \jmath_o\omega^2\sum_{\jmath=1}^{4}s_\jmath e^{s_\jmath\ell}c_\jmath$$

The system of four simultaneous equations can be solved for $c_1, \ldots, c_4$. Then the forced response solution corresponding to a foundation motion

$$\mathbf{Real}\left(y_o e^{\imath\omega t}\right) = y_o\cos(\omega t)$$

is given by

$$y(x,t) = \mathbf{Real}\left(f(x)e^{\imath\omega t}\right)$$

where $f(x)$ is complex valued.

The function **pvibs** evaluates the displacement, moment, and shear for $0 \le x \le \ell$, $0 \le t \le 2\pi/\omega$. Surface plots of these quantities are shown in Figures 9.7 through 9.9. Figure 9.10 is a single frame from an animation depicting how the pile and the attached mass move.

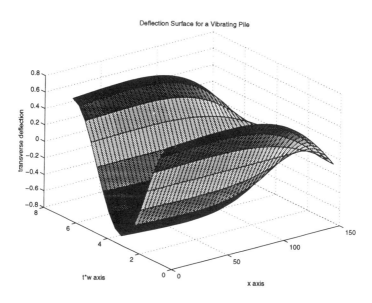

Figure 9.7. Deflection Surface for a Vibrating Pile

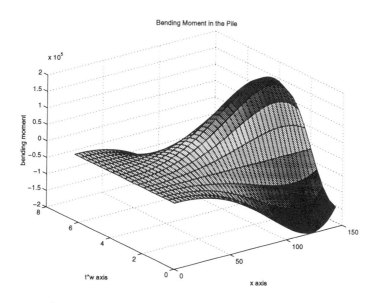

Figure 9.8. Bending Moment in a Vibrating Pile

Shear Force in the Pile

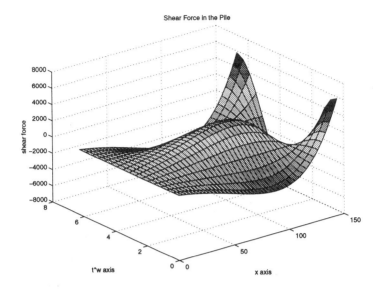

Figure 9.9. Shear Force in a Vibrating Pile

Forced Vibration of a Pile

Figure 9.10. Frame from Pile Animation

9.4.1 Program Output and Code

Script File runpvibs

```
 1: % Example: runpvibs
 2: % ~~~~~~~~~~~~~~~~~
 3: %
 4: % The routine is used to solve an example
 5: % problem using function pvibs. The example
 6: % involves a steel pile 144 inches long which
 7: % has a square cross section of 4 inch depth.
 8: % The pile is immersed in soil having an elastic
 9: % modulus of 200 psi. The attached mass weighs
10: % 736 lb. The foundation is shaken at an
11: % amplitude of 0.5 inch with a frequency of
12: % 20 cycles per second.
13: %
14: % User m functions required: pvibs, genprint
15: %-------------------------------------------------
16:
17: clear;
18: L=144; d=4; a=d^2; I=d^4/12; e=30e6; ei=e*I;
19: g=32.2*12; Density_steel=0.284;
20: rho=Density_steel/g;
21: Cap_w=36; Cap_h=18; Cap_t=4;
22: m0=Cap_w*Cap_h*Cap_t*rho;
23: j0=m0/12*(Cap_h^2+Cap_w^2);
24: h=Cap_h/2; arho=a*rho;
25: e_soil=200; k=e_soil*d; y0=0.5; w=40*pi;
26: nx=42; nt=17;
27:
28: [t,x,y,m,v]= ...
29:   pvibs(y0,ei,arho,L,k,w,h,m0,j0,nx,nt);
```

Function pvibs

```
 1: function [t,x,y,m,v]= ...
 2:             pvibs(y0,ei,arho,L,k,w,h,m0,j0,nx,nt)
 3: %
 4: % [t,x,y,m,v]=pvibs ...
 5: %            (y0,ei,arho,L,k,w,h,m0,j0,nx,nt)
 6: % ~~~~~~~~~~~~~~~~~~~~~~~~~~~~~~~~~~~~~~~~~~~
 7: %
 8: % This function computes the forced harmonic
 9: % response of a pile buried in an oscillating
10: % elastic medium. The lower end of the pile is
11: % free from shear and moment. The top of the
12: % pile carries an attached body having general
13: % mass and inertial properties. The elastic
14: % foundation is given a horizontal oscillation
15: % of the form
16: %
17: %   yf=real(y0*exp(i*w*t))
18: %
19: % The resulting transverse forced response of
20: % the pile is expressed as
21: %
22: %   y(x,t)=real(f(x)*exp(i*w*t))
23: %
24: % where f(x) is a complex valued function. The
25: % bending moment and shear force in the pile
26: % are also computed.
27: %
28: % y0   - amplitude of the foundation oscillation
29: % ei   - product of moment of inertia and
30: %        elastic modulus for the pile
31: % arho - mass per unit length of the pile
32: % L    - pile length
33: % k    - the elastic resistance constant for the
34: %        foundation described as force per unit
35: %        length per unit of transverse
36: %        deflection
37: % w    - the circular frequency of the
38: %        foundation oscillation which vibrates
39: %        like real(y0*exp(i*w*t))
40: % h    - the vertical distance above the pile
41: %        upper end to the gravity center of the
42: %        attached body
```

```
43: % m0    - the mass of the attached body
44: % j0    - the mass moment of inertia of the
45: %         attached body with respect to its
46: %         gravity center
47: % nx    - the number of equidistant values along
48: %         the pile at which the solution is
49: %         computed
50: % nt    - the number of values of t values at
51: %         which the solution is computed such
52: %         that 0 <= w*t <= 2*pi
53: %
54: % t     - a vector of time values such that the
55: %         pile moves through a full period of
56: %         motion. This means 0 <= t <= 2*pi/w
57: % x     - a vector of x values with 0 <= x <= L
58: % y     - the transverse deflection y(x,t) for
59: %         the pile with t varying from row to
60: %         row, and x varying from column to
61: %         column
62: % m,v   - matrices giving values bending moment
63: %         and shear force
64: %
65: % User m functions called: genprint
66: %----------------------------------------------
67:
68: % Default data for a steel pile 144 inches long
69: if nargin==0
70:    y0=0.5; ei=64e7; arho=0.0118; L=144; k=800;
71:    w=125.6637; h=9; m0=1.9051; j0=257.1876;
72:    nx=42; nt=17;
73: end
74:
75: w2=w^2; x=linspace(0,L,nx)';
76: t=linspace(0,2*pi/w,nt);
77:
78: % Evaluate characteristic roots and complex
79: % exponentials
80: s=((arho*w2-k)/ei)^(1/4)*[1,i,-1,-i];
81: s2=s.^2; s3=s2.*s;
82: c0=y0*k/(k-w2*arho); esl=exp(s*L);
83: esx=exp(x*s); eiwt=exp(i*w*t);
84:
85: % Solve for coefficients to satisfy the
86: % boundary conditions
87: c=[s2; s3; esl.*(h*s3+s2-j0*w2/ei*s); ...
```

```
88:    esl.*(s3+m0*w2/ei*(1+h*s))]\ ...
89:    [0;0;0;-c0*m0*w2/ei];
90:
91: % Compute the deflection, moment and shear
92: y=real((c0+esx*c)*eiwt)';
93: ype=real(s.*esl*c*eiwt)';
94: m=real(ei*s2(ones(nx,1),:).*esx*c*eiwt)';
95: v=real(ei*s3(ones(nx,1),:).*esx*c*eiwt)';
96: t=t'; x=x'; hold off; clf;
97:
98: % Make surface plots showing the deflection,
99: % moment, and shear over a complete period of
100: % the motion
101: surf(x,t*w,y);
102: xlabel('x axis'); ylabel('t*w axis');
103: zlabel('transverse deflection');
104: title('Deflection Surface for a Vibrating Pile');
105: grid on; figure(gcf); %genprint('defsurf');
106: dumy=input('Press [Enter] to continue','s');
107:
108: surf(x,t*w,m);
109: xlabel('x axis'); ylabel('t*w axis');
110: zlabel('bending moment');
111: title('Bending Moment in the Pile')
112: grid on; figure(gcf); %genprint('bendmom');
113: dumy=input('Press [Enter] to continue','s');
114:
115: surf(x,t*w,v);
116: xlabel('x axis'); ylabel('t*w axis');
117: zlabel('shear force');
118: title('Shear Force in the Pile');
119: grid on; figure(gcf); %genprint('shrforce');
120: dumy=input('Press [Enter] to continue','s');
121:
122: % Draw an animation depicting the pile response
123: % to the oscillation of the foundation
124: fu=.10/max(y(:)); p=[-0.70, 0.70, -.1, 1.3];
125: u=fu*y; upe=fu*L*ype; d=.15;
126: xm=[0,0,1,1,0,0]*d;
127: ym=[0,-1,-1,1,1,0]*d; zm=xm+i*ym;
128: close;
129: for jj=1:4
130:    for j=1:nt
131:       z=exp(i*atan(upe(j)))*zm;
132:       xx=real(z); yy=imag(z);
```

```
133:      ut=[u(j,:),u(j,nx)+yy]; xt=[x/L,1+xx];
134:      plot(ut,xt,'-'); axis(p); axis('square');
135:      title('Forced Vibration of a Pile');
136:      axis('off'); drawnow; figure(gcf);
137:    end
138: end
139: %genprint('pileanim');
140: fprintf('\nAll Done\n');
141: pause(2); axis('normal'); close;
```

9.5   Transient Heat Conduction in a One-Dimensional Slab

Let us analyze the temperature history in a slab which has the left side insulated while the right side temperature varies sinusoidally according to $U_0 \sin(\Omega T)$. The initial temperature in the slab is specified to be zero. The pertinent boundary value problem is

$$\alpha \frac{\partial^2 U}{\partial X^2}(X, T) = \frac{\partial U}{\partial T}(X, T) \qquad 0 < X < \ell \qquad T > 0$$

$$\frac{\partial U}{\partial X}(0, T) = 0 \qquad U(\ell, T) = U_0 \sin(\Omega T)$$

$$U(X, 0) = 0 \qquad 0 < X < \ell$$

where U, X, T, and α are, respectively, temperature, position, time, and thermal diffusivity.

The problem can be converted to dimensionless form by taking

$$u = U/U_0 \qquad x = X/\ell \qquad t = \alpha T/\ell^2 \qquad \omega = \Omega \ell^2/\alpha$$

Then we get

$$\frac{\partial^2 u}{\partial x^2} = \frac{\partial u}{\partial x} \qquad 0 < x < 1 \qquad t > 0$$

$$\frac{\partial u}{\partial x}(0, t) = 0 \qquad u(1, t) = \mathbf{imag}\left(e^{i\omega t}\right) \qquad u(x, 0) = 0$$

The solution consists of two parts as $u = w + v$ where w is a particular solution satisfying the differential equation and nonhomogeneous boundary conditions, and v is a solution satisfying homogeneous boundary conditions and specified to impose the desired zero initial temperature when combined with w. The appropriate form for the particular solution is

$$w = \mathbf{imag}\left[f(x)e^{i\omega t}\right]$$

Making w satisfy the heat equation requires

$$f''(x) = \imath w f(x).$$

Consequently

$$f(x) = c_1 \sin(\phi x) + c_2 \cos(\phi x)$$

where $\phi = \sqrt{-\imath w}$. The conditions of zero gradient at $x = 0$ and unit function value at $x = 1$ determine c_1 and c_2. We get the particular solution as

$$w = \mathbf{imag}\left[\frac{\cos(\phi x)}{\cos(\phi)} e^{\imath w t}\right]$$

This forced response solution evaluated at $t = 0$ yields

$$w(x,0) = \mathbf{imag}\left[\frac{\cos(\phi x)}{\cos(\phi)}\right]$$

The general solution of the heat equation satisfying zero gradient at $x = 0$ and zero function value at $x = 1$ is found to be

$$v(x,t) = \sum_{n=1}^{\infty} a_n \cos(\lambda_n x)e^{-\lambda_n^2 t}$$

when $\lambda_n = \pi(2n - 1)/2$. To make the initial temperature equal zero in the combined solution, the coefficients a_n are chosen to satisfy

$$\sum_{n=1}^{\infty} a_n \cos(\lambda_n x) = -\mathbf{imag}\left[\frac{\cos(\phi x)}{\cos(\phi)}\right]$$

The orthogonality of the functions $\cos(\lambda_n x)$ implies

$$a_n = -2 \int_0^1 \mathbf{imag}\left[\frac{\cos(\phi x)}{\cos(\phi)}\right]\cos(\lambda_n x)dx$$

which can be integrated to give

$$a_n = -\mathbf{imag}\left[\frac{(\sin(\lambda_n + \phi)/(\lambda_n + \phi) + \sin(\lambda_n - \phi)/(\lambda_n - \phi))}{\cos(\phi)}\right]$$

This completely determines the solution. Taking any finite number of terms in the series produces an approximate solution exactly satisfying the differential equation and boundary conditions. Exact satisfaction of the zero initial condition would theoretically require an infinite number of series terms. However, using a 250-term series produces initial temperature values not exceeding 10^{-6}. Thus, the finite series is satisfactory for practical purposes.

The above equations were evaluated in a function called **heat**. The script file **slabheat** was also written to plot numerical results. The code and resulting Figures (Figures 9.11 and 9.12) appear below. This example clearly illustrates how well MATLAB handles complex arithmetic and complex valued functions.

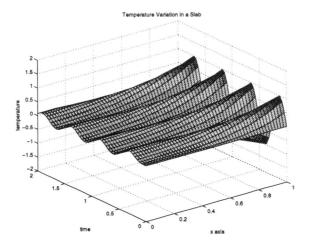

Figure 9.11. Temperature Variation in a Slab

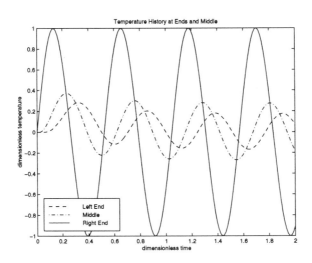

Figure 9.12. Temperature History at Ends and Middle

MATLAB Example

Script File slabheat

```
 1: % Example: slabheat
 2: % ~~~~~~~~~~~~~~~~~~
 3: % This program computes the temperature
 4: % variation in a one-dimensional slab with
 5: % the left end insulated and the right end
 6: % given a temperature variation sin(w*t).
 7: %
 8: % User m functions required:
 9: %    heat, genprint
10: %-------------------------------------------------
11:
12: [u1,t1,x1]=heat(12,0,2,50,0,1,51,250);
13: surf(x1,t1,u1); axis([0 1 0 2 -2 2]);
14: title('Temperature Variation in a Slab');
15: xlabel('x axis'); ylabel('time');
16: zlabel('temperature'); figure(gcf);
17: dumy=input(...
18:      '\nPress [Enter] to continue\n','s');
19: %genprint('tempsurf');
20:
21: [u2,t2,x2]=heat(12,0,2,150,0,1,3,250);
22: plot(t2,u2(:,1),'--',t2,u2(:,2),'-.', ...
23:      t2,u2(:,3),'-');
24: title(['Temperature History at Ends' ...
25:       ' and Middle']);
26: xlabel('dimensionless time');
27: ylabel('dimensionless temperature');
28: text1='Left End'; text2='Middle';
29: text3='Right End';
30: legend(text1,text2,text3,3); figure(gcf);
31: %genprint('tempplot');
32: fprintf('\All Done\n');
```

Function heat

```
 1: function [u,t,x]= ...
 2:             heat(w,tmin,tmax,nt,xmin,xmax,nx,nsum)
 3: %
 4: %[u,t,x]=heat(w,tmin,tmax,nt,xmin,xmax,nx,nsum)
```

```
 5: %~~~~~~~~~~~~~~~~~~~~~~~~~~~~~~~~~~~~~~~~~~~~~~~~~~~~~
 6: % This function evaluates transient heat
 7: % conduction in a slab which has the left end
 8: % (x=0) insulated and has the right end (x=1)
 9: % subjected to a temperature variation
10: % sin(w*t). The initial temperature of the slab
11: % is zero.
12: %
13: % w              - frequency of the right side
14: %                  temperature variation
15: % tmin,tmax      - time limits for solution
16: % nt             - number of uniformly spaced
17: %                  time values used
18: % xmin,xmax      - position limits for solution.
19: %                  Values should lie between zero
20: %                  and one.
21: % nx             - number of equidistant x values
22: % nsum           - number of terms used in the
23: %                  series solution
24: % u              - matrix of temperature values.
25: %                  Time varies from row to row.
26: %                  x varies from column to column.
27: % t,x            - vectors of time and x values
28: %
29: % User m functions called:  none.
30: %-------------------------------------------------
31:
32: t=tmin+(tmax-tmin)/(nt-1)*(0:nt-1);
33: x=xmin+(xmax-xmin)/(nx-1)*(0:nx-1)';
34: W=sqrt(-i*w); ln=pi*((1:nsum)-1/2);
35: v1=ln+W; v2=ln-W;
36: a=-imag((sin(v1)./v1+sin(v2)./v2)/cos(W));
37: u=imag(cos(W*x)*exp(i*w*t)/cos(W))+ ...
38:    (a(ones(nx,1),:).*cos(x*ln))* ...
39:    exp(-ln(:).^2*t);
40: u=u'; t=t(:);
```

Figure 9.13. Beam Geometry and Loading

9.6   Wave Propagation in a Beam with an Impact Moment Applied to One End

Analyzing the dynamic response caused when a time dependent moment acts on the end of an Euler beam involves a boundary value problem illustrating the use of Fourier series for solving linear partial differential equations. In the following example we consider a beam of uniform cross section which is pin-ended (hinged at the end) and is initially at rest. Suddenly, a harmonically varying moment $M_0 \cos(\Omega_0 T)$ is applied to the right end as shown in Figure 9.13. Determination of the resulting displacement and bending moment in the beam is desired.

Let U be transverse displacement, X longitudinal distance from the right end, and T time. The differential equation, boundary conditions and initial conditions characterizing the problem are

$$EI\frac{\partial^4 U}{\partial X^4} = -A\rho\frac{\partial^2 U}{\partial T^2} \qquad 0 < X < L \qquad T > 0$$

$$U(0,T) \;=\; 0 \qquad\qquad \tfrac{\partial^2 U}{\partial X^2}(0,T) \;=\; 0$$

$$U(L,T) \;=\; 0 \qquad\qquad \tfrac{\partial^2 U}{\partial X^2}(L,T) \;=\; M_0 \cos(\Omega_0 T)/(EI)$$

$$U(0,T) \;=\; 0 \qquad\qquad \tfrac{\partial U}{\partial T}(0,T) \;=\; 0$$

where L is beam length, EI is the product of elastic modulus and moment of inertia, and $A\rho$ is the product of cross section area and mass density.

This problem can be represented more conveniently by introducing dimensionless variables

x - dimensionless position $= X/L$

t - dimensionless time $= \left[\sqrt{(EI)/(A\rho)}\right] L^{-2}T$

u - dimensionless displacement $= (EI)/(M_0 L^2)U$

ω - dimensionless forcing frequency $= \left[\sqrt{(A\rho)/(EI)}\right] L^2\Omega_0$

m - dimensionless bending moment $= \frac{\partial^2 u}{\partial x^2}$

The new boundary value problem is then

$$\frac{\partial^4 u}{\partial x^4} = -\frac{\partial^2 u}{\partial t^2} \qquad 0 < x < 1 \quad \cdot \quad t > 0$$

$$u(0, t) = 0 \qquad \frac{\partial^2 u}{\partial x^2}(0, t) = 0$$

$$u(1, t) = 0 \qquad \frac{\partial^2 u}{\partial x^2}(1, t) = 0$$

$$u(x, 0) = 0 \qquad \frac{\partial u}{\partial t}(x, 0) = 0 \qquad 0 < x < 1$$

The problem can be solved by combining a particular solution w which satisfies the differential equation and nonhomogeneous boundary conditions with a homogeneous solution in series form which satisfies the differential equation and homogeneous boundary conditions. Thus we have $u = w + v$. The particular solution can be found in the form

$$u = f(x) \cos(\omega t)$$

where $f(x)$ satisfies

$$f''''(x) = \omega^2 f(x)$$

$$f(0) = f''(0) = f(1) \qquad f''(1) = 1$$

This ordinary differential equation is solvable as

$$f(x) = \sum_{k=1}^{4} c_k e^{s_k x}$$

where

$$s_k = \sqrt{\omega} \, e^{\pi \imath (k-1)/2} \qquad \imath = \sqrt{-1}$$

The boundary conditions require

$$\sum_{k=1}^{4} c_k = 0 \qquad \sum_{k=1}^{4} s_k^2 c_k = 1.$$

$$\sum_{k=1}^{4} c_k e^{s_k} = 0 \qquad \sum_{k=1}^{4} c_k s_k^2 e^{s_k} = 0$$

Solving these simultaneous equations determines the particular solution. The particular solution satisfies initial conditions

$$w(0, t) = f(x) = \sum_{k=-\infty}^{\infty} c_k e^{\imath \pi k x} \qquad \frac{\partial w}{\partial t}(0, t) = 0$$

where $f(x)$ is expandable in a complex Fourier series as an odd valued function such that

$$f(x) = f(x+2) = -f(2-x) \qquad 0 < x < 1$$

This implies that $f(x)$ is represented as a sine series, or

$$f(x) = \sum_{k=1}^{\infty} a_k \sin(k\pi x)$$

with

$$a_k = -2 * \mathbf{imag}(c_k)$$

The homogeneous solution is representable as

$$v(x,t) = -\sum_{k=1}^{\infty} a_k \cos(\pi^2 k^2 t) \sin(k\pi x)$$

so that $w + v$ combine to satisfy the desired initial conditions of zero displacement and velocity.

Of course, perfect satisfaction of the initial conditions cannot be achieved without taking an infinite number of terms in the Fourier series. However, the series converges very rapidly because the coefficients are of order n^{-3}. When a hundred or more terms are used, an approximate solution produces results which satisfy the differential equation and boundary conditions, and which insignificantly violates the initial displacement condition. It is important to remember the nature of this error when examining the bending moment results presented below. Effects of high frequency components are very evident in the moment. Despite the oscillatory character of the moments, these results are exact for the initial displacement conditions produced by the truncated series. These displacements agree closely with the exact solution.

A program was written to evaluate the series solution to compute displacements and moments as functions of position and time. Plots and surfaces showing these quantities are presented along with time-wise animations of the displacement and moment across the span. The computation involves the following steps:

1. evaluate $f(x)$

2. expand $f(x)$ using the FFT to get coefficients for the homogeneous series solution

3. combine the particular and homogeneous solution by summing the series for any number of terms desired by the user

4. plot u and m for selected times

5. plot surfaces showing $u(x, t)$ and $m(x, t)$

6. show animated plots of u and m

The principal parts of the program are shown in the table below.

rnbemimp	reads data and creates graphical output
beamresp	converts material property data to dimensionless form and calls **ndbemrsp**
ndbemrsp	construct the solution using Fourier series
sumser	sums the series for displacement and moment
animate	animates the time history of displacement and moment
genprint	saves file for printing

The numerical results show the response for a beam subjected to a moment close to the natural frequencies of the beam. It can be shown that, in the dimensionless problem, the system of equations defining the particular solution becomes singular when ω assumes values of the form $k^2\pi^2$ for integer k. In that instance the series solution provided here will fail. However, ω values near to resonance can be used to show how the displacements and moments quickly become large. In our example we let EI, $A\rho$, l, and M_0 all equal unity, and $\omega = 0.95\pi^2$. Figures 9.14 and 9.15 show displacement and bending moment patterns shortly after motion is initiated. The surfaces in Figures 9.16 and 9.17 also show how the displacement and moment grow quickly with increasing time. The reader may find it interesting to run the program for various choices of ω and observe how dramatically the chosen forcing frequency affects results.

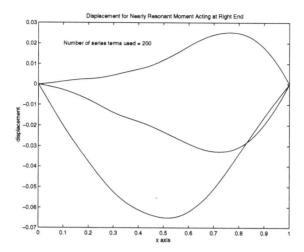

Figure 9.14. Displacement Due to Impact Moment at Right End

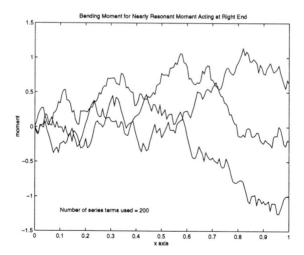

Figure 9.15. Bending Moment in the Beam

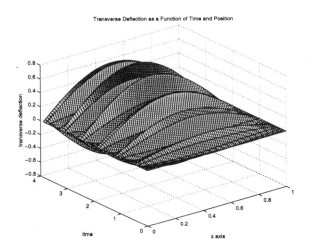

Figure 9.16. Displacement Growth Near Resonance

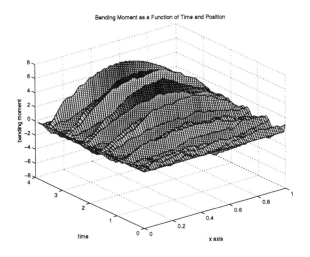

Figure 9.17. Moment Growth Near Resonance

MATLAB Example

Script File rnbemimp

```
 1: % Example: rnbemimp
 2: % ~~~~~~~~~~~~~~~~~~~
 3: % This program analyzes an impact dynamics
 4: % problem for an elastic Euler beam of
 5: % constant cross section which is simply
 6: % supported at each end. The beam is initially
 7: % at rest when a harmonically varying moment
 8: % m0*cos(w0*t) is applied to the right end.
 9: % The resulting transverse displacements and
10: % bending moments are computed.  The
11: % displacement and moment are plotted as
12: % functions of x for the first several
13: % successive time steps. Animated plots of the
14: % entire displacement and moment history are
15: % also given.
16: %
17: % User m functions required:
18: %    beamresp, animate, sumser,
19: %    ndbemrsp, rnbemimp, genprint
20: %------------------------------------------------
21:
22: fprintf('\nDYNAMICS OF A BEAM WITH AN ');
23: fprintf('OSCILLATORY END MOMENT\n');
24: ei=1; arho=1; len=1; m0=1; w0=.90*pi^2;
25: tmin=0; tmax=5; nt=101;
26: xmin=0; xmax=len; nx=151; ntrms=200;
27: [t,x,displ,mom]=beamresp(ei,arho,len,m0,w0,...
28:               tmin,tmax,nt,xmin,xmax,nx,ntrms);
29:
30: np=[3 5 8]; clf;
31: dip=displ(np,:); mop=mom(np,:);
32: plot(x,dip(1,:),'-r',x,dip(2,:),'-r',...
33:     x,dip(3,:),'-r');
34: xlabel('x axis'); ylabel('displacement');
35: hh=gca;
36: r(1:2)=get(hh,'XLim'); r(3:4)=get(hh,'YLim');
37: xp=r(1)+(r(2)-r(1))/10;
38: dp=r(4)-(r(4)-r(3))/10;
39: tstr=['Displacement for Nearly Resonant' ...
40:     ' Moment Acting at Right End'];
```

```
41: title(tstr);
42: text(xp,dp,['Number of series terms ' ...
43:            'used = ',int2str(ntrms)]);
44: estr='\nPress [Enter] to continue';
45: figure(gcf); pauz(estr);
46: %genprint('3positns');
47:
48: clf;
49: plot(x,mop(1,:),'-r',x,mop(2,:),'-r',...
50:      x,mop(3,:),'-r');
51: h=gca;
52: r(1:2)=get(h,'XLim'); r(3:4)=get(h,'YLim');
53: mp=r(3)+(r(4)-r(3))/10;
54: xlabel('x axis'); ylabel('moment');
55: tstr=['Bending Moment for Nearly Resonant' ...
56:        ' Moment Acting at Right End'];
57: title(tstr);
58: text(xp,mp,['Number of series terms ' ...
59:             'used = ',int2str(ntrms)]);
60: figure(gcf); pauz(estr);
61: %genprint('3moments');
62:
63: inct=2; incx=2;
64: ht=0.75; it=1:inct:.8*nt; ix=1:incx:nx;
65: tt=t(it); xx=x(ix);
66: dd=displ(it,ix); mm=mom(it,ix);
67: % a=mesh(xx,tt,dd);
68: % set(a,'edge','b'); set(a,'face','w');
69: a=surf(xx,tt,dd);
70: tstr=['Transverse Deflection as a ' ...
71:        'Function of Time and Position'];
72: title(tstr);
73: xlabel('x axis'); ylabel('time');
74: zlabel('transverse deflection');
75: figure(gcf); %genprint('dispsrf');
76: pauz(estr);
77:
78: a=surf(xx,tt,mm);
79: title(['Bending Moment as a Function ' ...
80:         'of Time and Position'])
81: xlabel('x axis'); ylabel('time');
82: zlabel('bending moment'); figure(gcf);
83: fprintf('\nPress [Enter] to animate ');
84: fprintf('beam deflection\n');
85: %genprint('momsrf'); pauz;
```

```
86: animate(x,displ,.1,'Transverse Deflection', ...
87:        'x axis','deflection'); pauz;
88: fprintf('\nPress [Enter] to animate ');
89: fprintf('bending moment\n'); pauz;
90: animate(x,mom,.1,'Bending Moment History', ...
91:        'x axis','moment');
92: fprintf('\nAll Done\n'); close;
```

Function beamresp

```
 1: function [t,x,displ,mom]= ...
 2:         beamresp(ei,arho,len,m0,w0,tmin,tmax, ...
 3:                 nt,xmin,xmax,nx,ntrms)
 4: %
 5: % [t,x,displ,mom]=beamresp(ei,arho,len,m0, ...
 6: %         w0,tmin,tmax,nt,xmin,xmax,nx,ntrms)
 7: % ~~~~~~~~~~~~~~~~~~~~~~~~~~~~~~~~~~~~~~~~~~~~~~~
 8: % This function evaluates the time dependent
 9: % displacement and moment in a constant
10: % cross-section simply-supported beam which
11: % is initially at rest when a harmonically
12: % varying moment is suddenly applied at the
13: % right end.  The resulting time histories of
14: % displacement and moment are computed.
15: %
16: % ei       - modulus of elasticity times
17: %            moment of inertia
18: % arho     - mass per unit length of the
19: %            beam
20: % len      - beam length
21: % m0,w0    - amplitude and frequency of the
22: %            harmonically varying right end
23: %            moment
24: % tmin,tmax - minimum and maximum times for
25: %            the solution
26: % nt       - number of evenly spaced
27: %            solution times
28: % xmin,xmax - minimum and maximum position
29: %            coordinates for the solution.
30: %            These values should lie between
31: %            zero and len (x=0 and x=len at
32: %            the left and right ends).
33: % nx       - number of evenly spaced solution
```

```
34: %                   positions
35: % ntrms         - number of terms used in the
36: %                   Fourier sine series
37: % t             - vector of nt equally spaced time
38: %                   values varying from tmin to tmax
39: % x             - vector of nx equally spaced
40: %                   position values varying from
41: %                   xmin to xmax
42: % displ         - matrix of transverse
43: %                   displacements with time varying
44: %                   from row to row, and position
45: %                   varying from column to column
46: % mom           - matrix of bending moments with
47: %                   time varying from row to row,
48: %                   and position varying from column
49: %                   to column
50: %
51: % User m functions called:  ndbemrsp
52: %-----------------------------------------------------
53:
54: tcof=sqrt(arho/ei)*len^2; dcof=m0*len^2/ei;
55: tmin=tmin/tcof; tmax=tmax/tcof; w=w0*tcof;
56: xmin=xmin/len; xmax=xmax/len;
57: [t,x,displ,mom]=...
58: ndbemrsp(w,tmin,tmax,nt,xmin,xmax,nx,ntrms);
59: t=t*tcof; x=x*len;
60: displ=displ*dcof; mom=mom*m0;
```

Function ndbemrsp

```
1: function [t,x,displ,mom]= ...
2:     ndbemrsp(w,tmin,tmax,nt,xmin,xmax,nx,ntrms)
3: %
4: % [t,x,displ,mom]=ndbemrsp(w,tmin,tmax,nt,...
5: %                            xmin,xmax,nx,ntrms)
6: % ~~~~~~~~~~~~~~~~~~~~~~~~~~~~~~~~~~~~~~~~~~~~~~~~
7: % This function evaluates the nondimensional
8: % displacement and moment in a constant
9: % cross-section simply-supported beam which
10: % is initially at rest when a harmonically
11: % varying moment of frequency w is suddenly
12: % applied at the right end. The resulting
13: % time history is computed.
```

```
14: %
15: % w              - frequency of the harmonically
16: %                  varying end moment
17: % tmin,tmax      - minimum and maximum
18: %                  dimensionless times
19: % nt             - number of evenly spaced
20: %                  solution times
21: % xmin,xmax      - minimum and maximum
22: %                  dimensionless position
23: %                  coordinates. These values
24: %                  should lie between zero and
25: %                  one (x=0 and x=1 give the
26: %                  left and right ends).
27: % nx             - number of evenly spaced
28: %                  solution positions
29: % ntrms          - number of terms used in the
30: %                  Fourier sine series
31: % t              - vector of nt equally spaced
32: %                  time values varying from
33: %                  tmin to tmax
34: % x              - vector of nx equally spaced
35: %                  position values varying
36: %                  from xmin to xmax
37: % displ          - matrix of dimensionless
38: %                  displacements with time
39: %                  varying from row to row,
40: %                  and position varying from
41: %                  column to column
42: % mom            - matrix of dimensionless
43: %                  bending moments with time
44: %                  varying from row to row, and
45: %                  position varying from column
46: %                  to column
47: %
48: % User m functions called:  sumser
49: %-----------------------------------------------
50:
51: if nargin < 8, w=0; end; nft=512; nh=nft/2;
52: xft=1/nh*(0:nh)';
53: x=xmin+(xmax-xmin)/(nx-1)*(0:nx-1)';
54: t=tmin+(tmax-tmin)/(nt-1)*(0:nt-1)';
55: cwt=cos(w*t);
56:
57: % Get particular solution for nonhomogeneous
58: % end condition
```

```
59: if w ==0 % Case for a constant end moment
60:   cp=[1 0 0 0; 0 0 2 0; 1 1 1 1; 0 0 2 6]\ ...
61:     [0;0;0;1];
62:   yp=[ones(x), x, x.^2, x.^3]*cp; yp=yp';
63:   mp=[zeros(nx,2), 2*ones(nx,1), 6*x]*cp;
64:   mp=mp';
65:   ypft=[ones(xft), xft, xft.^2, xft.^3]*cp;
66:
67: % Case where end moment oscillates
68: % with frequency w
69: else
70:   s=sqrt(w)*[1, i, -1, -i]; es=exp(s);
71:   cp=[ones(1,4); s.^2; es; es.*s.^2]\ ...
72:     [0; 0; 0; 1];
73:   yp=real(exp(x*s)*cp); yp=yp';
74:   mp=real(exp(x*s)*(cp.*s(:).^2)); mp=mp';
75:   ypft=real(exp(xft*s)*cp);
76: end
77:
78: % Fourier coefficients for
79: % particular solution
80: yft=-fft([ypft;-ypft(nh:-1:2)])/nft;
81:
82: % Sine series coefficients for
83: % homogeneous solution
84: acof=-2*imag(yft(2:ntrms+1));
85: ccof=pi*(1:ntrms)'; bcof=ccof.^2;
86:
87: % Sum series to evaluate Fourier
88: % series part of solution. Then combine
89: % with the particular solution.
90: displ=sumser(acof,bcof,ccof,'cos','sin',...
91:                 tmin,tmax,nt,xmin,xmax,nx);
92: displ=displ+cwt*yp; acof=acof.*bcof;
93: mom=sumser(acof,bcof,ccof,'cos','sin',...
94:                 tmin,tmax,nt,xmin,xmax,nx);
95: mom=-mom+cwt*mp;
```

Function sumser

```
1: function [u,t,x] = sumser(a,b,c,funt,funx, ...
2:                 tmin,tmax,nt,xmin,xmax,nx)
3: %
```

```
 4: % [u,t,x] = sumser(a,b,c,funt,funx,tmin, ...
 5: %                 tmax,nt,xmin,xmax,nx)
 6: % ~~~~~~~~~~~~~~~~~~~~~~~~~~~~~~~~~~~~~~~~~~~
 7: % This function evaluates a function U(t,x)
 8: % which is defined by a finite series. The
 9: % series is evaluated for t and x values taken
10: % on a rectangular grid network. The matrix u
11: % has elements specified by the following
12: % series summation:
13: %
14: % u(i,j)   =    sum(    a(k)*funt(t(i)* ...
15: %            k=1:nsum
16: %                      b(k))*funx(c(k)*x(j))
17: %
18: % where nsum is the length of each of the
19: % vectors a, b, and c.
20: %
21: % a,b,c         - vectors of coefficients in
22: %                 the series
23: % funct,funx    - functions which accept a
24: %                 matrix argument. funct is
25: %                 evaluated for an argument of
26: %                 the form func(t*b) where t is
27: %                 a column and b is a row. funx
28: %                 is evaluated for an argument
29: %                 of the form funx(c*x) where
30: %                 c is a column and x is a row.
31: % tmin,tmax,nt - produces vector t with nt
32: %                 evenly spaced values between
33: %                 tmin and tmax
34: % xmin,xmax,nx - produces vector x with nx
35: %                 evenly spaced values between
36: %                 xmin and xmax
37: % u             - the nt by nx matrix
38: %                 containing values of the
39: %                 series evaluated at t(i),x(j),
40: %                 for i=1:nt and j=1:nx
41: % t,x           - column vectors containing t
42: %                 and x values. These output
43: %                 values are optional.
44: %
45: % User m functions called:  none.
46: %-------------------------------------------------
47:
48: tt=(tmin:(tmax-tmin)/(nt-1):tmax)';
```

```
49: xx=(xmin:(xmax-xmin)/(nx-1):xmax); a=a(:).';
50: u=a(ones(nt,1),:).*feval(funt,tt*b(:).')*...
51:   feval(funx,c(:)*xx);
52: if nargout>1, t=tt; x=xx'; end
```

Function animate

```
 1: function animate(x,u,tpause,titl,xlabl,ylabl)
 2: %
 3: % animate(x,u,tpause,titl,xlabl,ylabl)
 4: % ~~~~~~~~~~~~~~~~~~~~~~~~~~~~~~~~~~~~~~~~
 5: % This function draws an animated plot of data
 6: % values stored in array u.  The different
 7: % columns of u correspond to position values
 8: % in vector x.  The successive rows of u
 9: % correspond to different times. Parameter
10: % tpause controls the speed of animation.
11: %
12: % u       - matrix of values to animate plots
13: %             of u versus x
14: % x       - spatial positions for different
15: %             columns of u
16: % tpause  - clock seconds between output of
17: %             frames. The default is .1 secs
18: %             when tpause is left out. When
19: %             tpause=0, a new frame appears
20: %             when the user presses any key.
21: % titl    - graph title
22: % xlabl   - label for horizontal axis
23: % ylabl   - label for vertical axis
24: %
25: % User m functions called:  genprint
26: %-------------------------------------------------
27:
28: if nargin<6, ylabl=''; end;
29: if nargin<5, xlabl=''; end
30: if nargin<4, titl=''; end;
31: if nargin<3, tpause=.1; end;
32:
33: [ntime,nxpts]=size(u);
34: umin=min(u(:)); umax=max(u(:));
35: udif=umax-umin;
36: uavg=.5*(umin+umax);
```

```
37: xmin=min(x); xmax=max(x);
38: xdif=xmax-xmin; xavg=.5*(xmin+xmax);
39: xwmin=xavg-.55*xdif; xwmax=xavg+.55*xdif;
40: uwmin=uavg-.55*udif; uwmax=uavg+.55*udif; clf;
41: axis([xwmin,xwmax,uwmin,uwmax]); title(titl);
42: xlabel(xlabl); ylabel(ylabl); hold on;
43:
44: for j=1:ntime
45:   ut=u(j,:);
46:   plot(x,ut,'-'); axis('off'); figure(gcf);
47:   if tpause==0
48:     pause;
49:   else
50:     pause(tpause);
51:   end
52:   if j==ntime, break, else, cla; end
53: end
54: %genprint('cntltrac');
55: hold off; clf;
```

9.7   Torsional Stresses in a Beam of Rectangular Cross Section

Elastic beams of uniform cross section are commonly used structural members. Evaluation of the stresses caused when beams undergo torsional moments depends on finding a particular type of complex valued function. This function is analytic inside the beam cross section and has its imaginary part known on the boundary [71]. The shear stresses τ_{XZ} and τ_{YZ} are obtained from the stress function $f(z)$ of the complex variable $z = x + \imath y$ according to

$$\frac{\tau_{ZX} - \imath\tau_{ZY}}{\mu\alpha} = f'(z) - \imath\bar{z}$$

where μ is the shear modulus and α is the twist per unit length. In the case for a simply connected cross section, such as a rectangle or a semicircle, the necessary boundary condition is

$$\mathbf{imag}[f(z)] = \frac{1}{2}|z|^2$$

at all boundary points. It can also be shown that the torsional moment causes the beam cross section to warp. The warped shape is given by the real part of $f(z)$.

The geometry we will analyze is rectangular. As long as the ratio of side length remains fairly close to unity, $f(z)$ can be approximated well in series form as

$$f(z) = \imath \sum_{j=1}^{n} c_j \left(\frac{z}{s}\right)^{2j-2}$$

where $c_1, \ldots, c_n$ are real coefficients computed to satisfy the boundary conditions in the least square sense. Parameter s is used for scaling to prevent occurrence of large numbers when n becomes large. We take a rectangle with sides parallel to the coordinate axes and assume side lengths of $2a$ and $2b$ for the horizontal and vertical directions, respectively. The scaling parameter will be chosen as the larger of a and b. The boundary conditions state that for any point z_i on the boundary we should have

$$\sum_{j=1}^{n} c_j \, \mathbf{real}\left[(\frac{z_i}{s})^{2j-2}\right] = \frac{1}{2}|z_i|^2$$

Once the series coefficients are found, then shear stresses are computed as

$$\frac{\tau_{XZ} - \imath\tau_{YZ}}{\mu\alpha} = -\imath\bar{z} + 2\imath s^{-1} \sum_{j=2}^{n} (j-1)c_j \left(\frac{z}{s}\right)^{2j-3}$$

A program was written to compute stresses in a rectangular beam and to show graphically the cross section warping and the dimensionless

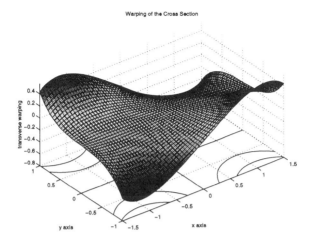

Figure 9.18. Warping of the Cross Section

stress values. The program is short and the necessary calculations are almost self explanatory. It is worthwhile to observe, however, the ease with which MATLAB handles complex functions. Note how intrinsic function **linspace** is used to generate boundary data and **meshgrid** is used to generate a grid of complex values (see lines 50, 51, 72, 73, and 74 of function **recstrs**). The sample problem employs a rectangle of dimension 2 units by 4 units. The maximum stress occurs at the middle of the longest side. Figures 9.18–9.21 plot the results of this analysis.

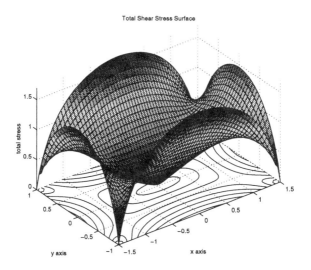

Figure 9.19. Total Shear Stress Surface

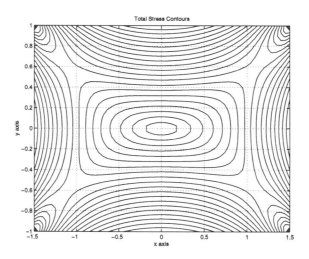

Figure 9.20. Total Stress Contours

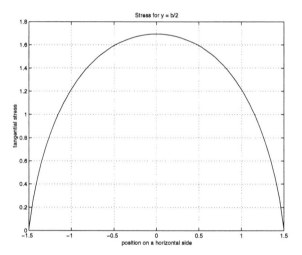

Figure 9.21. Stress for $y = b/2$

MATLAB Example

Output from Example

```
>> rector


===   TORSIONAL STRESS CALCULATION IN A RECTANGULAR   ===
===       BEAM USING LEAST SQUARE APPROXIMATION       ===


Input the lengths of the horizontal and the vertical sides
(make the long side horizontal)
? 3,2


To approximate the boundary conditions, give the number of
least square points on the horizontal side and the number of
least square points on the vertical side
(40,40 is typical for a square)
? 60,40


Input the number of terms used in the stress function
(20 terms is usually enough)
? 30


To display the stress and displacement quantities on
a grid of the cross section, input the number of grid
lines for the X direction and for the Y direction.
(25,25 is usually adequate)
? 60,40


Press [Enter] to plot warping surface
Press [Enter] to show total stress surface
Press [Enter] to display stress contours
Press [Enter] to plot maximum
  stress on rectangle side

The Maximum Shear Stress = 1.6949
at x = -0.025424 and y = -1
```

Script File rector

```
 1: % Example:  rector
 2: % ~~~~~~~~~~~~~~~~~
 3: % This program employs point matching to obtain
 4: % an approximate solution for torsional
 5: % stresses in a Saint Venant beam of
 6: % rectangular cross section. The complex stress
 7: % function is analytic inside the rectangle and
 8: % has its real part equal to abs(z*z)/2 on the
 9: % boundary. The problem is solved approximately
10: % using a polynomial stress function which fits
11: % the boundary condition in the least square
12: % sense. Surfaces and contour curves describing
13: % the stress and deformation pattern in the
14: % beam cross section are drawn.
15: %
16: % User m functions required:
17: %    recstrs, read, genprint
18: %-------------------------------------------------
19:
20: clear;
21: fprintf('\n===  TORSIONAL STRESS CALCULATION');
22: fprintf(' IN A RECTANGULAR  ===');
23: fprintf('\n===     BEAM USING LEAST SQUARE ');
24: fprintf('APPROXIMATION     ===\n');
25: fprintf('\nInput the lengths of the ');
26: fprintf('horizontal and the vertical sides\n');
27: fprintf('(make the long side horizontal)\n');
28: [a,b]=read; a=a/2; b=b/2;
29: fprintf('\nTo approximate the boundary ');
30: fprintf('conditions, give the number of');
31: fprintf('\nleast square points on the ');
32: fprintf('horizontal side and the number of');
33: fprintf('\nleast square points on the ');
34: fprintf('vertical side');
35: fprintf('\n(40,40 is typical for a square)\n');
36: [nsega,nsegb]=read;
37: nsega=fix(nsega/2); nsegb=fix(nsegb/2);
38: fprintf('\nInput the number of terms ');
39: fprintf('used in the stress function');
40: fprintf('\n(20 terms is usually enough)\n');
41: ntrms=input('? ');
42: fprintf('\nTo display the stress and ');
```

```
43: fprintf('displacement quantities on');
44: fprintf('\na grid of the cross section,');
45: fprintf(' input the number of grid');
46: fprintf('\nlines for the X direction and ');
47: fprintf('for the Y direction.');
48: fprintf('\n(25,25 is usually adequate)\n');
49: [nxout,nyout]=read;
50:
51: [c,phi,stres,z] = ...
52:    recstrs(a,nsega,b,nsegb,ntrms,nxout,nyout);
53: [smax,k]=max(abs(stres(:))); zmax=z(:);
54: zmax=zmax(k); xmax=real(zmax); ymax=imag(zmax);
55: disp(...
56: ['The Maximum Shear Stress = ',num2str(smax)]);
57: disp(['at x = ',num2str(xmax),' and y = ',...
58:       num2str(ymax)]);
59: disp(' '); disp('All Done');
```

Function recstrs

```
 1: function [c,phi,stres,z]=...
 2:    recstrs(a,nsega,b,nsegb,ntrms,nxout,nyout)
 3: %
 4: % [c,phi,stres,z]=...
 5: %    recstrs(a,nsega,b,nsegb,ntrms,nxout,nyout)
 6: % ~~~~~~~~~~~~~~~~~~~~~~~~~~~~~~~~~~~~~~~~~~~~~~
 7: % This function employs point matching to
 8: % obtain an approximate solution for torsional
 9: % stresses in a Saint Venant beam of
10: % rectangular cross section. The complex stress
11: % function is analytic inside the rectangle
12: % and has its real part equal to abs(z*z)/2 on
13: % the boundary. The problem is solved
14: % approximately using a polynomial stress
15: % function which fits the boundary condition
16: % in the least square sense. The beam is 2*a
17: % wide along the x axis and 2*b deep along
18: % the y axis. The shear stresses in the beam
19: % are given by the complex stress formula:
20: %
21: % (tauzx-i*tauzy)/(mu*alpha) = -i*conj(z)+f'(z)
22: %
23: % where
```

```
24: %
25: %    f(z)=i*sum( c(j)*z^(2*j-2), j=1:ntrms )
26: %
27: % and c(j) are real.
28: %
29: % a,b       - half the side lengths of the
30: %               horizontal and vertical sides
31: % nsega,    - numbers of subintervals used in
32: % nsegb       formation of least square
33: %             equations
34: % ntrms     - number of polynomial terms used in
35: %             polynomial stress function
36: % nxout,    - number of grid points used to
37: % nyout       evaluate output
38: % c         - coefficients defining the stress
39: %             function
40: % phi       - values of the membrane function
41: % stres     - array of complex stress values
42: % z         - complex point array at which
43: %             stresses are found
44: %
45: % User m functions called:  genprint
46: %-------------------------------------------------
47:
48: % Generate vector zbdry of boundary points
49: % for point matching.
50: db=b/nsegb; zvert=linspace(a,a+i*(b-db),nsegb);
51: zhoriz=linspace(a+i*b,i*b,nsega+1);
52: zbdry=[zvert(:);zhoriz(:)];
53:
54: % Determine a scaling parameter used to
55: % prevent occurrence of large numbers when
56: % high powers of z are used
57: s=max(a,b);
58:
59: % Form the least square equations to impose
60: % the boundary conditions.
61: neq=length(zbdry); amat=ones(neq,ntrms);
62: ztmp=(zbdry/s).^2; bvec=.5*abs(zbdry).^2;
63: for j=2:ntrms
64:   amat(:,j)=amat(:,j-1).*ztmp;
65: end
66:
67: % Solve the least square equations.
68: amat=real(amat); c=pinv(amat)*bvec;
```

```
69:
70: % Generate grid points to evaluate
71: % the solution.
72: xsid=linspace(-a,a,nxout);
73: ysid=linspace(-b,b,nyout);
74: [xg,yg]=meshgrid(xsid,ysid);
75: z=xg+i*yg; zz=(z/s).^2;
76:
77: % Evaluate the warping function
78: phi=-imag(polyval(flipud(c),zz));
79:
80: % Evaluate stresses and plot results
81: cc=(2*(1:ntrms)-2)'.*c;
82: stres=-i*conj(z)+i* ...
83:       polyval(flipud(cc),zz)./(z+eps*(z==0));
84: am=num2str(-a);ap=num2str(a);
85: bm=num2str(-b);bp=num2str(b);
86:
87: % Plot results
88: fprintf('\nPress [Enter] to plot ');
89: fprintf('warping surface\n');
90: pause; hold off; clf; surfc(xg,yg,phi);
91: title('Warping of the Cross Section')
92: xlabel('x axis'); ylabel('y axis');
93: zlabel('transverse warping');
94: axis('equal'), figure(gcf);
95: fprintf('\nPress [Enter] to show total ');
96: fprintf('stress surface\n'); pauz;
97: %genprint('warpsurf');
98:
99: surfc(xg,yg,abs(stres));
100: title('Total Shear Stress Surface');
101: xlabel('x axis'); ylabel('y axis');
102: zlabel('total stress'); axis('equal');
103: figure(gcf);
104: fprintf('Press [Enter] to display ');
105: fprintf('stress contours\n');
106: %genprint('torstrsu'); pauz;
107:
108: contour(xg,yg,abs(stres),20);
109: title('Total Stress Contours');
110: xlabel('x axis'); ylabel('y axis');
111: figure(gcf);
112: fprintf('Press [Enter] to plot maximum');
113: fprintf('\n  stress on rectangle side\n');
```

```
114: %genprint('torcontu'); pauz;
115:
116: plot(xsid,abs(stres(1,:)));
117: grid; ylabel('tangential stress');
118: xlabel('position on a horizontal side');
119: title('Stress for y = b/2'); figure(gcf);
120: %genprint('torstsid');
```

9.8   Accuracy Comparison for Euler Beam Natural Frequencies Obtained by Finite Element and Finite Difference Methods

This chapter concludes with an example involving natural frequency computation for a cantilever beam. Accuracy comparisons are made among results from the following three methods: a) solution of the frequency equation for the true continuum model, b) approximation of the equations of motion using finite difference to replace spatial derivatives, and c) use of finite element methods implying a piecewise cubic spatial interpolation of the displacement field. The first method is less appealing as a general tool than the last two methods because the frequency equation would be awkward to obtain for geometries of variable cross section. Frequencies found using finite difference and finite element methods are compared with results from the exact model; and it is observed that the finite element method produces results which are superior to those from finite differences for comparable degrees of freedom. In addition, the natural frequencies and mode shapes given by finite elements are used to compute and animate system response produced when a beam, initially at rest, is suddenly subjected to two concentrated loads.

9.8.1   Mathematical Formulation

The differential equation governing transverse vibrations of an elastic beam of constant depth is [68]

$$EI\frac{\partial^4 Y}{\partial X^4} = -\rho\frac{\partial^2 Y}{\partial T^2} + W(X,T) \qquad 0 \le X \le \ell \qquad T \ge 0$$

where

$$
\begin{aligned}
Y(X,T) &\quad-\quad \text{transverse displacement}\\
X &\quad-\quad \text{horizontal position along the beam length}\\
T &\quad-\quad \text{time}\\
EI &\quad-\quad \text{product of moment of inertia and Young's modulus}\\
\rho &\quad-\quad \text{mass per unit length of the beam}\\
W(X,T) &\quad-\quad \text{external applied force per unit length}
\end{aligned}
$$

In the present study, we consider the cantilever beam shown in Figure 9.22, having end conditions which are

$$Y(0,T) = 0 \qquad\qquad \frac{\partial Y(0,T)}{\partial X} = 0$$

and

$$EI\frac{\partial^2 Y(\ell,T)}{\partial X^2} = M_E(T) \qquad\qquad EI\frac{\partial^3 Y(\ell,T)}{\partial X^3} = V_E(T)$$

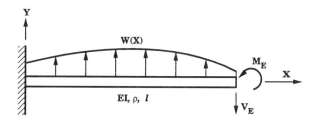

Figure 9.22. Cantilever Beam Subjected to Impact Loading

This problem can be expressed more concisely using dimensionless variables

$$x = \frac{X}{\ell} \qquad y = \frac{Y}{\ell} \qquad t = \sqrt{\frac{EI}{\rho}}\left(\frac{T}{\ell^2}\right)$$

Then the differential equations and boundary conditions become

$$\frac{\partial^4 y}{\partial x^4} = -\frac{\partial^2 y}{\partial t^2} + w(x,t) \qquad 0 \le x \le 1 \qquad t \ge 0$$

$$y(0,t) = 0 \qquad \frac{\partial y}{\partial x}(0,t) = 0$$

$$\frac{\partial^2 y}{\partial x^2}(1,t) = m_e(t) \qquad \frac{\partial^3 y}{\partial x^3}(1,t) = v_e(t)$$

where

$$w = (W\ell^3)/(EI) \qquad m_e = (M_E\ell)/(EI)$$

and

$$v_e = (V_E\ell^2)/(EI)$$

The natural frequencies of the system are obtained by computing homogeneous solutions of the form $y(x,t) = f(x)\sin(\omega t)$ which exist when $w = m_e = v_e = 0$. This implies

$$\frac{d^4 f}{dx^4} = \lambda^4 f \qquad \lambda = \sqrt{\omega}$$

subject to

$$f(0) = 0 \qquad f'(0) = 0 \qquad f''(1) = 0 \qquad f'''(1) = 0$$

The solution satisfying this fourth order differential equation with homogeneous boundary conditions has the form

$$\begin{aligned} f = \;& [\cos(\lambda x) - \cosh(\lambda x)][\sin(\lambda) + \sinh(\lambda)] - \\ & [\sin(\lambda x) - \sinh(\lambda x)][\cos(\lambda) + \cosh(\lambda)] \end{aligned}$$

where λ values must be roots of the frequency equation

$$p(\lambda) = \cos(\lambda) + 1/\cosh(\lambda) = 0$$

Although the roots cannot be obtained explicitly, asymptotic approximations valid for large n are evidently

$$\lambda_n = (2k - 1)\pi/2$$

These estimates can be used as the starting points for a rapidly convergent Newton iteration indicated by

$$\lambda_{NEW} = \lambda_{OLD} - p(\lambda_{OLD})/p'(\lambda_{OLD})$$

with

$$p'(\lambda) = -\sin(\lambda) - [\sinh(\lambda)/\cosh^2(\lambda)]$$

The exact solution will be used to compare related results produced by finite difference and finite element methods. First we consider finite differences. The following difference formulas have a quadratic truncation error derivable from Taylor's series [1]:

$$\frac{\partial^4 y}{\partial x^4}(x) = [y(x - 2h) - 4y(x - h) + 6y(x) -$$
$$4y(x + h) + y(x + 2h)]/h^4$$
$$y'(x) = [-y(x - h) + y(x + h)]/(2h)$$
$$y''(x) = [y(x - h) - 2y(x) + y(x + h)]/h^2$$
$$y'''(x) = [-y(x - 2h) + 2y(x - h) - 2y(x + h) +$$
$$y(x + 2h)]/(2h^3)$$

The step-size will be taken as $h = 1/n$ so $x_j = jh$, $0 \le j \le n$. Therefore, x_0 is at the left end and x_n is at the right end of the beam. It is desirable to include additional fictitious points x_{-1}, x_{n+1}, and x_{n+2}. Then the left end conditions imply

$$y_0 = y_1 \qquad \text{and} \qquad y_{-1} = y_1$$

and the right end conditions imply

$$y_{n+1} = -y_{n-1} + 2y_n$$

$$y_{n+2} = y_{n-2} - 4y_{n-1} + 4y_n$$

Using these relations, the algebraic eigenvalue problem derived from the difference approximation is

$$\begin{aligned}
7y_1 - 4y_2 + y_3 &= \tilde{\lambda}y_1 & \tilde{\lambda} = h^4\lambda \\
-4y_1 + 6y_2 - 4y_3 + y_4 &= \tilde{\lambda}y_2 & \\
y_{j-2} - 4y_{j-1} + 6y_j - 4y_{j+1} + y_{j+2} &= \tilde{\lambda}y_j & 2 < j < (n - 1) \\
y_{n-3} - 4y_{n-2} + 5y_{n-1} - 2y_n &= \tilde{\lambda}y_{n-1} & \\
2y_{n-2} - 4y_{n-1} + 2y_n &= \tilde{\lambda}y_n &
\end{aligned}$$

This can be solved for the natural frequencies provided satisfactory eigenvalue calculation software is available.

The finite element method leads to a similar problem involving global mass and stiffness matrices [53]. When we consider a single beam element of mass m and length ℓ, the elemental mass and stiffness matrices found using a cubically varying displacement approximation are

$$M_e = \frac{m}{420} \begin{bmatrix} 156 & 22\ell & 54 & -13\ell \\ 22\ell & 4\ell^2 & 13\ell & -3\ell^2 \\ 54 & 13\ell & 156 & -22\ell \\ -13\ell & -3\ell^2 & -22\ell & 4\ell^2 \end{bmatrix}$$

$$K_e = \frac{EI}{\ell^3} \begin{bmatrix} 6 & 3\ell & -6 & 3\ell \\ 3\ell & 2\ell^2 & -3\ell & \ell^2 \\ -6 & -3\ell & 6 & -3\ell \\ 3\ell & \ell^2 & -3\ell & 2\ell^2 \end{bmatrix}$$

and the elemental equation of motion has the form

$$M_e Y_e'' + K_e Y_e = F_e$$

where

$$Y_e = [Y_1, Y_1', Y_2, Y_2']^T$$
$$F_e = [F_1, M_1, F_2, M_2]^T$$

are generalized elemental displacement and force vectors. The global equation of motion is obtained as an assembly of element matrices and has the form

$$MY'' + KY = F$$

A system with N elements involves $N+1$ nodal points. For the cantilever beam studied here both Y_0 and Y_0' are zero, so removing these two variables leaves a system of $n = 2N$ unknowns. The solution of this equation in the case of a non-resonant harmonic forcing function will be discussed further. The matrix analog of the simple harmonic equation is

$$M\ddot{Y} + KY = F_1 \cos(\omega t) + F_2 \sin(\omega t)$$

with initial conditions

$$Y(0) = Y_0 \qquad \dot{Y}(0) = V_0$$

Solution of this differential equation is facilitated by using a particular solution and a homogeneous solution

$$Y = Y_P + Y_H$$

where

$$Y_H = Y_1 \cos(\omega t) + Y_2 \sin(\omega t)$$

with

$$Y_j = (K - \omega^2 M)^{-1} F_j \qquad j = 1, 2$$

This assumes, of course, that $K - \omega^2 M$ is nonsingular. The homogeneous equation satisfies

$$M\ddot{Y}_H + KY_H = 0$$

with the initial conditions

$$Y_H(0) = Y_0 - Y_1 \qquad \dot{Y}_H(0) = V_0 - \omega Y_2$$

The homogeneous solution components have the form

$$Y_{jH} = U_j \cos(\omega_j t + \phi_j)$$

where ω_j and U_j are natural frequencies and modal vectors satisfying the eigenvalue equation

$$KU_j = \omega_j^2 M U_j$$

Consequently, the homogeneous solution completing the modal response is

$$Y_H(t) = \sum_{j=1}^{n} U_j[\cos(\omega_j t)c_j + \sin(\omega_j t)d_j/\omega_j]$$

where c_j and d_j are computed to satisfy the initial conditions which require

$$C = U^{-1}(Y_0 - Y_1) \qquad D = U^{-1}(V_0 - \omega Y_2)$$

The next section presents the MATLAB program. Natural frequencies from finite difference and finite element matrices are compared and modal vectors from the finite element method are used to analyze a time response problem.

9.8.2   Discussion of the Code

A program was written to compare exact frequencies from the original continuous beam model with approximations produced using finite differences and finite elements. The finite element results were also employed to calculate a time response by modal superposition for any structure which has general mass and stiffness matrices, and is subjected to loads which are constant or harmonically varying.

The code below is longer than earlier problem solutions because various MATLAB capabilities are applied to three different solution methods. The following function summary involves eight segments, several of which were used earlier in the text.

cbfreq	driver to input data, call computation modules, and print results
cbfrqnwm	function to compute exact natural frequencies by Newton's method for root calculation
cbfrqfdm	forms equations of motion using finite difference and calls **eig** to compute natural frequencies
cbfrqfem	uses the finite element method to form the equation of motion and calls **eig** to compute natural frequencies and modal vectors
frud	function which solves the structural dynamics equation by methods developed in Chapter 7
examplmo	evaluates the response caused when a downward load at the middle and an upward load at the free end are applied
animate	plots successive positions of the beam to animate the motion
plotsave	plots the beam frequencies for the three methods. Also plots percent errors showing how accurate finite element and finite difference methods are
read	reads a sequence of numbers

Several characteristics of the functions assembled for this program are worth examining in detail. The next table contains remarks relevant to the code.

Routine	Line	Operation
Output		Natural frequencies are printed along with error percentages. The output shown here has been extracted from the actual output to show only the highest and lowest frequencies.
cbfrqnwm	23	Asymptotic estimates are used to start a Newton method iteration.
	26-32	Root corrections are carried out for all roots until the correction to any root is sufficiently small.
cbfrqfdm	23-24	The equations of motion are formed without corrections for end conditions.
	26-33	End conditions are applied.
	37	**eig** computes the frequencies.
cbfrqfem	28-33	Form elemental mass matrix.
	35-39	Form elemental stiffness matrix.
	44-48	Global equations of motion are formed using an element by element loop.
	52	Boundary conditions are applied requiring zero displacement and slope at the left end, and zero moment and shear at the right end.
	54-61	Frequencies and modal vectors are computed. Note that modal vector computation is made optional since this takes longer than only computing frequencies.
frud		Compute time response by modal superposition. Theoretical details pertaining to this function appear in Chapter 7.
examplmo	23-29	The time step and maximum time for response calculation is selected.
	31-32	Function **frud** is used to compute displacement and rotation response. Only displacement is saved.
	36-43	Free end displacement is plotted.
	45-53	A surface showing displacement as a function of position and time is shown.
		continued on next page

		continued from previous page
Routine	Line	Operation
	57	Function **animate** is called.
animate	34-40	Window limits are determined.
	44-53	Each position is plotted. Then it is erased before proceeding to the next position.
plotsave		Plot and save graphs showing the frequencies and error percentages.

Table 9.1. Description of Code in Example

9.8.3 Numerical Results

The dimensionless frequency estimates from finite difference and finite element methods were compared for various numbers of degrees-of-freedom. Typical program output for $n = 100$ is shown at the end of this section. The frequency results and error percentages are shown in Figures 9.23 and 9.24. It is evident that the finite difference frequencies are consistently low and the finite element results are consistently high. The finite difference estimates degrade smoothly with increasing order. The finite element frequencies are surprisingly accurate for ω_k when $k < n/2$. At $k = n/2$ and $k = n$, the finite element error jumps sharply. This peculiar error jump halfway through the spectrum has also been observed in [53]. The most important and useful result seen from Figure 9.24 is that in order to obtain a particular number of frequencies, say N, which are accurate within 3.5%, it is necessary to employ more than $2N$ elements and keep only half of the predicted values.

The final result presented is the time response of a beam which is initially at rest when a concentrated downward load of five units is applied at the middle and a one unit upward load is applied at the free end. The time history was computed using function **frud**. Figure 9.25 shows the time history of the free end. Figure 9.26 is a surface plot illustrating how the deflection pattern changes with time. Finally, Figure 9.27 shows successive deflection positions produced by function **animate**. The output was obtained by suppressing the graph clearing option for successive configurations.

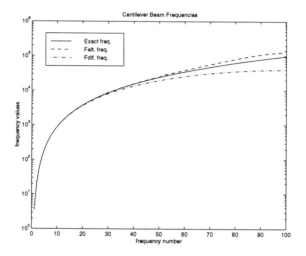

Figure 9.23. Cantilever Beam Frequencies

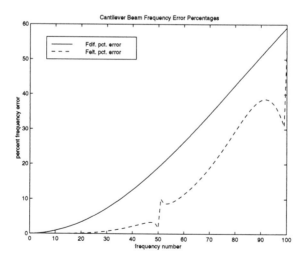

Figure 9.24. Cantilever Beam Frequency Error Percentages

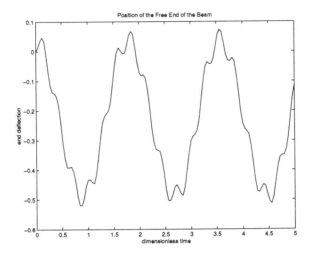

Figure 9.25. Position of the Free End of the Beam

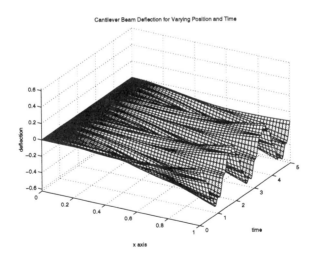

Figure 9.26. Beam Deflection History for Varying Time

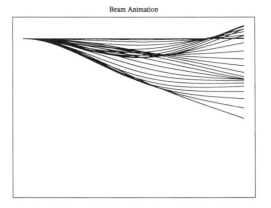

Beam Animation

Figure 9.27. Beam Animation

MATLAB Example

Output from Example

>> cbfreq

CANTILEVER BEAM FREQUENCIES BY FINITE DIFFERENCE AND
        FINITE ELEMENT APPROXIMATION

Give the number of frequencies to be computed
(use an even number greater than 2)
? > 100

freq. number	exact. freq.	fdif. freq.	fd. pct. error	felt. freq.	fe. pct. error
1	3.51602e+00	3.51572e+00	-0.008	3.51602e+00	0.000
2	2.20345e+01	2.20250e+01	-0.043	2.20345e+01	0.000
3	6.16972e+01	6.16414e+01	-0.090	6.16972e+01	0.000
4	1.20902e+02	1.20714e+02	-0.155	1.20902e+02	0.000
5	1.99860e+02	1.99386e+02	-0.237	1.99860e+02	0.000
6	2.98556e+02	2.97558e+02	-0.334	2.98558e+02	0.001
7	4.16991e+02	4.15123e+02	-0.448	4.16999e+02	0.002
8	5.55165e+02	5.51957e+02	-0.578	5.55184e+02	0.003
9	7.13079e+02	7.07918e+02	-0.724	7.13119e+02	0.006
10	8.90732e+02	8.82842e+02	-0.886	8.90809e+02	0.009
11	1.08812e+03	1.07655e+03	-1.064	1.08826e+03	0.013
12	1.30526e+03	1.28884e+03	-1.257	1.30550e+03	0.019
13	1.54213e+03	1.51950e+03	-1.467	1.54252e+03	0.026
14	1.79874e+03	1.76830e+03	-1.692	1.79937e+03	0.035
15	2.07508e+03	2.03497e+03	-1.933	2.07605e+03	0.047
16	2.37117e+03	2.31926e+03	-2.189	2.37261e+03	0.061
17	2.68700e+03	2.62088e+03	-2.461	2.68908e+03	0.077
18	3.02257e+03	2.93951e+03	-2.748	3.02551e+03	0.098
19	3.37787e+03	3.27486e+03	-3.050	3.38197e+03	0.121
20	3.75292e+03	3.62657e+03	-3.367	3.75851e+03	0.149

        ====== INTERMEDIATE LINES OF OUTPUT DELETED ======

90	7.90580e+04	3.88340e+04	-50.879	1.09328e+05	38.288
91	8.08345e+04	3.90347e+04	-51.710	1.11989e+05	38.541
92	8.26308e+04	3.92169e+04	-52.540	1.14512e+05	38.582
93	8.44468e+04	3.93804e+04	-53.367	1.16860e+05	38.384
94	8.62825e+04	3.95250e+04	-54.191	1.18999e+05	37.917
95	8.81380e+04	3.96507e+04	-55.013	1.20889e+05	37.159
96	9.00133e+04	3.97572e+04	-55.832	1.22496e+05	36.086
97	9.19082e+04	3.98445e+04	-56.648	1.23786e+05	34.684
98	9.38229e+04	3.99125e+04	-57.460	1.24730e+05	32.941
99	9.57574e+04	3.99611e+04	-58.268	1.25305e+05	30.857
100	9.77116e+04	3.99903e+04	-59.073	1.49694e+05	53.200

Evaluate the time response from two
concentrated loads. One downward at the
middle and one upward at the free end.

input the time step and the maximum time
(0.04 and 5.0) are typical. Use 0,0 to stop

? .04,5

Evaluate the time response resulting from a
concentrated downward load at the middle and
an upward end load.

input the time step and the maximum time
(0.04 and 5.0) are typical. Use 0,0 to stop

? 0,0

Script File cbfreq

```
 1: % Example:  cbfreq
 2: % ~~~~~~~~~~~~~~~~~
 3: % This program computes approximate natural
 4: % frequencies of a uniform depth cantilever
 5: % beam using finite difference and finite
 6: % element methods. Error results are presented
 7: % which demonstrate that the finite element
 8: % method is much more accurate than the finite
 9: % difference method when the same matrix orders
10: % are used in computation of the eigenvalues.
11: %
12: % User m functions required:
13: %    cbfrqnwm, cbfrqfdm, cbfrqfem, frud,
14: %    examplmo, animate, plotsave, read,
15: %    genprint
16: %------------------------------------------------
17:
18: clear;
19: fprintf('\n\n');
20: fprintf('CANTILEVER BEAM FREQUENCIES BY ');
21: fprintf('FINITE DIFFERENCE AND');
22: fprintf(...
23: '\n          FINITE ELEMENT APPROXIMATION\n');
24:
25: fprintf('\nGive the number of frequencies ');
26: fprintf('to be computed');
27: fprintf('\n(use an even number greater ');
28: fprintf('than 2)\n'), n=input('? > ');
29: if rem(n,2) ~= 0, n=n+1; end
30:
31: % Exact frequencies from solution of
32: % the frequency equation
33: wex = cbfrqnwm(n,1e-12);
34:
35: % Frequencies for the finite
36: % difference solution
37: wfd = cbfrqfdm(n);
38:
39: % Frequencies, modal vectors, mass matrix,
40: % and stiffness matrix from the finite
41: % element solution.
42: nelts=n/2; [wfe,mv,mm,kk] = cbfrqfem(nelts);
```

```
43: pefdm=(wfd-wex)./(.01*wex);
44: pefem=(wfe-wex)./(.01*wex);
45:
46: nlines=17; nloop=round(n/nlines);
47: v=[(1:n)',wex,wfd,pefdm,wfe,pefem];
48: disp(' '); lo=1;
49: t1=[' freq.      exact.           fdif.' ...
50:    '          fd. pct.'];
51: t1=[t1,'    felt.        fe. pct.'];
52: t2=['number      freq.            freq.' ...
53:    '          error  '];
54: t2=[t2,'    freq.          error  '];
55: while lo < n
56:   disp(t1),disp(t2);
57:   hi=min(lo+nlines-1,n);
58:   for j=lo:hi
59:     s1=sprintf('\n %4.0f %13.5e %13.5e', ...
60:                v(j,1),v(j,2),v(j,3));
61:     s2=sprintf(' %9.3f %13.5e %9.3f', ...
62:                v(j,4),v(j,5),v(j,6));
63:     fprintf([s1,s2]);
64:   end
65:   fprintf('\n\nPress [Enter] to continue\n\n');
66:   pause;
67:   lo=lo+nlines;
68: end
69: plotsave(wex,wfd,pefdm,wfe,pefem);
70: nfe=length(wfe); nmidl=nfe/2;
71: if rem(nmidl,2)==0, nmidl=nmidl+1; end
72: x0=zeros(nfe,1); v0=x0; w=0;
73: f1=zeros(nfe,1); f2=f1; f1(nfe-1)=1;
74: f1(nmidl)=-5;
75: xsav=examplmo(mm,kk,f1,f2,x0,v0,wfe,mv);
76: close; fprintf('\nAll Done\n');
```

Function cbfrqnwm

```
1: function z=cbfrqnwm(n,tol)
2: %
3: % z=cbfrqnwm(n,tol)
4: % ~~~~~~~~~~~~~~~~~~
5: % Cantilever beam frequencies by Newton's
6: % method.  Zeros of
```

```
 7: %          f(z) = cos(z) + 1/cosh(z)
 8: % are computed.
 9: %
10: % n   - Number of frequencies required
11: % tol - Error tolerance for terminating
12: %       the iteration
13: % z   - Dimensionless frequencies are the
14: %       squares of the roots of f(z)=0
15: %
16: % User m functions called:  none
17: %-----------------------------------------------
18:
19: if nargin ==1, tol=1.e-5; end
20:
21: % Base initial estimates on the asymptotic
22: % form of the frequency equation
23: zbegin=((1:n)-.5)'*pi; zbegin(1)=1.875; big=10;
24:
25: % Start Newton iteration
26: while big > tol
27:   t=exp(-zbegin); tt=t.*t;
28:   f=cos(zbegin)+2*t./(1+tt);
29:   fp=-sin(zbegin)-2*t.*(1-tt)./(1+tt).^2;
30:   delz=-f./fp;
31:   z=zbegin+delz; big=max(abs(delz)); zbegin=z;
32: end
33: z=z.*z;
```

Function cbfrqfdm

```
 1: function [wfindif,mat]=cbfrqfdm(n)
 2: %
 3: % [wfindif,mat]=cbfrqfdm(n)
 4: % ~~~~~~~~~~~~~~~~~~~~~~~~~~
 5: % This function computes approximate cantilever
 6: % beam frequencies by the finite difference
 7: % method. The truncation error for the
 8: % differential equation and boundary
 9: % conditions are of order h^2.
10: %
11: % n       - Number of frequencies to be
12: %           computed
13: % wfindif - Approximate frequencies in
```

```
14: %              dimensionless form
15: % mat      - Matrix having eigenvalues which
16: %            are the square roots of the
17: %            frequencies
18: %
19: % User m functions called:  none
20: %-----------------------------------------------
21:
22: % Form the primary part of the frequency matrix
23: mat=3*diag(ones(n,1))-4*diag(ones(n-1,1),1)+...
24:     diag(ones(n-2,1),2); mat=(mat+mat');
25:
26: % Impose left end boundary conditions
27: % y(0)=0 and y'(0)=0
28: mat(1,[1:3])=[7,-4,1]; mat(2,[1:4])=[-4,6,-4,1];
29:
30: % Impose right end boundary conditions
31: % y''(1)=0 and y'''(1)=0
32: mat(n-1,[n-3:n])=[1,-4,5,-2];
33: mat(n,[n-2:n])=[2,-4,2];
34:
35: % Compute approximate frequencies and
36: % sort these values
37: w=eig(mat); w=sort(w); h=1/n;
38: wfindif=sqrt(w)/(h*h);
```

Function cbfrqfem

```
1: function [wfem,modvecs,mm,kk]= ...
2:                   cbfrqfem(nelts,mas,len,ei)
3: %
4: % [wfem,modvecs,mm,kk]=
5: %                   cbfrqfem(nelts,mas,len,ei)
6: % ~~~~~~~~~~~~~~~~~~~~~~~~~~~~~~~~~~~~~~~~~~~~~~~
7: % Determination of natural frequencies of a
8: % uniform depth cantilever beam by the Finite
9: % Element Method.
10: %
11: %   nelts   - number of elements in the beam
12: %   mas     - total beam mass
13: %   len     - total beam length
14: %   ei      - elastic modulus times moment
15: %              of inertia
```

```
16: %   wfem    - dimensionless circular frequencies
17: %   modvecs - modal vector matrix
18: %   mm,kk   - reduced mass and stiffness
19: %             matrices
20: %
21: % User m functions called: none
22: %-----------------------------------------------
23:
24: if nargin==1, mas=1; len=1; ei=1; end
25: n=nelts; le=len/n; me=mas/n;
26: c1=6/le^2; c2=3/le; c3=2*ei/le;
27:
28: % element mass matrix
29: masselt=me/420* ...
30:          [   156,    22*le,      54,   -13*le
31:            22*le,  4*le^2,   13*le,  -3*le^2
32:               54,   13*le,     156,   -22*le
33:           -13*le, -3*le^2,  -22*le,   4*le^2];
34:
35: % element stiffness matrix
36: stifelt=c3*[ c1,   c2,   -c1,   c2
37:              c2,    2,   -c2,    1
38:             -c1,  -c2,    c1,  -c2
39:              c2,    1,   -c2,    2];
40:
41: ndof=2*(n+1); jj=0:3;
42: mm=zeros(ndof);  kk=zeros(ndof);
43:
44: % Assemble equations
45: for i=1:n
46:   j=2*i-1+jj; mm(j,j)=mm(j,j)+masselt;
47:   kk(j,j)=kk(j,j)+stifelt;
48: end
49:
50: % Remove degrees of freedom for zero
51: % deflection and zero slope at the left end.
52: mm=mm(3:ndof,3:ndof); kk=kk(3:ndof,3:ndof);
53:
54: % Compute frequencies
55: if nargout ==1
56:   wfem=sqrt(sort(real(eig(mm\kk))));
57: else
58:   [modvecs,wfem]=eig(mm\kk);
59:   [wfem,id]=sort(diag(wfem));
60:   wfem=sqrt(wfem); modvecs=modvecs(:,id);
```

```
61: end
```

Function frud

```
 1: function [t,x]= ...
 2:          frud(m,k,f1,f2,w,x0,v0,wn,modvc,h,tmax)
 3: %
 4: % [t,x]=frud(m,k,f1,f2,w,x0,v0,wn,modvc,h,tmax)
 5: % ~~~~~~~~~~~~~~~~~~~~~~~~~~~~~~~~~~~~~~~~~~~~~~~~
 6: % This function employs modal superposition
 7: % to solve
 8: %
 9: %    m*x'' + k*x = f1*cos(w*t) + f2*sin(w*t)
10: %
11: % m,k    - mass and stiffness matrices
12: % f1,f2  - amplitude vectors for the forcing
13: %          function
14: % w      - forcing frequency not matching any
15: %          natural frequency component in wn
16: % wn     - vector of natural frequency values
17: % x0,v0  - initial displacement and velocity
18: %          vectors
19: % modvc  - matrix with modal vectors as its
20: %          columns
21: % h,tmax - time step and maximum time for
22: %          evaluation of the solution
23: % t      - column of times at which the
24: %          solution is computed
25: % x      - solution matrix in which row j
26: %          is the solution vector at
27: %          time t(j)
28: %
29: % User m functions called:  none
30: %-------------------------------------------------
31:
32: t=0:h:tmax; nt=length(t); nx=length(x0);
33: wn=wn(:); wnt=wn*t;
34:
35: % Evaluate the particular solution.
36: x12=(k-(w*w)*m)\[f1,f2];
37: x1=x12(:,1); x2=x12(:,2);
38: xp=x1*cos(w*t)+x2*sin(w*t);
39:
```

```
40: % Evaluate the homogeneous solution.
41: cof=modvc\[x0-x1,v0-w*x2];
42: c1=cof(:,1)'; c2=(cof(:,2)./wn)';
43: xh=(modvc.*c1(ones(1,nx),:))*cos(wnt)+...
44:    (modvc.*c2(ones(1,nx),:))*sin(wnt);
45:
46: % Combine the particular and
47: % homogeneous solutions.
48: t=t(:); x=(xp+xh)';
```

Function examplmo

```
 1: function x=examplmo(mm,kk,f1,f2,x0,v0,wfe,mv)
 2: %
 3: % x=examplmo(mm,kk,f1,f2,x0,v0,wfe,mv)
 4: % ~~~~~~~~~~~~~~~~~~~~~~~~~~~~~~~~~~~~~~~
 5: % Evaluate the response caused when a downward
 6: % load at the middle and an upward load at the
 7: % free end is applied.
 8: %
 9: % mm, kk - mass and stiffness matrices
10: % f1, f2 - forcing function magnitudes
11: % x0, v0 - initial position and velocity
12: % wfe    - forcing function frequency
13: % mv     - matrix of modal vectors
14: %
15: % User m functions called:  frud, animate, read
16: %-----------------------------------------------
17:
18: w=0; n=length(x0); t0=0;
19: s1=['\nEvaluate the time response from two',...
20:    '\nconcentrated loads. One downward at the',...
21:    '\nmiddle and one upward at the free end.'];
22: while 1
23:    fprintf(s1); fprintf('\n\n');
24:    fprintf('Input the time step and ');
25:    fprintf('the maximum time ');
26:    fprintf('\n(0.04 and 5.0) are typical.');
27:    fprintf(' Use 0,0 to stop\n');
28:    [h,tmax]=read; disp(' ');
29:    if norm([h,tmax])==0, return; end
30:
31:    [t,x]= ...
```

```
32:        frud(mm,kk,f1,f2,w,x0,v0,wfe,mv,h,tmax);
33:   x=x(:,1:2:n-1); x=[zeros(length(t),1),x];
34:   [nt,nc]=size(x); hdist=linspace(0,1,nc);
35:
36:   clf; plot(t,x(:,nc),'-');
37:   title('Position of the Free End of the Beam');
38:   xlabel('dimensionless time');
39:   ylabel('end deflection'); figure(gcf);
40:   disp('Press [Enter] for a surface plot of');
41:   disp('  transverse deflection versus x and t');
42:   pauz; %genprint('endpos1');
43:   xc=linspace(0,1,nc); zmax=1.2*max(abs(x(:)));
44:
45:   clf; surf(xc,t,x); view(30,35);
46:   axis([0,1,0,tmax,-zmax,zmax]);
47:   xlabel('x axis'); ylabel('time');
48:   zlabel('deflection');
49:   title(['Cantilever Beam Deflection ' ...
50:           'for Varying Position and Time']);
51:   figure(gcf); %genprint('endpos2');
52:   disp('Press [Enter] to animate');
53:   disp('  the beam motion'); pauz;
54:
55:   titl='Beam Animation';
56:   xlab='x axis'; ylab='displacement';
57:   animate(hdist,x,0.1,titl,xlab,ylab); close;
58: end
```

Function animate

```
1: function animate(x,u,tpause,titl,xlabl,ylabl)
2: %
3: % animate(x,u,tpause,titl,xlabl,ylabl)
4: % ~~~~~~~~~~~~~~~~~~~~~~~~~~~~~~~~~~~~~~
5: % This function draws an animated plot of data
6: % values stored in array u.  The different
7: % columns of u correspond to position values
8: % in vector x.  The successive rows of u
9: % correspond to different times. Parameter
10: % tpause controls the speed of animation.
11: %
12: % u       - matrix of values to animate plots
13: %             of u versus x
```

```
14: %   x        - spatial positions for different
15: %              columns of u
16: %   tpause   - clock seconds between output of
17: %              frames. The default is .1 secs
18: %              when tpause is left out. When
19: %              tpause=0, a new frame appears
20: %              when the user presses any key.
21: %   titl     - graph title
22: %   xlabl    - label for horizontal axis
23: %   ylabl    - label for vertical axis
24: %
25: % User m functions called:  genprint
26: %------------------------------------------------
27:
28: if nargin<6, ylabl=''; end;
29: if nargin<5, xlabl=''; end
30: if nargin<4, titl=''; end;
31: if nargin<3, tpause=.1; end;
32:
33: [ntime,nxpts]=size(u);
34: umin=min(u(:)); umax=max(u(:));
35: udif=umax-umin;
36: uavg=.5*(umin+umax);
37: xmin=min(x); xmax=max(x);
38: xdif=xmax-xmin; xavg=.5*(xmin+xmax);
39: xwmin=xavg-.55*xdif; xwmax=xavg+.55*xdif;
40: uwmin=uavg-.55*udif; uwmax=uavg+.55*udif; clf;
41: axis([xwmin,xwmax,uwmin,uwmax]); title(titl);
42: xlabel(xlabl); ylabel(ylabl); hold on;
43:
44: for j=1:ntime
45:   ut=u(j,:);
46:   plot(x,ut,'-'); axis('off'); figure(gcf);
47:   if tpause==0
48:     pause;
49:   else
50:     pause(tpause);
51:   end
52:   if j==ntime, break, else, cla; end
53: end
54: %genprint('cntltrac');
55: hold off; clf;
```

Function plotsave

```
 1: function plotsave(wex,wfd,pefd,wfe,pefem)
 2: %
 3: % function plotsave(wex,wfd,pefd,wfe,pefem)
 4: % ~~~~~~~~~~~~~~~~~~~~~~~~~~~~~~~~~~~~~~~~~~~
 5: % This function plots errors in frequencies
 6: % computed by two approximate methods.
 7: %
 8: % wex        - exact frequencies
 9: % wfd        - finite difference frequencies
10: % wfe        - finite element frequencies
11: % pefd,pefem - percent errors by both methods
12: %
13: % User m functions called:  genprint
14: %-------------------------------------------------
15:
16: % plot results comparing accuracy
17: % of both frequency methods
18: w=[wex(:);wfd(:);wfd];
19: wmin=min(w); wmax=max(w);
20: n=length(wex);wht=wmin+.001*(wmax-wmin);
21: j=1:n;
22:
23: semilogy(j,wex,'-r',j,wfe,'--g',j,wfd,'-.m')
24: title('Cantilever Beam Frequencies');
25: xlabel('frequency number');
26: ylabel('frequency values');
27: legend('Exact freq.','Felt. freq.', ...
28:        'Fdif. freq.'); figure(gcf);
29: disp('Press [Enter] for a frequency');
30: disp('  error plot'); pauz;
31: %genprint('beamfrq1');
32:
33: plot(j,abs(pefd),'-r',j,abs(pefem),'--g');
34: title(['Cantilever Beam Frequency ' ...
35:        'Error Percentages']);
36: xlabel('frequency number');
37: ylabel('percent frequency error');
38: legend('Fdif. pct. error','Felt. pct. error');
39: figure(gcf);
40: disp('Press [Enter] for a transient')
41: disp('  response calculation'); pauz;
42: %genprint('beamfrq2');
```

Chapter 10

Stress Analysis and Eigenvalue Analysis

10.1 Introduction

Eigenvalue problems occur often in mechanics applications. One of the most important concerns determining natural frequencies and mode shapes associated with the matrix differential equation

$$M\ddot{X} + KX = 0$$

which is studied in Chapters 7 and 9. The present chapter discusses some other important applications involving principal axes of stress and inertia tensors, buckling of variable depth beams, and modal response of pin-connected trusses. Stress and inertia tensors lead to symmetric matrices of order three. The column buckling problem involves a homogeneous differential equation and homogeneous boundary conditions which are dealt with by finite differences. The truss vibration problem is solved for two-dimensional structures of general shape. Analysis of all these problems is facilitated by the important MATLAB function **eig**.

10.2 Stress Transformation and Principal Coordinates

The state of stress at a point in a three-dimensional continuum is described in terms of a symmetric three by three matrix $t = [t(\imath, \jmath)]$ where $t(\imath, \jmath)$ denotes the stress component in the direction of the $x_\imath$ axis on the plane with it normal in the direction of the $x_\jmath$ axis [9]. Suppose we introduce a rotation of axes defined by matrix b such that row $b(\imath, :)$ represents the components of a unit vector along the new $\tilde{x}_\imath$ axis measured relative to the initial reference state. It can be shown that the stress matrix $\tilde{t}$ corresponding to the new axis system can be computed by the transformation

$$\tilde{t} = btb^T$$

Sometimes it is desirable to locate a set of reference axes such that $\tilde{t}$ is diagonal, in which case the diagonal components of $\tilde{t}$ represent extremal values of normal stress. This means that seeking maximum or minimum normal stress on a plane leads to the same condition as requiring zero shear stress on the plane. The eigenfunction operation

```
[eigvecs,eigvals]=eig(t);
```

applied to a symmetric matrix t produces an orthonormal set of eigenvectors stored in the columns of **eigvecs**, and a diagonal matrix **eigvals** having the eigenvalues on the diagonal. These matrices satisfy

$$\text{eigvecs}^T \text{ t eigvecs} = \text{eigvals}$$

Consequently, the rotation matrix b needed to transform to principal axes is simply the transpose of the matrix of orthonormalized eigenvectors. In other words, the eigenvectors of the stress tensor give the unit normals to the planes on which the normal stresses are extreme and the shear stresses are zero. The function **prnstres** performs the principal axis transformation.

10.2.1 Program Output and Code

Function prnstres

```
 1: function [pstres,pvecs]=prnstres(stress)
 2: %
 3: % [pstres,pvecs]=prnstres(stress)
 4: % ~~~~~~~~~~~~~~~~~~~~~~~~~~~~~~~~~
 5: %
 6: % This function computes principal stresses
 7: % and principal stress directions for a three-
 8: % dimensional stress state.
 9: %
10: % stress - a vector defining the stress
11: %           components in the order
12: %           [sxx,syy,szz,sxy,sxz,syz]
13: %
14: % pstres - the principal stresses arranged in
15: %           ascending order
16: % pvecs  - the transformation matrix defining
17: %           the orientation of the principal
18: %           axis system.  The rows of this
19: %           matrix define the surface normals to
20: %           the planes on which the extremal
21: %           normal stresses act
22: %
23: % User m functions called:  none
24: %------------------------------------------------
25:
26: s=stress(:)';
27: s=([s([1 4 5]); s([4 2 6]); s([5 6 3])]);
28: [pvecs,pstres]=eig(s);
29: [pstres,k]=sort(diag(pstres));
30: pvecs=pvecs(:,k)';
31: if det(pvecs)<0, pvecs(3,:)=-pvecs(3,:); end
```

10.3   Principal Axes of the Inertia Tensor

Consider the problem of computing the kinetic energy of a rigid body which rotates with angular velocity $\omega = [\omega_x; \omega_y; \omega_z]$ about the reference origin [47]. The kinetic energy of the body can be obtained by

$$K = \frac{1}{2}\omega^T J\omega$$

with the inertia tensor J computed as

$$J = \int \int \int \rho \left[Ir^T r - rr^T \right] d(\text{vol})$$

where ρ is the mass per unit volume, I is the identity matrix, and r is the cartesian radius vector. The inertia tensor is characterized by a symmetric matrix expressed in component form as

$$J = \int \int \int \begin{bmatrix} y^2 + z^2 & -xy & -xz \\ -xy & x^2 + z^2 & -yz \\ -xz & -yz & x^2 + y^2 \end{bmatrix} d(\text{mass})$$

Under the rotation transformation

$$\tilde{r} = br \quad \text{with} \quad b^T b = I$$

we can see that the inertia tensor transforms as

$$\tilde{J} = bJb^T$$

which is identical to the transformation law for the stress component matrix discussed earlier. Consequently, the inertia tensor will also possess principal axes which make the off-diagonal components zero. The kinetic energy is expressed more simply as

$$K = \frac{1}{2} \left(\omega_1^2 J_{11} + \omega_2^2 J_{22} + \omega_3^2 J_{33} \right)$$

where the components of ω and J must be referred to the principal axes. The function **prnstres** can also be used to locate principal axes of the inertia tensor since the same transformations apply. As an example of principal axis computation consider the inertia tensor for a cube of side length A and mass M which has a corner at $(0,0,0)$ and edges along the coordinate axes. The inertia tensor is found to be

$$J = MA^2 \begin{bmatrix} 2/3 & -1/4 & -1/4 \\ -1/4 & 2/3 & -1/4 \\ -1/4 & -1/4 & 2/3 \end{bmatrix}$$

The computation

```
[pvl,pvc]=prnstres([2/3,2/3,2/3,-1/4,-1/4,-1/4]);
```

produces the results

$$\mathbf{pvl} = \begin{bmatrix} 0.1667 \\ 0.9167 \\ 0.9167 \end{bmatrix} \qquad \mathbf{pvc} = \begin{bmatrix} 0.5574 & 0.5574 & 0.5574 \\ 0.4342 & -0.8159 & 0.3817 \\ 0.6915 & 0.0303 & -0.7218 \end{bmatrix}$$

This shows that the smallest possible inertial component equals $1/6 \approx 0.1667$ about the diagonal line through the origin while the maximal inertial moments of $11/12 \approx 0.9167$ occur about the axes normal to the diagonal.

10.4 Vibration of Truss Structures

Trusses are a familiar type of structure used in diverse applications such as bridges, roof supports, and power transmission towers. These structures can be envisioned as a series of nodal points among which various axially loaded members are connected. These members act like linearly elastic springs supporting tension or compression. Typically, displacement constraints apply at one or more points to prevent movement of the truss from its supports. The present article computes the natural frequencies and mode shapes of two-dimensional trusses when the member properties are known and the loads of interest arise from inertial forces occurring during vibration. A similar loading analysis pertaining to statically loaded trusses has been published recently [100].

Consider an axially loaded member of constant cross section connected between nodes i and j which have displacement components (u_i, v_i) and (u_j, v_j) as indicated in Figure 10.1. The member length is given by

$$\ell = \sqrt{(x_j - x_i)^2 + (y_j - y_i)^2}$$

and the member inclination is quantified by the trigonometric functions

$$c = \cos\theta = \frac{x_j - x_i}{\ell} \quad \text{and} \quad s = \sin\theta = \frac{y_j - y_i}{\ell}$$

The axial extension for small deflections is

$$\Delta = (u_j - u_i)c + (v_j - v_i)s$$

The axial force needed to extend a member having length ℓ, elastic modulus E and cross section area A is given by

$$P_{ij} = \frac{AE}{\ell}\Delta = \frac{AE}{\ell}[-c, \; -s, \; c, \; s] \, u_{ij}$$

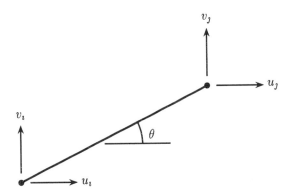

Figure 10.1. Typical Truss Element

where
$$u_{ij} = [u_i;\ v_i;\ u_j;\ v_j]$$

is a column matrix describing the nodal displacements of the member ends. The corresponding end forces are represented by

$$F_{ij} = [F_{ix};\ F_{iy};\ F_{jx};\ F_{jy}] = P_{ij}\,[-c,\ -s,\ c,\ s]$$

so the end forces and end displacements are related by the matrix equation
$$F_{ij} = K_{ij}U_{ij}$$

where the element stiffness matrix is

$$K_{ij} = \frac{AE}{\ell}\,[-c;\ -s;\ c;\ s]\,[-c,\ -s,\ c,\ s]$$

In regard to mass effects in a member, we will assume that any transverse motion is negligible and half of the mass of each member can be lumped at each end. Hence the mass placed at each end would be $A\rho\ell/2$ where ρ is the mass per unit volume.

The deflection of a truss with n nodal points can be represented using a generalized displacement vector

$$U = [u_1;\ v_1;\ u_2;\ v_2;\ \ldots;\ u_n;\ v_n]$$

and a generalized nodal force vector

$$F = [F_{1x};\ F_{1y};\ F_{2x};\ F_{2y};\ \ldots;\ F_{nx};\ F_{ny}]$$

When the contributions of all members in the network are assembled together, a global matrix relation results in the form

$$F = KU$$

where K is called the global stiffness matrix. Before we formulate procedures for assembling the global stiffness matrix, dynamical aspects of the problem will be discussed.

In the current application, the applied nodal forces are attributable to the acceleration of masses located at the nodes and to support reactions at points where displacement constraints occur. The mass concentrated at each node will equal half the sum of the masses of all members connected to the node. According to the principle of D'Alembert [47] a particle having mass m and acceleration $\ddot{u}$ is statically equivalent to a force $-m\ddot{u}$ so, the equation of motion for the truss, without accounting for support reactions, is

$$KU = -M\ddot{U}$$

where M is a global mass matrix given by

$$M = \mathbf{diag}\left([m_1; \, m_1; \, m_2; \, m_2; \, \ldots; \, m_n; \, m_n]\right)$$

with m_i denoting the mass concentrated at the i'th node. The equation of motion

$$M\ddot{U} + KU = 0$$

will also be subjected to constraint equations arising when some points are fixed or have roller supports. This type of support implies a matrix equation of the form

$$CU = 0$$

Natural frequency analysis investigates states-of-motion where each node of the structure simultaneously moves with simple harmonic motion of the same frequency. This means solutions are sought of the form

$$U = X \cos(\omega t)$$

where ω denotes a natural frequency and X is a modal vector describing the deflection pattern for the corresponding frequency. The assumed mode of motion implies $\ddot{U} = -\lambda U$ where $\lambda = \omega^2$, so we are led to an eigenvalue problem of the form

$$KX = \lambda M X$$

with a side constraint

$$CX = 0$$

needed to satisfy support conditions.

MATLAB provides the excellent intrinsic functions **eigen** and **null** which deal with this problem effectively. Using function **null** we can write

$$X = QY$$

where Q has columns which are an orthonormal basis for the null space of matrix C. Expressing the eigenvalue equation in terms of Y and multiplying both sides by Q^T gives

$$K_o Y = \lambda M_o Y$$

where

$$K_o = Q^T K Q \qquad M_o = Q^T M Q$$

It can be shown from physical considerations that, in general, K and M are symmetric matrices such that K has real non-negative eigenvalues and M has real positive eigenvalues. This implies that M_o can be factored as

$$M_o = N^T N$$

where N is an upper triangular matrix. Then the eigenvalue problem can be rewritten as

$$K_1 Z = \lambda Z \qquad Y = NZ \qquad K_1 = \left(N^T\right)^{-1} K_o N^{-1}$$

Because matrix K_1 will be real and symmetric, intrinsic function **eig** generates orthonormal eigenvectors. Function **eigsym** used by program **trusvibs** produces a set of eigenvectors in the columns of X which satisfy generalized orthogonality conditions of the form

$$X^T M X = I \quad \text{and} \quad X^T K X = \Lambda$$

where Λ is a diagonal matrix containing the squares of the natural frequencies arranged in ascending order. The authors believe the calculations performed in function **eigsym** illustrate especially well the excellent matrix manipulative features that MATLAB embodies.

Before we discuss a physical example, the problem of assembling the global stiffness matrix will be addressed. It is helpful to think of all nodal displacements as if they were known and then compute the nodal forces by adding the stiffness contributions of all elements. Although the total force at each node results only from the forces in members touching the node, it is better to accumulate force contributions on an element-by-element basis instead of working node by node. For example, a member connecting node i and node j will involve displacement components at row positions $2i - 1, 2i, 2j - 1$ and $2j$ in the global displacement vector and force components at similar positions in the generalized force matrix. Because principles of superposition apply, the stiffness contributions of individual members can be added, one member at a time, into the global stiffness matrix. This process is implemented in function **assemble** which also forms the mass matrix. First, selected points constrained to have zero displacement components are specified. Next the global stiffness and mass matrices are formed. This is followed by an eigenvalue analysis which yields the natural frequencies and the modal vectors. Finally the motion associated with each vibration mode is described by superimposing on the coordinates of each nodal point a multiple of the corresponding modal vector varying sinusoidally with time. Redrawing the structure produces an appearance of animated motion.

The complete program has several functions which should be studied individually for complete understanding of the methods developed. These functions and their purposes are summarized in the table below.

trusvibs	reads data and guides interactive input to animate the various vibration modes
crossdat	function typifying the nodal and element data to define a problem
assemble	assembles the global stiffness and mass data matrices
elmstf	forms the stiffness matrix and calculates the volume of an individual member
eigc	forms the constraint equations implied when selected displacement components are set to zero
eigsym	solves the constrained eigenvalue problem pertaining to the global stiffness and mass matrices
trifacsm	factors a positive definite matrix into upper and lower global triangular parts
drawtrus	draws the truss in deflected positions
cubrange	a utility routine to determine a window for drawing the truss without scale distortion

The data in function **crossdat** contains the information for node points, element data, and constraint conditions needed to define a problem. Once the data values are read, mode shapes and frequencies are computed and the user is allowed to observe the animation of modes ordered from the lowest to the highest frequency. The number of modes produced equals twice the number of nodal points minus the number of constraint conditions. The plot in Figure 10.2 shows mode eleven for the sample problem. This mode has no special significance aside from the interesting deflection pattern produced. The reader may find it instructive to run the program and select several modes by using input such as 3:5 or a single mode by specifying a single mode number.

Figure 10.2. Truss Vibration Mode Number 11

10.4.1   Program Output and Code

Script File trusvibs

```
1:  % Example: trusvibs
2:  % ~~~~~~~~~~~~~~~~~~
3:  %
4:  % This program analyzes natural vibration modes
5:  % for a general plane pin-connected truss. The
6:  % direct stiffness method is employed in
7:  % conjunction with eigenvalue calculation to
8:  % evaluate the natural frequencies and mode
9:  % shapes. The truss is defined in terms of a
10: % set of nodal coordinates and truss members
11: % connected to different nodal points. Global
12: % stiffness and mass matrices are formed. Then
13: % the frequencies and mode shapes are computed
14: % with provision for imposing zero deflection
15: % at selected nodes. The user is then allowed
16: % to observe animated motion of the various
17: % vibration modes.
18: %
19: % User m functions called:
20: %         eigsym, crossdat, drawtrus, eigc,
21: %         assemble, elmstf, cubrange,
22: %         trifacsm, genprint
23: %-------------------------------------------------
24:
25: clear; disp(' '); kf=1; idux=[]; iduy=[];
26: disp(['Modal Vibrations for a Pin ', ...
27:       'Connected Truss']); disp(' ');
28:
29: % A sample data file defining a problem is
30: % given in crossdat.m
31: disp(['Give the name of an m-file ', ...
32:       'containing your data']);
33: disp(['Do not include .m in the name ', ...
34:       '(try file crossdat)']);
35: filename=input('>? ','s');
36: eval(filename); disp(' ');
37:
38: % Assemble the global stiffness and
39: % mass matrices
40: [stiff,masmat]= ...
```

```
41:    assemble(x,y,inode,jnode,area,elast,rho);
42:
43: % Compute natural frequencies and modal vectors
44: % accounting for the fixed nodes
45: ifixed=[2*idux(:)-1; 2*iduy(:)];
46: [modvcs,eigval]=eigc(stiff,masmat,ifixed);
47: natfreqs=sqrt(eigval);
48:
49: % Set parameters used in modal animation
50: nsteps=31; s=sin(linspace(0,6.5*pi,nsteps));
51: x=x(:); y=y(:); np=2*length(x);
52: bigxy=max(abs([x;y])); scafac=.04*bigxy;
53: highmod=size(modvcs,2); hm=num2str(highmod);
54:
55: % Show animated plots of the vibration modes
56: while 1
57:    disp('Give the mode numbers to be animated?');
58:    disp(['Do not exceed a total of ',hm, ...
59:          ' modes.']); disp('Input 0 to stop');
60:    if kf==1, disp(['Try 1:',hm]); kf=kf+1; end
61:    str=input('>? ','s');
62:    nmode=eval(['[',str,']']);
63:    if sum(nmode)==0; break; end
64:    % Animate the various vibration modes
65:    hold off; clf; ovrsiz=1.1;
66:    w=cubrange([x(:),y(:)],ovrsiz);
67:    axis(w); axis('square'); axis('off'); hold on;
68:    for kk=1:length(nmode)  % Loop over each mode
69:      kkn=nmode(kk);
70:      titl=['Truss Vibration Mode Number ', ...
71:            num2str(kkn)];
72:      dd=modvcs(:,kkn); mdd=max(abs(dd));
73:      dx=dd(1:2:np); dy=dd(2:2:np);
74:      clf; pause(1);
75:      % Loop through several cycles of motion
76:      for jj=1:nsteps
77:        sf=scafac*s(jj)/mdd;
78:        xd=x+sf*dx; yd=y+sf*dy; clf;
79:        axis(w); axis('square'); axis('off');
80:        drawtrus(xd,yd,inode,jnode); title(titl);
81:        drawnow; figure(gcf);
82:      end
83:    end
84: end
85: disp(' ');
```

Function crossdat

```
 1: % Data set: crossdat
 2: % ~~~~~~~~~~~~~~~~~~
 3: %
 4: % Data specifying a cross shaped truss.
 5: %
 6: %-----------------------------------------------
 7:
 8: % Nodal point data are defined by:
 9: %   x - a vector of x coordinates
10: %   y - a vector of y coordinates
11: x=10*[.5 2.5 1 2 0 1 2 3 0 1 2 3 1 2];
12: y=10*[ 0   0 1 1 2 2 2 2 3 3 3 3 4 4];
13:
14: % Element data are defined by:
15: %    inode - index vector defining the I-nodes
16: %    jnode - index vector defining the J-nodes
17: %    elast - vector of elastic modulus values
18: %    area  - vector of cross section area values
19: %    rho   - vector of mass per unit volume
20: %            values
21: inode=[1 1 2 2 3 3 4 3 4 5 6 7 5 6 6 6 7 7 7 ...
22:        8 9 10 11 10 11 10 11 13];
23: jnode=[3 4 3 4 4 6 6 7 7 6 7 8 9 9 10 11 10 ...
24:        11 12 12 10 11 12 13 13 14 14 14];
25: elast=3e7*ones(1,28);
26: area=ones(1,28); rho=ones(1,28);
27:
28: % Any points constrained against displacement
29: % are defined by:
30: %    idux - indices of nodes having zero
31: %           x-displacement
32: %    iduy - indices of nodes having zero
33: %           y-displacement
34: idux=[1 2]; iduy=[1 2];
```

Function assemble

```
 1: function [stif,masmat]= ...
 2:   assemble(x,y,id,jd,a,e,rho)
 3: %
```

```
 4: % [stif,masmat]=assemble(x,y,id,jd,a,e,rho)
 5: % ~~~~~~~~~~~~~~~~~~~~~~~~~~~~~~~~~~~~~~~~~~~~~
 6: %
 7: % This function assembles the global
 8: % stiffness matrix and mass matrix for a
 9: % plane truss structure. The mass density of
10: % each element equals unity.
11: %
12: % x,y    - nodal coordinate vectors
13: % id,jd - nodal indices of members
14: % a,e    - areas and elastic moduli of members
15: % rho    - mass per unit volume of members
16: %
17: % stif   - global stiffness matrix
18: % masmat- global mass matrix
19: %
20: % User m functions called: elmstf
21: %-------------------------------------------------
22:
23: numnod=length(x); numelm=length(a);
24: id=id(:); jd=jd(:);
25: stif=zeros(2*numnod); masmat=stif;
26: ij=[2*id-1,2*id,2*jd-1,2*jd];
27: for k=1:numelm, kk=ij(k,:);
28:    [stfk,volmk]= ...
29:      elmstf(x,y,a(k),e(k),id(k),jd(k));
30:    stif(kk,kk)=stif(kk,kk)+stfk;
31:    masmat(kk,kk)=masmat(kk,kk)+ ...
32:                   rho(k)*volmk/2*eye(4,4);
33: end
```

Function elmstf

```
 1: function [k,vol]=elmstf(x,y,a,e,i,j)
 2: %
 3: % [k,vol]=elmstf(x,y,a,e,i,j)
 4: % ~~~~~~~~~~~~~~~~~~~~~~~~~~~~~
 5: %
 6: % This function forms the stiffness matrix for
 7: % a truss element. The member volume is also
 8: % obtained.
 9: %
10: % User m functions called: none
```

```
11: %-------------------------------------------------
12:
13: xx=x(j)-x(i); yy=y(j)-y(i);
14: L=norm([xx,yy]); vol=a*L;
15: c=xx/L; s=yy/L; k=a*e/L*[-c;-s;c;s]*[-c,-s,c,s];
```

Function eigc

```
1: function [vecs,eigvals]=eigc(k,m,idzero)
2: %
3: % [vecs,eigvals]=eigc(k,m,idzero)
4: % ~~~~~~~~~~~~~~~~~~~~~~~~~~~~~~~~
5: % This function computes eigenvalues and
6: % eigenvectors for the problem
7: %              k*x=eigval*m*x
8: % with some components of x constrained to
9: % equal zero. The imposed constraint is
10: %              x(idzero(j))=0
11: % for each component identified by the index
12: % matrix idzero.
13: %
14: % k        - a real symmetric stiffness matrix
15: % m        - a positive definite symmetric mass
16: %              matrix
17: % idzero   - the vector of indices identifying
18: %              components to be made zero
19: %
20: % vecs     - eigenvectors for the constrained
21: %              problem. If matrix k has dimension
22: %              n by n and the length of idzero is
23: %              m (with m<n), then vecs will be a
24: %              set on n-m vectors in n space
25: % eigvals  - eigenvalues for the constrained
26: %              problem. These are all real.
27: %
28: % User m functions called:  eigsym
29: %-------------------------------------------------
30:
31: n=size(k,1); j=1:n; j(idzero)=[];
32: c=eye(n,n); c(j,:)=[];
```

33: `[vecs,eigvals]=eigsym((k+k')/2, (m+m')/2, c);`

Function eigsym

```
 1: function [evecs,eigvals]=eigsym(k,m,c)
 2: %
 3: % [evecs,eigvals]=eigsym(k,m,c)
 4: % ~~~~~~~~~~~~~~~~~~~~~~~~~~~~~~~~~~~
 5: % This function solves the eigenvalue of the
 6: % constrained eigenvalue problem
 7: %    k*x=(lambda)*m*x, with c*x=0.
 8: % Matrix k must be real symmetric and matrix
 9: % m must be symmetric and positive definite,
10: % otherwise computed results will be wrong.
11: %
12: % k       - a real symmetric matrix
13: % m       - a real symmetric positive
14: %           definite matrix
15: % c       - a matrix defining the constraint
16: %           condition c*x=0. This matrix is
17: %           omitted if no constraint exists.
18: %
19: % evecs   - matrix of eigenvectors orthogonal
20: %           with respect k and m. The
21: %           following relations apply:
22: %           evecs'*m*evecs=identity_matrix
23: %           evecs'*k*evecs=diag(eigvals).
24: % eigvals - a vector of the eigenvalues
25: %           sorted in increasing order
26: %
27: % User m functions called: trifacsm
28: %-------------------------------------------------
29:
30: if nargin==3
31:   q=null(c); m=q'*m*q; k=q'*k*q;
32: end
33: u=trifacsm(m); k=u'\k/u; k=(k+k')/2;
34: [evecs,eigvals]=eig(k);
35: [eigvals,j]=sort(diag(eigvals));
36: evecs=evecs(:,j); evecs=u\evecs;
```

37: if nargin==3, evecs=q*evecs; end

Function trifacsm

```
1: function u=trifacsm(a)
2: %
3: % u=trifacsm(a)
4: % ~~~~~~~~~~~~~
5: %
6: % This function determines an upper triangular
7: % matrix u such that u'*u=a where a must be
8: % symmetric and positive definite.
9: %
10: % User m functions called: none
11: %-----------------------------------------------
12:
13: [L,u]=lu(a); d=1./sqrt(diag(u));
14: u=d(:,ones(length(d),1)).*u;
```

Function drawtrus

```
1: function drawtrus(x,y,i,j)
2: %
3: % drawtrus(x,y,i,j)
4: % ~~~~~~~~~~~~~~~~~
5: %
6: % This function draws a truss defined by nodal
7: % coordinates defined in x,y and member indices
8: % defined in i,j.
9: %
10: % User m functions called: none
11: %-----------------------------------------------
12:
13: hold on;
14: for k=1:length(i)
15:   plot([x(i(k)),x(j(k))],[y(i(k)),y(j(k))]);
16: end
```

10.5   Buckling of Axially Loaded Columns

Computing the buckling load and deflection curve for a slender axially loaded column leads to an interesting type of eigenvalue problem. Let us analyze a column of length L subjected to a critical value of axial load P just large enough to hold the column in a deflected configuration. Reducing the load below the critical value will allow the column to straighten out, whereas increasing the load above the buckling value will result in a structural failure. To prevent sudden collapse of structures using axially loaded members, designers must be able to calculate buckling loads corresponding to various end constraints. We will present an analysis allowing the flexural rigidity EI to vary along the length. Four common types of end conditions of interest are shown in Figure 10.3. For each of these systems we will assume that the coordinate origin is at the left end of the column[1] with $y(0) = 0$. Cases I and II involve statically determinate columns. Cases III and IV are different because unknown end reactions occur in the boundary conditions.

All four problems lead to a homogeneous linear differential equation subjected to homogeneous boundary conditions. All of these cases possess a trivial solution where $y(x)$ vanishes identically. However, the solutions of practical interest involve a nonzero deflection configuration which is only possible when P equals the buckling load. Finite difference methods can be used to accurately approximate the differential equation and boundary conditions. In this manner we obtain a linear algebraic eigenvalue problem subjected to side constraints characterized by an underdetermined system of linear simultaneous equations. This type of problem is readily solved using the intrinsic functions in MATLAB.

Consider a beam element relating the bending moment m, the transverse shear v, the axial load P, and the transverse deflection y as shown in Figure 10.4. Equilibrium considerations imply

$$dv = 0 \qquad dm + Py'\, dx = v\, dx$$

$$v'(x) = 0 \qquad m'(x) + Py'(x) = v$$

Since no transverse external loading acts on the column between the end supports, the shear v will be constant. Differentiating the moment equation gives

$$m''(x) + Py''(x) = 0$$

Furthermore, flexural deformation theory of slender elastic beams implies

$$EIy''(x) = m(x)$$

---

[1] Although columns are usually positioned vertically, we show them as horizontal for convenience.

$$y = 0 \qquad\qquad\qquad y = 0$$
$$m = 0 \qquad\qquad\qquad m = 0$$

I) Pinned-Pinned

$$y = 0 \qquad\qquad\qquad y = y(\ell)$$
$$m = 0 \qquad\qquad\qquad y' = 0$$

II) Free-Fixed

$$y = 0 \qquad\qquad\qquad y = 0$$
$$m = 0 \qquad\qquad\qquad y' = 0$$

III) Pinned-Fixed

$$y = 0 \qquad\qquad\qquad y = 0$$
$$y' = 0 \qquad\qquad\qquad y' = 0$$

IV) Fixed-Fixed

Figure 10.3. Buckling Configurations

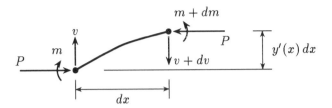

Figure 10.4. Beam Element Subjected to Axial Load

which leads to the following homogeneous differential equation governing the bending moment

$$EIm''(x) + Pm(x) = 0$$

We need to find values of P allowing nontrivial solutions of this differential equation subject to the required homogeneous boundary conditions.

The four types of end conditions shown in Figure 10.3 impose both deflection and moment conditions at the ends. Case I and II can be formulated completely in terms of displacements because moment conditions evidently imply

$$EIy''(x) = m = -Py$$

To handle cases III and IV, we need to relate the displacement and slope conditions at the ends to the bending moment. Let us denote the function $1/(EI)$ as $k(x)$ so that

$$y''(x) = k(x)m(x)$$

Integration gives

$$y'(x) = y'(0) + \int_0^x k(\xi)m(\xi)\,d\xi$$

and

$$y(x) = y(0) + y'(0)x + \int_0^x \int_0^x k(\xi)m(\xi)\,d\xi\,d\xi$$

The last integral can be transformed by integrating by parts to yield

$$y(x) = y(0) + y'(0)x + \int_0^x (x - \xi)k(\xi)m(\xi)\,d\xi$$

The boundary conditions for the pinned-fixed case require that

     a) $y(0) = 0$ b) $y'(L) = 0$ c) $y(L) = 0$

Case		Differential Equation	Boundary Conditions
I:	pinned-pinned	$EIy''(x) + Py(x) = 0$	$y(0) = 0$
			$y(L) = 0$
II:	free-fixed	$EIy''(x) + Py(x) = 0$	$y(0) = 0$
			$y'(L) = 0$
III:	pinned-fixed	$EIm''(x) + Pm(x) = 0$	$m(0) = 0$
			$\int_0^L k(x)m(x)\,dx = 0$
IV:	fixed-fixed	$EIm''(x) + Pm(x) = 0$	$\int_0^L k(x)m(x)\,dx = 0$
			$\int_0^L xk(x)m(x)\,dx = 0$

Table 10.1. Buckling Problem Summary

Condition b requires

$$y'(0) = -\int_0^L k(\xi)m(\xi)\,d\xi$$

whereas a and c combined lead to

$$y(L) = y(0) - L\int_0^L km\,d\xi + \int_0^L (L-\xi)km\,d\xi$$

Consequently, the conditions characterizing Case III are

$$EIm''(x) + Pm(x) = 0 \qquad 0 < x < L$$

$$m(0) = 0 \qquad \int_0^L xk(x)m(x)\,dx = 0$$

The boundary conditions for Case IV are handled similarly. Since we must have $y'(0) = y'(L) = 0$ and $y(0) = y(L) = 0$, this problem is formulated as

$$EIm''(x) + Pm(x) = 0$$

$$\int_0^L k(x)m(x)\,dx = 0 \qquad \int_0^L xk(x)m(x)\,dx = 0$$

The results for each case can now be summarized. Each problem requires a nontrivial solution of a homogeneous differential equation satisfying homogeneous boundary conditions as indicated in Table 10.1.

Each of these boundary value problems can be transformed to linear algebraic form by choosing a set of evenly spaced grid points across the span and approximating $y''(x)$ by finite differences. It follows from Taylor's series that

$$y''(x) = \frac{y(x-h) - 2y(x) + y(x+h)}{h^2} + O(h^2)$$

For sufficiently small h we can neglect the truncation error and write

$$y''(x_j) = \frac{y(x_{j-1}) - 2y(x_j) + y(x_{j+1})}{h^2}$$

$$x_j = jh \qquad h = \frac{L}{n+1} \qquad 1 \le j \le n$$

Employing the notation $y_j = y(x_j)$, $m_j = m(x_j)$ we have

$$\frac{(EI)_j[y_{j-1} - 2y_j + y_{j+1}]}{h^2} + Py_j = 0$$

for Cases I or II and

$$\frac{(EI)_j[m_{j-1} - 2m_j + m_{j+1}]}{h^2} + Pm_j = 0$$

for Cases III or IV. At the left end, either y or m is zero in all cases. Case I also has $y(L) = y_{n+1} = 0$. Case II requires $y'(L) = 0$. This is approximated in finite difference form as

$$y_{n+1} = \frac{4y_n - y_{n-1}}{3}$$

which implies for Case II that

$$y_n'' = \frac{2(y_{n-1} - y_n)}{3h^2}$$

Cases III and IV are slightly more involved than I and II. The condition that

$$\int_0^L \frac{mx}{EI}\, dx = 0$$

can be formulated using the trapezoidal rule to give

$$b_1 * [m_1, \ldots, m_n, m_{n+1}]^T = 0$$

where the asterisk indicates matrix multiplication involving a row matrix b_1 defined by

$$b_1 = [1, 1, \ldots, 1, 1/2] \;.*\; [x_1, x_2, \ldots, x_n, L] \;./\; [EI_1, \ldots, EI_n, EI_{n+1}]$$

Similarly, the condition

$$\int_0^L \frac{m}{EI}\, dx = 0$$

leads to

$$b_2 * [m_1, \ldots, m_n]^T + \frac{1}{2} \left[\frac{m_0}{EI_0} + \frac{m_{n+1}}{EI_{n+1}} \right] = 0$$

with

$$b_2 = \left[\frac{1}{EI_1}, \ldots, \frac{1}{EI_n} \right]$$

The first of these equations involving b_1 allows m_{n+1} to be eliminated in Case III, whereas the two equations involving b_1 and b_2 allow elimination of m_0 and m_{n+1} (the moments at $x = 0$ and $x = L$) for Case IV. Hence, in all cases, we are led to an eigenvalue problem typified as

$$EI_j(-m_{j-1} + 2m_j - m_{j+1}) = \lambda m_j$$

with $\lambda = h^2 P$ and we understand that the equations for $j = 1$ and $j = n$ may require modification to account for pertinent boundary conditions. We are led to solve

$$Am = \lambda m$$

where the desired buckling loads are associated with the smallest positive eigenvalue of matrix A. Case I and II lead directly to the deflection curve forms. However, III and IV require that the deflection curve be computed from the trapezoidal rule as

$$y'(x) = y'(0) + \int_0^x \frac{m}{EI} \, dx$$

and

$$y(x) = y(0) + y'(0) + x \int_0^x \frac{m}{EI} \, dx - \int_0^x \frac{mx}{EI} \, dx$$

The deflection curves can be normalized to make $y_{\max}$ equal unity. This completes the formulation needed in the buckling analysis for all four cases studied. These solutions have been implemented in the program described later in this section. An example, which is solvable exactly, will be discussed next to demonstrate that the finite difference formulation actually produces good results.

10.5.1   Example for a Linearly Tapered Circular Cross Section

Consider a column with circular cross section tapered linearly from diameter h_1 at $x = 0$ to diameter h_2 at $x = L$. The moment of inertia is given by

$$I = \frac{\pi d^4}{64}$$

which leads to

$$EI = E_o I_o \left(1 + \frac{sx}{L} \right)^4$$

where

$$s = \frac{h_2 - h_1}{h_1} \qquad I_o = \frac{\pi h_1^4}{64}$$

and E_o is the elastic modulus assumed to have a constant value. The differential equation governing the moment in all cases (and for y in cases I or II) is

$$\left(1 + \frac{sx}{L}\right)^4 m''(x) + \frac{P}{E_o I_o} m(x) = 0$$

This equation can be reduced to a simpler form by introducing variable changes. Let us replace x and $m(x)$ by t and $g(t)$ defined by

$$t = \left(1 + \frac{sx}{L}\right)^{-1} \qquad g(t) = t\, m(x)$$

The differential equation for $g(t)$ is found to be

$$g''(t) + \lambda^2 g(t) = 0 \qquad \lambda = \frac{L}{|s|}\sqrt{\frac{P}{E_o I_o}}$$

Therefore,

$$m(x) = \left(1 + \frac{sx}{L}\right)\left[c_1 \sin\left(\frac{\lambda}{1 + \frac{sx}{L}}\right) + c_2 \cos\left(\frac{\lambda}{1 + \frac{sx}{L}}\right)\right]$$

where c_1 and c_2 are arbitrary constants which can be found by imposing the boundary conditions. We will determine the constants for Cases I, II, and III. Case IV can be solved similarly and is left as an exercise for the reader.

To deal with Cases I, II, and III it is convenient to begin with a solution which vanishes at $x = 0$. A function satisfying this requirement has the form

$$m(x) = \left(1 + \frac{sx}{L}\right)\left[\sin\left(\frac{\lambda}{1 + \frac{sx}{L}}\right)\cos\lambda - \cos\left(\frac{\lambda}{1 + \frac{sx}{L}}\right)\sin\lambda\right]$$

which can be written as

$$m(x) = \left(1 + \frac{sx}{L}\right)\sin\left(\frac{\lambda}{1 + \frac{sx}{L}} - \lambda\right)$$

This equation can also represent the deflection curve for Cases I and II or the moment curve for Case III. Imposition of the remaining boundary conditions leads to an eigenvalue equation which is used to determine λ and the buckling load P. The deflection curve for Case I is taken as

$$y(x) = \left(1 + \frac{sx}{L}\right)\sin\left(\frac{\lambda}{1 + \frac{sx}{L}} - \lambda\right)$$

and the requirement that $y(L) = 0$ yields

$$\sin\left(\frac{\lambda}{1+s} - \lambda\right) = \sin\left(\frac{-\lambda s}{1+s}\right)$$

Hence, we need

$$\frac{\lambda s}{1+s} = \left(\frac{s}{1+s}\right)\left(\frac{L}{s}\sqrt{\frac{P}{E_o I_o}}\right) = \pi$$

which means that the buckling load is

$$P = \frac{\pi^2 E_o I_o}{L^2}(1+s)^2 \qquad s = \frac{h_2 - h_1}{h_1}$$

It is interesting to note that the buckling load for the tapered column ($s \neq 0$) is simply obtained by multiplying the buckling load for the constant cross section column ($s = 0$) by a factor

$$(1+s)^2 = \left(\frac{h_2}{h_1}\right)^2$$

This is also true for Cases III and IV, but is not true for Case II. Let us derive the characteristic equation for Case III. The constraint condition for the pinned-fixed case requires

$$\int_0^L \frac{x\, m(x)}{EI}\, dx = 0$$

so we need

$$\int_0^L x\left(1 + \frac{sx}{L}\right)^{-3} \sin\left(\frac{\lambda}{1+\frac{sx}{L}} - \lambda\right) dx = 0$$

This equation can be integrated using the substitution

$$\left(1 + \frac{sx}{L}\right)^{-1} = t$$

This leads to a characteristic equation of the form

$$\theta = \tan\theta \qquad \theta = \frac{\lambda s}{1+s} = \frac{L}{1+s}\sqrt{\frac{P}{E_o I_o}}$$

The smallest positive root of this equation is $\theta = 4.4934$, which yields

$$P = \frac{20.1906 E_o I_o}{L^2}(1+s)^2 \qquad \text{for Case III.}$$

Further analysis produces

$$P = \frac{4\pi^2 E_o I_o}{L^2} (1+s)^2 \qquad \text{for Case IV.}$$

The characteristic equation for Case II can be obtained by starting with the Case I deflection equation and imposing the condition $y'(L) = 0$. This leads to

$$s \sin \theta + \theta \cos \theta = 0 \qquad \theta = \frac{L}{1+s} \sqrt{\frac{P}{E_o I_o}}$$

When $s = 0$, the smallest positive root of this equation is $\theta = \pi/2$. Therefore, buckling load (when $s = 0$) is

$$P = \frac{\pi^2 E_o I_o}{4L^2} \qquad \text{for Case II}$$

and the dependence on s found in the other cases does not hold for the free-fixed problem.

10.5.2   Numerical Results

The function **colbuc**, which uses the above relationships, was written to analyze variable depth columns using any of the four types of end conditions discussed. The program allows a piecewise linear variation of EI. The program employs function **lintrp** for interpolation and function **trapsum** to perform trapezoidal rule integration. Comparisons were made with results presented by Beer and Johnston [9] and a comprehensive handbook on stability [19]. We will present some examples to show how well the program works. It is known that a column of length L and constant cross section stiffness $E_o I_o$ has buckling loads of

$$\frac{\pi^2 E_o I_o}{L^2} \qquad \frac{\pi^2 E_o I_o}{(2L)^2} \qquad \frac{\pi^2 E_o I_o}{(0.6992L)^2} \qquad \frac{\pi^2 E_o I_o}{(0.5L)^2}$$

for pinned-pinned, free-fixed, pinned-fixed, and fixed-fixed end conditions respectively. These cases were verified using the program **colbuc**. Let us illustrate the capability of the program to handle approximately a discontinuous cross section change. We analyze a column twenty inches long consisting of a ten inch section pinned at the outer end and joined to a ten inch long section which is considered rigid and fixed at the outer end. We use $E_o I_o = 1$ for the flexible section and $E_o I_o = 10000$ for the rigid section. This configuration should behave much like a pinned-fixed column of length 100 with a buckling load of

$$\frac{\pi^2}{6.992^2} = 0.2019$$

Using 100 segments (**nseq=100**) the program yields a value of 0.1975, which agrees within 2.2% of the expected value. A graph of the computed deflection configuration is shown in Figure 10.5. The code necessary to solve this problem is:

```
ei=[1 0; 1 10; 10000 10; 10000 20];
nseg=100; endc=3; len=20;
[p,y,x]=colbuc(len,ei,nseg,endc)
```

For a second example we consider a ten inch long column of circular cross section which is tapered from a one inch diameter at one end to a two inch diameter at the other end. We employ a fixed-fixed end condition and use $E_o = 1$. The theoretical results for this configuration indicate a buckling load of

$$\frac{\pi^3}{400} = 0.07752$$

Using 100 segments (**nseq=100**) the program produces a value of 0.07728, which agrees within 0.3% of the exact result. The code to generate this result utilizes function **eilt** and can be summarized as:

```
ei=eilt(1,2,10,101,1);
[p,y,x]=colbuc(10,ei,100,4);
```

The examples presented illustrate well the effectiveness of using finite difference methods in conjunction with the intrinsic eigenvalue solver in MATLAB to compute buckling loads. Furthermore, the provision for piecewise linear EI variation provided in the program is adequate to handle various column shapes.

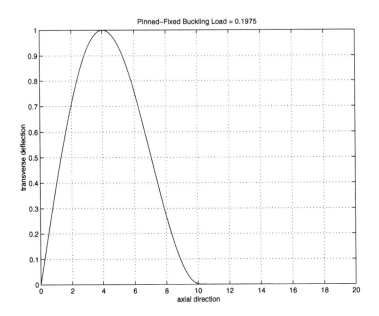

Figure 10.5. Analysis of Discontinuous Pinned-Fixed Column

10.5.3 Program Output and Code

Function colbuc

```
 1: function [p,y,x]=colbuc(len,ei,nseg,endc)
 2: %
 3: % [p,y,x]=colbuc(len,ei,nseg,endc)
 4: % ~~~~~~~~~~~~~~~~~~~~~~~~~~~~~~~~~~
 5: %
 6: % This function determines the Euler buckling
 7: % load for a slender column of variable cross
 8: % section which can have any one of four
 9: % constraint conditions at the column ends.
10: %
11: % len  - the column length
12: % ei   - the product of Young's modulus and the
13: %         cross section moment of inertia. This
14: %         quantity is defined as a piecewise
15: %         linear function specified at one or
16: %         more points along the length.  ei(:,1)
17: %         contains ei values at points
18: %         corresponding to x values given in
19: %         ei(:,2). Values at intermediate points
20: %         are computed by linear interpolation
21: %         using function lintrp which allows
22: %         jump discontinuities in ei.
23: % nseg - the number of segments into which the
24: %         column is divided to perform finite
25: %         difference calculations.The stepsize h
26: %         equals len/nseg.
27: % endc - a parameter specifying the type of end
28: %         condition chosen.
29: %            endc=1, both ends pinned
30: %            endc=2, x=0 free, x=len fixed
31: %            endc=3, x=0 pinned, x=len fixed
32: %            endc=4, both ends fixed
33: %
34: % p    - the Euler buckling load of the column
35: % x,y  - vectors describing the shape of the
36: %         column in the buckled mode. x varies
37: %         between 0 and len. y is normalized to
38: %         have a maximum value of one.
39: %
40: % User m functions called:  lintrp,
```

```
41: %                              trapsum, genprint
42: %-----------------------------------------------
43:
44: % If the column has constant cross section,
45: % then ei can be given as a single number. Also,
46: % use at least 20 segments to assure computed
47: % results will reasonable.
48: if size(ei,1) < 2
49:   ei=[ei(1,1),0; ei(1,1),len];
50: end
51: nseg=max(nseg,30);
52:
53: if endc==1
54: % pinned-pinned case (y=0 at x=0 and x=len)
55:   str='Pinned-Pinned Buckling Load = ';
56:   h=len/nseg; n=nseg-1; x=linspace(h,len-h,n);
57:   eiv=lintrp(ei(:,2),ei(:,1),x);
58:   a=-diag(ones(n-1,1),1);
59:   a=a+a'+diag(2*ones(n,1));
60:   [yvecs,pvals]=eig(diag(eiv/h^2)*a);
61:   pvals=diag(pvals);
62:   % Discard any spurious nonpositive eigenvalues
63:   j=find(pvals<=0);
64:   if length(j)>0, pvals(j)=[]; yvecs(:,j)=[]; end
65:   [p,k]=min(pvals); y=[0;yvecs(:,k);0];
66:   [ym,j]=max(abs(y)); y=y/y(j); x=[0;x(:);len];
67: elseif endc==2
68: % free-fixed case (y=0 at x=0 and y'=0 at x=len)
69:   str='Free-Fixed Buckling Load = ';
70:   h=len/nseg; n=nseg-1; x=linspace(h,len-h,n);
71:   eiv=lintrp(ei(:,2),ei(:,1),x);
72:   a=-diag(ones(n-1,1),1);
73:   a=a+a'+diag(2*ones(n,1));
74:   % Zero slope at x=len implies
75:   % y(n+1)=4/3*y(n)-1/3*y(n-1). This
76:   % leads to y''(n)=(y(n-1)-y(n))*2/(3*h^2).
77:   a(n,[n-1,n])=[-2/3,2/3];
78:   [yvecs,pvals]=eig(diag(eiv/h^2)*a);
79:   pvals=diag(pvals);
80:   % Discard any spurious nonpositive eigenvalues
81:   j=find(pvals<=0);
82:   if length(j)>0, pvals(j)=[]; yvecs(:,j)=[]; end
83:   [p,k]=min(pvals); y=yvecs(:,k);
84:   y=[0;y;4*y(n)/3-y(n-1)/3]; [ym,j]=max(abs(y));
85:   y=y/y(j); x=[0;x(:);len];
```

```
86: elseif endc==3
87: % pinned-fixed case
88: % (y=0 at x=0 and x=len, y'=0 at x=len)
89:   str='Pinned-Fixed Buckling Load = ';
90:   h=len/nseg; n=nseg; x=linspace(h,len,n);
91:   eiv=lintrp(ei(:,2),ei(:,1),x);
92:   a=-diag(ones(n-1,1),1);
93:   a=a+a'+diag(2*ones(n,1));
94: % Use a five point backward difference
95: % approximation for the second derivative
96: % at x=len.
97:   v=-[35/12,-26/3,19/2,-14/3,11/12];
98:   a(n,n:-1:n-4)=v; a=diag(eiv/h^2)*a;
99: % Form the equation requiring zero deflection
100: %    at x=len.
101:   b=x(:)'.*[ones(1,n-1),1/2]./eiv(:)';
102: % Impose the homogeneous boundary condition
103:   q=null(b); [z,pvals]=eig(q'*a*q);
104:   pvals=diag(pvals);
105: % Discard any spurious nonpositive eigenvalues
106:   k=find(pvals<=0);
107:   if length(k)>0, pvals(k)=[]; z(:,k)=[]; end;
108:   vecs=q*z; [p,k]=min(pvals); mom=[0;vecs(:,k)];
109: % Compute the slope and deflection from
110: %    moment values.
111:   yp=trapsum(0,len,mom./[1;eiv(:)]);
112:   yp=yp-yp(n+1);  y=trapsum(0,len,yp);
113:   [ym,j]=sort(abs(y)); y=y/y(j); x=[0;x(:)];
114: else
115: % fixed-fixed case
116: % (y and y' both zero at each end)
117:   str='Fixed-Fixed Buckling Load = ';
118:   h=len/nseg; n=nseg+1; x=linspace(0,len,n);
119:   eiv=lintrp(ei(:,2),ei(:,1),x);
120:   a=-diag(ones(n-1,1),1);
121:   a=a+a'+diag(2*ones(n,1));
122: % Use five point forward and backward
123: % difference approximations for the second
124: % derivatives at each end.
125:   v=-[35/12,-26/3,19/2,-14/3,11/12];
126:   a(1,1:5)=v; a(n,n:-1:n-4)=v;
127:   a=diag(eiv/h^2)*a;
128: % Write homogeneous equations to make the
129: % slope and deflection vanish at x=len.
130:   b=[1/2,ones(1,n-2),1/2]./eiv(:)';
```

```
131:    b=[b;x(:)'.*b];
132:    % Impose the homogeneous boundary conditions
133:    q=null(b); [z,pvals]=eig(q'*a*q);
134:    pvals=diag(pvals);
135:    % Discard any spurious nonpositive eigenvalues
136:    k=find(pvals<=0);
137:    if length(k>0), pvals(k)=[]; z(:,k)=[]; end;
138:    vecs=q*z; [p,k]=min(pvals); mom=vecs(:,k);
139:    % Compute the moment and slope from moment
140:    % values.
141:    yp=trapsum(0,len,mom./eiv(:));
142:    y=trapsum(0,len,yp);
143:    [ym,j]=max(abs(y)); y=y/y(j);
144: end
145:
146: close;
147: plot(x,y); grid on;
148: xlabel('axial direction');
149: ylabel('transverse deflection');
150: title([str,num2str(p)]); figure(gcf);
151: %genprint('buck');
```

Function eilt

```
1:  function ei=eilt(h1,h2,L,n,E)
2:  %
3:  % ei=eilt(h1,h2,L,n,E)
4:  % ~~~~~~~~~~~~~~~~~~~~
5:  %
6:  % This function computes the moment of inertia
7:  % along a linearly tapered circular cross
8:  % section and then uses that value to produce
9:  % the product EI.
10: %
11: % h1,h2 - column diameters at each end
12: % L     - column length
13: % n     - number of points at which ei is
14: %         computed
15: % E     - Young's modulus
16: %
17: % ei    - vector of EI values along column
18: %
19: % User m functions called:  none
```

```
20: %-------------------------------------------------
21:
22: if nargin<5, E=1; end;
23: x=linspace(0,L,n)';
24: ei=E*pi/64*(h1+(h2-h1)/L*x).^4;
25: ei=[ei(:),x(:)];
```

Function trapsum

```
 1: function v=trapsum(a,b,y,n)
 2: %
 3: % v=trapsum(a,b,y,n)
 4: % ~~~~~~~~~~~~~~~~~~~
 5: %
 6: % This function evaluates:
 7: %
 8: %    integral(a=>x, y(x)*dx) for a<=x<=b
 9: %
10: % by the trapezoidal rule (which assumes linear
11: % function variation between successive function
12: % values).
13: %
14: % a,b - limits of integration
15: % y   - integrand which can be a vector valued
16: %        function returning a matrix such that
17: %        function values vary from row to row.
18: %        It can also be input as a matrix with
19: %        the row size being the number of
20: %        function values and the column size
21: %        being the number of components in the
22: %        in the vector function.
23: % n   - the number of function values used to
24: %        perform the integration. When y is a
25: %        matrix then n is computed as the number
26: %        of rows in matrix y.
27: %
28: % v   - integral value
29: %
30: % User m functions called:  none
31: %-------------------------------------------------
32:
33: if isstr(y)
34:    % y is an externally defined function
```

```
35:    x=linspace(a,b,n)'; h=x(2)-x(1);
36:    Y=feval(y,x); % Function values must vary in
37:                  % row order rather than column
38:                  % order or computed results
39:                  % will be wrong.
40:    m=size(Y,2);
41: else
42:    % y is column vector or a matrix
43:    Y=y; [n,m]=size(Y); h=(b-a)/(n-1);
44: end
45: v=[zeros(1,m); ...
46:    h/2*cumsum(Y(1:n-1,:)+Y(2:n,:))];
```

Chapter 11

Bending Analysis of Beams of General Cross Section

11.1 Introduction

Elastic beams are important components in many types of structures. Consequently methods to analyze the shear, moment, slope, and deflection in beams with complex loading and general cross section variation are of significant interest. A typical beam of the type considered is shown in Figure 11.1. The study of Euler beam theory is generally regarded as an elementary topic dealt with in undergraduate engineering courses. However, simple analyses presented in standard textbooks usually do not reveal difficulties encountered with statically indeterminate problems and general geometries [113]. Finite element approximations intended to handle arbitrary problems typically assume a piecewise constant depth profile and a piecewise cubic transverse deflection curve. This contradicts even simple instances such as a constant depth beam subjected to a linearly varying distributed load which actually leads to a deflection curve which is a fifth order polynomial. Exact solutions of more involved problems where the beam depth changes linearly, for example, are more complicated. Therefore, an exact analysis of the beam problem is desirable to handle depth variation, a combination of concentrated and distributed loads, and static indeterminacy providing for general end conditions and multiple in-span supports. The current formulation considers a beam carrying any number of concentrated loads and linearly varying distributed loads. The equations for the shear and moment in the beam are obtained explicitly. Expressions for slope and deflection are formulated for evaluation by numerical integration allowing as many integration steps as necessary to achieve high accuracy. A set of simultaneous equations imposing desired constraints at the beam ends and at supports is solved for support reactions and any unknown end conditions. Knowledge of these quantities then allows evaluation of internal load and deformation quantities throughout the beam. The analytical formulation is implemented in a program using a concise prob-

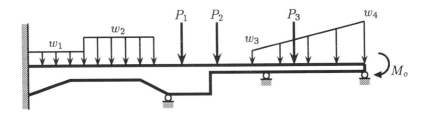

Figure 11.1. General Beam

lem definition specifying all loading, geometry, and constraint conditions without reference to beam elements or nodal points as might be typical in a finite element formulation. The program and example problem are discussed next.

11.1.1   Analytical Formulation

Solution of beam problems utilizes some mathematical idealizations such as a concentrated load, which implies infinite load intensity acting over an infinitesimal area. Also of importance are linearly varying distributed loads, or ramp loads. Treatment of these entities is facilitated by use of singularity functions [43]. The singularity function of order n is denoted by $< x - x_0 >^n$ and is defined as

$$< x - x_0 >^n \; = \; \begin{cases} 0 & x < x_0 \\ (x - x_0)^n & x \geq x_0 \end{cases}$$

For $n \geq 0$, the function satisfies

$$\int_0^x < x - x_0 >^n \; dx = \frac{< x - x_0 >^{n+1}}{n+1}$$

The special case where $n = -1$ is appropriate for describing a concentrated load. The term $< x - x_0 >^{-1}$ means the limit as $\epsilon \to 0$ of the following function

$$< x - x_0 >^{-1} \; = \; \begin{cases} 0 & x < x_0 \\ \frac{1}{\epsilon} & x_0 \leq x \leq (x_0 + \epsilon) \\ 0 & x > (x_0 + \epsilon) \end{cases}$$

Consequently, in the limit as ϵ approaches zero the integral becomes

$$\int_0^x < x - x_0 >^{-1} \; dx = < x - x_0 >^0$$

Analyzing the loads and deformations in the beam requires computation of the shear, moment, slope, and deflection designated as $v(x)$, $m(x)$, $y'(x)$, and $y(x)$. The beam lies in the range $0 \leq x \leq L$. A total of four end conditions are imposed at $x = 0$ and $x = L$. Normally, two conditions will be specified at each end, so two unknown conditions applicable at $x = 0$ need to be found during the solution process. Along with the end conditions, interior supports may exist at $x = r_j$, $1 \leq j \leq N_s$. Displacements y_j will occur at supports, and the reactions R_j, as well as four end conditions, needed to cause the deflections will have to be determined during the analysis. Within the beam span, the applied loading will consist of known external loads described as $w_e(x)$ and the support reactions. Fundamentals of Euler beam theory developed in standard textbooks [9, 100] imply the following differential and integral relations:

I) **Load**

$$\frac{dv}{dx} = w_e(x) + \sum_{j=1}^{N_s} R_j < x - r_j >^{-1}$$

II) **Shear**

$$v(x) = v_0 + v_e(x) + \sum_{j=1}^{N_s} R_j < x - r_j >^0$$

$$v_e(x) = \int_0^x w_e(x)\, dx$$

III) **Moment and Second Derivative**

$$\frac{dm}{dx} = v$$

$$m(x) = m_0 + v_0 x + m_e(x) + \sum_{j=1}^{N_s} R_j < x - r_j >^1$$

$$m_e(x) = \int_0^x v_e(x)\, dx$$

$$y''(x) = k(x) \left[m_0 + v_0 x + m_e(x) + \sum_{j=1}^{N_s} R_j < x - r_j >^1 \right]$$

$$k(x) = \frac{1}{E(x)\, I(x)}$$

IV) Slope

$$y'(x) \;=\; y_0' + m_0 \int_0^x k(x)\,dx + v_0 \int_0^x x\,k(x)\,dx +$$

$$\int_0^x k(x)\,m_e(x)\,dx + \sum_{j=1}^{N_s} R_j \int_0^x <x - r_j>^1 k(x)\,dx$$

V) Deflection

$$y(x) \;=\; y_0 + y_0'x + m_0 \int_0^x \int_0^x k(x)\,dx\,dx +$$

$$v_0 \int_0^x \int_0^x x\,k(x)\,dx\,dx + \int_0^x \int_0^x k(x)\,m_e(x)\,dx\,dx +$$

$$\sum_{j=1}^{N_s} R_j \int_0^x \int_0^x <x - r_j>^1 k(x)\,dx\,dx$$

where $E(x)\,I(x)$ means the product of Young's modulus and the cross section moment of inertia. The quantities y_0, y_0', v_0, m_0 mean left-end values of deflection, slope, shear, and moment. The property $k(x)$ will be spatially variable unless EI is constant which yields the following simple formulas

$$EI\,y'(x) \;=\; EI\,y_0' + m_0 x + \frac{v_0 x^2}{2} + \int_0^x m_e(x)\,dx +$$

$$\frac{1}{2}\sum_{j=1}^{N_s} R_j <x - r_j>^2$$

$$EI\,y(x) \;=\; EI\,y_0 + EI\,y_0'x + \frac{m_0 x^2}{2} + \frac{v_0 x^3}{6} +$$

$$\int_0^x \int_0^x m_e(x)\,dx\,dx + \frac{1}{6}\sum_{j=1}^{N_s} R_j <x - r_j>^3$$

The external loading conditions employed here can handle most practical situations. It is assumed that several concentrated loads F_j act at positions f_j, $1 \le j \le N_f$. Distributed loads are described by linearly varying ramp loads. A typical ramp load starts at position p_j with intensity P_j and varies linearly to magnitude Q_j at position q_j. The ramp load is zero unless $p_j \le x \le q_j$. A total of N_r ramp loads may be present. Instances where $P_j = Q_j$ can also occur, implying a uniformly distributed load. The general external loading chosen can be represented as

$$w_e(x) = \sum_{j=1}^{N_f} F_j < x - f_j >^{-1} +$$

$$\sum_{j=1}^{N_r} [P_j < x - p_j >^0 -Q_j < x - q_j >^0 +$$

$$S_j \left(< x - p_j >^1 - < x - q_j >^1 \right)]$$

where

$$S_j = \frac{Q_j - P_j}{q_j - p_j}$$

and each summation extends over the complete range of pertinent values. Similiarly, integration using the properties of singularity functions yields

$$v_e(x) = \sum_{j=1}^{N_f} F_j < x - f_j >^0 +$$

$$\sum_{j=1}^{N_r} [P_j < x - p_j >^1 -Q_j < x - q_j >^1 +$$

$$\frac{S_j}{2} \left(< x - p_j >^2 - < x - q_j >^2 \right)]$$

and

$$m_e(x) = \sum_{j=1}^{N_f} F_j < x - f_j >^1 +$$

$$\sum_{j=1}^{N_r} \left[\frac{P_j}{2} < x - p_j >^2 -\frac{Q_j}{2} < x - q_j >^2 + \right.$$

$$\left. \frac{S_j}{6} \left(< x - p_j >^3 - < x - q_j >^3 \right) \right]$$

The single and double integrals given earlier involving $m_e(x)$ and $k(x)$ can easily be evaluated exactly when EI is constant, but these are not needed here. Since $k(x)$ will generally be spatially variable in the target problem set, the integrations to compute $y'(x)$ and $y(x)$ are best performed numerically. Leaving the number of integration increments as an independent parameter allows high accuracy evaluation of all integrals whenever this is desirable. Typically, problems using several hundred integration points only require a few seconds to solve using a personal computer.

Completing the problem solution requires formulations and solution of a system of simultaneous equations involving v_0, m_0, y_0', y_0, R_1, ..., R_{N_s}. The desired equations are created by specifying the displacement constraints at the supports, as well as four of eight possible end conditions. To present the equations more concisely the following notation is adopted:

$$\int_0^x k(x)\,dx = K_1(x) \qquad \int_0^x \int_0^x k(x)\,dx\,dx = K_2(x)$$

$$\int_0^x x\,k(x)\,dx = L_1(x) \qquad \int_0^x \int_0^x x\,k(x)\,dx\,dx = L_2(x)$$

$$\int_0^x m_e(x)\,k(x)\,dx = I_1(x) \qquad \int_0^x \int_0^x m_e(x)\,k(x)\,dx\,dx = I_2(x)$$

$$\int_0^x <x - r_j>^1 k(x)\,dx = J_1(x, r_j)$$

$$\int_0^x \int_0^x <x - r_j>^1 k(x)\,dx\,dx = J_2(x, r_j)$$

and it is evident from their definitions that both $J_1(x, r_j)$ and $J_2(x, r_j)$ both equal zero for $x \leq r_j$.

At a typical support location r_i, the deflection will have an imposed value y_i. Consequently, the displacement constraints require

$$y_0 + r_i y_0' + K_2(r_i)\,m_0 + L_2(r_i)\,v_0 + \sum_{j=i+1}^{N_s} J_2(r_i, r_j)\,R_j =$$

$$y_i - I_2(r_i) \qquad 1 \leq i \leq N_s$$

The remaining four end conditions can specify any legitimate combination of conditions yielding a unique solution. For example, a beam cantilevered at $x = 0$ and pin supported at $x = L$ would require $y(0) = 0$, $y'(0) = 0$, $m(L) = 0$, and $y(L) = 0$. In general, conditions imposed at $x = 0$ have an obvious form since only v_0, m_0, y_0, or y_0' are explicitly involved. To illustrate a typical right end condition, let us choose slope, for example. This yields

$$y_0' + y_0' + K_1(L)\,m_0 + L_1(L)\,v_0 + \sum_{j=1}^{N_s} J_1(L, r_j)\,R_j = y'(L) - I_1(L)$$

Equations for other end conditions have similar form and all eight possibilities are implemented in the computer program listed at the end of the chapter. Once the reactions and any initially unknown left-end conditions have been determined, load and deformation quantities anywhere in the beam can be readily found.

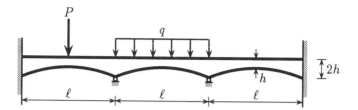

Figure 11.2. Parabolic Beam from Arbabi and Li

11.1.2 Program to Analyze Beams of General Cross Section

A program to solve general beam problems was written which tabulates and plots the shear, moment, slope, and deflection. The driver program **vdb** defines the data, calls the analysis functions, and outputs the results. Six functions which implement the methods given in this section were written. Understanding the program details can best be achieved by studying the code closely. The program was checked extensively using examples from several texts and reference books. The three span beam having parabolically tapered haunches shown in Figure 11.2 was analyzed previously by Arbabi and Li [5]. The program **vdb** was used to analyze the same problem and produces results which agree well with the paper.

We believe that the computer program is general enough to handle a wide variety of practical problems. Some readers may want to extend the program by adding interactive input or input from a data file. Such a modification is straightforward.

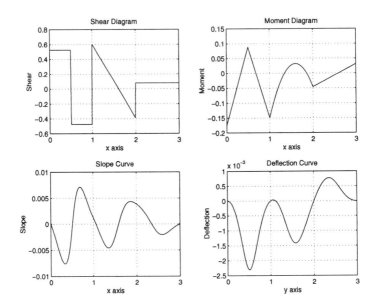

Figure 11.3. Results for Arbabi and Li Example

11.1.3 Program Output and Code

Output from Arbabi and Li Example

```
Analysis of a Variable Depth Elastic Beam
-----------------------------------------

Title: Problem from Arbabi and Li

Beam Length:               3
Number of integration segments: 301
Print frequency for results:    10

Interior Supports: (2)
   |   #    X-location    Deflection
   |  ---  ------------  ------------
   |   1   1.0000e+000   0.0000e+000
   |   2   2.0000e+000   0.0000e+000


Concentrated Forces: (1)
   |   #    X-location        Force
   |  ---  ------------  ------------
   |   1   5.0000e-001  -1.0000e+000


Ramp loads: (1)
   |   #       X-start         Load       X-end          Load
   |  ---  ------------  ------------  ------------  ------------
   |   1   1.0000e+000  -1.0000e+000   2.0000e+000  -1.0000e+000


End conditions:
   | End    Function          Value
   | ------ ----------   ------------
   | left   slope         0.0000e+000
   | left   deflection    0.0000e+000
   | right  slope         0.0000e+000
   | right  deflection    0.0000e+000


EI values are specified
   |   #       X-start     EI-value
   |  ---  ------------  ------------
   |   1   0.0000e+000   7.9976e+000
   |   2   1.0101e-002   7.5273e+000
   |   3   2.0202e-002   7.0848e+000
   |   4   3.0303e-002   6.6688e+000
   |   5   4.0404e-002   6.2776e+000

   Material deleted for publication

   | 296   2.9596e+000   6.2776e+000
   | 297   2.9697e+000   6.6688e+000
   | 298   2.9798e+000   7.0848e+000
   | 299   2.9899e+000   7.5273e+000
   | 300   3.0000e+000   7.9976e+000


Solution time was 0.55 secs.
```

```
Reactions at Internal Supports:
  |   X-location        Reaction
  |  ------------   ------------
  |            1    1.0782e+000
  |            2    4.7506e-001

Table of Results:
  |  X-location         Shear        Moment         Theta         Delta
  |  -----------   ------------  ------------  ------------  ------------
  |            0    5.2494e-001  -1.7415e-001   0.0000e+000   0.0000e+000
  |          0.1    5.2494e-001  -1.2166e-001  -2.4859e-003  -1.1943e-004
  |          0.2    5.2494e-001  -6.9164e-002  -5.3262e-003  -5.0996e-004
  |          0.3    5.2494e-001  -1.6670e-002  -7.4251e-003  -1.1612e-003
  |          0.4    5.2494e-001   3.5824e-002  -6.5761e-003  -1.8965e-003
  |          0.5   -4.7506e-001   8.8318e-002  -5.5680e-004  -2.3003e-003
  |          0.6   -4.7506e-001   4.0812e-002   5.6976e-004  -1.9998e-003
  |          0.7   -4.7506e-001  -6.6940e-003   7.1119e-003  -1.3258e-003
  |          0.8   -4.7506e-001  -5.4200e-002   5.6441e-003  -6.7385e-004
  |          0.9   -4.7506e-001  -1.0171e-001   3.3302e-003  -2.2402e-004
  |            1    6.0309e-001  -1.4921e-001   1.2242e-003  -2.4286e-017
  |          1.1    5.0309e-001  -9.3903e-002  -7.9439e-004   2.3707e-005
  |          1.2    4.0309e-001  -4.8593e-002  -2.8814e-003  -1.6165e-004
  |          1.3    3.0309e-001  -1.3284e-002  -4.3574e-003  -5.3250e-004
  |          1.4    2.0309e-001   1.2025e-002  -4.2883e-003  -9.8078e-004
  |          1.5    1.0309e-001   2.7334e-002  -2.3015e-003  -1.3242e-003
  |          1.6    3.0918e-003   3.2643e-002   6.5694e-004  -1.4078e-003
  |          1.7   -9.6908e-002   2.7953e-002   3.0625e-003  -1.2125e-003
  |          1.8   -1.9691e-001   1.3262e-002   4.1954e-003  -8.3907e-004
  |          1.9   -2.9691e-001  -1.1429e-002   4.2843e-003  -4.0860e-004
  |            2    7.8151e-002  -4.6120e-002   3.8358e-003  -1.1102e-016
  |          2.1    7.8151e-002  -3.8305e-002   3.1202e-003   3.5021e-004
  |          2.2    7.8151e-002  -3.0490e-002   2.0801e-003   6.1308e-004
  |          2.3    7.8151e-002  -2.2675e-002   7.2881e-004   7.5555e-004
  |          2.4    7.8151e-002  -1.4860e-002  -6.9898e-004   7.5597e-004
  |          2.5    7.8151e-002  -7.0445e-003  -1.7447e-003   6.2865e-004
  |          2.6    7.8151e-002   7.7058e-004  -2.0539e-003   4.3228e-004
  |          2.7    7.8151e-002   8.5857e-003  -1.7105e-003   2.4008e-004
  |          2.8    7.8151e-002   1.6401e-002  -1.0840e-003   9.9549e-005
  |          2.9    7.8151e-002   2.4216e-002  -4.7454e-004   2.2493e-005
  |            3    7.8151e-002   3.2031e-002  -4.4409e-016  -2.2204e-016
```

Script File vdb

```
 1: % Example: vdb
 2: % ~~~~~~~~~~~~
 3: %
 4: % This program calculates the shear, moment,
 5: % slope, and deflection of a variable depth
 6: % indeterminate beam subjected to complex
 7: % loading and general end conditions. The
 8: % input data is defined in the program
 9: % statements below.
10: %
11: % User m functions required:
12: %    bmvardep, extload, lintrp, oneovrei,
13: %    sngf, trapsum, genprint
14: %-------------------------------------------------
15:
16: clear all; Problem=1;
17: if Problem == 1
18:    Title=['Problem from Arbabi and Li'];
19:    Printout=10; % Output frequency
20:    BeamLength=3; % Beam length
21:    NoSegs=301;   % # of beam divisions for
22:                  % integration
23:    % External concentrated loads and location
24:    ExtForce= [-1]; ExtForceX=[.5];
25:    % External ramp loads and range
26:    %       q1  q2  x1  x2
27:    ExtRamp=[-1  -1   1   2];
28:    % Interior supports: initial displacement
29:    % and location
30:    IntSupX=    [1; 2]; IntSupDelta=[0; 0];
31:    % End (left and right) conditions
32:    EndCondVal= [0; 0; 0; 0];   % magnitude
33:    % 1=shear,2=moment,3=slope,4=delta
34:    EndCondFunc=[3; 4; 3; 4];
35:    % 1=left end,2=right end
36:    EndCondEnd= [1; 1; 2; 2];
37:    % EI or beam depth specification
38:    EIorDepth=1; % 1=EI values specified
39:                 % 2=depth values specified
40:    if EIorDepth == 1
41:       % Discretize the parabolic haunch for the
42:       % three spans
```

```
43:     Width=1; E=1; a=0.5^2; Npts=100;
44:     h1=0.5; k1=1; x1=linspace(0,1,Npts);
45:     h2=1.5; k2=1; x2=linspace(1,2,Npts);
46:     h3=2.5; k3=1; x3=linspace(2,3,Npts);
47:     y1=(x1-h1).^2/a+k1; y2=(x2-h2).^2/a+k2;
48:     y3=(x3-h3).^2/a+k3;
49:     EIx=[x1 x2 x3]'; h=[y1 y2 y3]';
50:     EIvalue=E*Width/12*h.^3;
51:     mn=min(EIvalue); EIvalue=EIvalue./mn;
52:   else
53:     % Beam width and Young's modulus
54:     BeamWidth=[]; BeamE=[]; Depth=[]; DepthX=[];
55:   end
56: elseif Problem == 2
57:   Title=['From Timoshenko and Young,', ...
58:           ' pp 434, haunch beam'];
59:   Printout=12; NoSegs=144*4+1; BeamLength=144;
60:   ExtForce=[]; ExtForceX=[];
61:   ExtRamp=[-1 -1 0 108];
62:   IntSupX=[36; 108]; IntSupDelta=[0; 0];
63:   EndCondVal=[0; 0; 0; 0];
64:   EndCondFunc=[2; 4; 2; 4];
65:   EndCondEnd= [1; 1; 2; 2]; EIorDepth=2;
66:   if EIorDepth == 1
67:     EIvalue=[]; EIx=[];
68:   else
69:     BeamWidth=[1]; BeamE=[1];
70:     % Discretize the parabolic sections
71:     a=36^2/5; k=2.5; h1=0; h2=72; h3=144;
72:     N1=36; N2=72; N3=36;
73:     x1=linspace(  0, 36,N1); y1=(x1-h1).^2/a+k;
74:     x2=linspace( 36,108,N2); y2=(x2-h2).^2/a+k;
75:     x3=linspace(108,144,N3); y3=(x3-h3).^2/a+k;
76:     Depth=[y1 y2 y3]'; DepthX=[x1 x2 x3]';
77:     % Comparison values
78:     I=BeamWidth*Depth.^3/12; Imin=min(I); L1=36;
79:     k1=BeamE*Imin/L1; k2=k1/2; k3=k1;
80:     t0=10.46/k1; t1=15.33/k1; t2=22.24/k1;
81:     t3=27.95/k1;
82:     fprintf('\n\nValues from reference');
83:     fprintf('\n  Theta (x=  0): %12.4e',t0);
84:     fprintf('\n  Theta (x= 36): %12.4e',t1);
85:     fprintf('\n  Theta (x=108): %12.4e',t2);
86:     fprintf('\n  Theta (x=144): %12.4e\n',t3);
87:   end
```

```matlab
88: end
89:
90: % Load input parameters into matrices
91: Force=[ExtForce,ExtForceX];
92: NoExtForce=length(ExtForce);
93: [NoExtRamp,ncol]=size(ExtRamp);
94: IntSup=[IntSupDelta,IntSupX];
95: NoIntSup=length(IntSupX);
96: EndCond=[EndCondVal,EndCondFunc,EndCondEnd];
97: if EIorDepth == 1
98:    BeamProp=[]; NoEIorDepths=length(EIx);
99:    EIdata=[EIvalue EIx];
100: else
101:    BeamProp=[BeamWidth BeamE];
102:    NoEIorDepths=length(DepthX);
103:    EIdata=[Depth DepthX];
104: end
105:
106: % Output input data
107: label1=['shear      ';'moment     '; ...
108:         'slope      ';'deflection'];
109: label2=['left  ';'right '];
110: fprintf('\n\nAnalysis of a Variable Depth ');
111: fprintf('Elastic Beam');
112: fprintf('\n--------------------------------');
113: fprintf('---------');
114: fprintf('\n\n');
115: disp(['Title: ' Title]);
116: fprintf ...
117:    ('\nBeam Length:                    %g', ...
118:    BeamLength);
119: fprintf ...
120:    ('\nNumber of integration segments: %g', ...
121:    NoSegs);
122: fprintf ...
123:    ('\nPrint frequency for results:    %g', ...
124:    Printout);
125: fprintf('\n\nInterior Supports: (%g)', ...
126:    NoIntSup);
127: if NoIntSup > 0
128:    fprintf('\n |   #   X-location   Deflection');
129:    fprintf('\n | --- ------------ ------------');
130:    for i=1:NoIntSup
131:      fprintf('\n |%4.0f %12.4e %12.4e', ...
132:              i,IntSup(i,2),IntSup(i,1));
```

```
133:   end
134: end
135: fprintf('\n\nConcentrated Forces: (%g)', ...
136:   NoExtForce);
137: if NoExtForce > 0
138:    fprintf('\n |  #   X-location         Force');
139:    fprintf('\n | --- ------------ ------------');
140:    for i=1:NoExtForce
141:       fprintf('\n |%4.0f %12.4e %12.4e', ...
142:             i,Force(i,2),Force(i,1));
143:    end
144: end
145: fprintf('\n\nRamp loads: (%g)', NoExtRamp);
146: if NoExtRamp > 0
147:    fprintf('\n |  #      X-start        Load');
148:    fprintf('       X-end        Load');
149:    fprintf('\n | --- ------------ ------------');
150:    fprintf(' ------------ ------------');
151:    for i=1:NoExtRamp
152:       fprintf('\n |%4.0f %12.4e %12.4e ', ...
153:             i,ExtRamp(i,3),ExtRamp(i,1));
154:       fprintf('%12.4e %12.4e', ...
155:             ExtRamp(i,4),ExtRamp(i,2));
156:    end
157: end
158: fprintf('\n\nEnd conditions:');
159: fprintf('\n | End     Function          Value');
160: fprintf('\n ');
161: fprintf('| ------ ----------  ------------\n');
162: for i=1:4
163:    j=EndCond(i,3); k=EndCond(i,2);
164:    strg=sprintf(' %12.4e',EndCond(i,1));
165:    disp([' | ' label2(j,:) label1(k,:) strg]);
166: end
167: if EIorDepth == 1
168:    fprintf('\nEI values are specified');
169:    fprintf('\n |  #      X-start      EI-value')
170:    fprintf('\n | --- ------------ ------------');
171:    for i=1:NoEIorDepths
172:       fprintf('\n |%4.0f %12.4e %12.4e', ...
173:             i,EIdata(i,2),EIdata(i,1));
174:    end
175: else
176:    fprintf('\nDepth values are specified for ');
177:    fprintf('rectangular cross section');
```

```
178:   fprintf('\n  | Beam width:        %12.4e', ...
179:           BeamProp(1));
180:   fprintf('\n  | Young''s modulus: %12.4e', ...
181:           BeamProp(2));
182:   fprintf('\n  |');
183:   fprintf('\n  |   #       X-start        Depth')
184:   fprintf('\n  | --- ------------ ------------');
185:   for i=1:NoEIorDepths
186:      fprintf('\n  |%4.0f %12.4e %12.4e', ...
187:             i,EIdata(i,2),EIdata(i,1));
188:   end
189: end
190: disp(' ');
191:
192: % Begin analysis
193: x=linspace(0,BeamLength,NoSegs)'; t=clock;
194: [V,M,Theta,Delta,Reactions]= ...
195:   bmvardep(NoSegs,BeamLength,Force,ExtRamp, ...
196:            EndCond,IntSup,EIdata,BeamProp);
197: t=etime(clock,t);
198:
199: % Output results
200: disp(' ');
201: disp(['Solution time was ',num2str(t),' secs.']);
202: if NoIntSup > 0
203:   fprintf('\nReactions at Internal Supports:');
204:   fprintf('\n  |   X-location      Reaction');
205:   fprintf('\n  | ------------ -------------');
206:   for i=1:NoIntSup
207:      fprintf('\n  | %12.8g %12.4e', ...
208:             IntSup(i,2),Reactions(i));
209:   end
210: end
211: fprintf('\n\nTable of Results:');
212: fprintf('\n  | X-location        Shear');
213: fprintf('       Moment');
214: fprintf('        Theta        Delta');
215: fprintf('\n  | ----------- ------------ ');
216: fprintf('------------');
217: fprintf(' ------------ ------------');
218: if Printout > 0
219:   for i=1:Printout:NoSegs
220:      fprintf('\n  |%12.4g %12.4e %12.4e', ...
221:             x(i),V(i),M(i));
222:      fprintf(' %12.4e %12.4e',Theta(i),Delta(i));
```

```
223:    end
224:    disp(' ');
225: else
226:    i=1; j=NoSegs;
227:    fprintf('\n  |%12.4g %12.4e %12.4e', ...
228:            x(i),V(i),M(i));
229:    fprintf(' %12.4e %12.4e',Theta(i),Delta(i));
230:    fprintf('\n  |%12.8g %12.4e %12.4e', ...
231:            x(j),V(j),M(j));
232:    fprintf(' %12.4e %12.4e',Theta(j),Delta(j));
233: end
234: fprintf('\n\n');
235: subplot(2,2,1);
236:    plot(x,V); grid; xlabel('x axis');
237:    ylabel('Shear'); title('Shear Diagram');
238: subplot(2,2,2);
239:    plot(x,M); grid; xlabel('x axis');
240:    ylabel('Moment'); title('Moment Diagram')
241: subplot(2,2,3);
242:    plot(x,Theta); grid; xlabel('x axis');
243:    ylabel('Slope'); title('Slope Curve');
244: subplot(2,2,4);
245:    plot(x,Delta); grid; xlabel('y axis');
246:    ylabel('Deflection');
247:    title('Deflection Curve');
248: drawnow; %genprint('vdb');
```

Function bmvardep

```
1: function [V,M,Theta,Delta,Reactions]= ...
2:   bmvardep(NoSegs,BeamLength,Force,ExtRamp, ...
3:   EndCond,IntSup,EIdata,BeamProp)
4: % [V,M,Theta,Delta,Reactions]=bmvardep ...
5: % (NoSegs,BeamLength,Force,ExtRamp,EndCond, ...
6: % IntSup,EIdata,BeamProp)
7: % ~~~~~~~~~~~~~~~~~~~~~~~~~~~~~~~~~~~~~~~~~~~~~~~~
8: %
9: % This function computes the shear, moment,
10: % slope and deflection in a variable depth
11: % elastic beam having specified end conditions,
12: % intermediate supports with given
13: % displacements, and general applied loading
14: % allowing concentrated loads and linearly
15: % varying ramp loads.
16: %
17: % NoSegs     - number of beam divisions for
18: %                integration
19: % BeamLength - beam length
20: % Force      - matrix containing the magnitudes
21: %                and locations for concentrated
22: %                loads
23: % ExtRamp    - matrix containing the end
24: %                magnitudes and end locations
25: %                for ramp loads
26: % EndCond    - matrix containing the type of
27: %                end conditions, the magnitudes,
28: %                and the whether left or right
29: %                end
30: % IntSup     - matrix containing the location
31: %                and delta for interior supports
32: % EIdata     - either EI or depth values
33: % BeamProp   - either null or beam widths
34: %
35: % V          - vector of shear values
36: % M          - vector of moment values
37: % Theta      - vector of slope values
38: % Delta      - vector of deflection values
39: % Reactions  - reactions at interior supports
40: %
41: % User m functions required:
42: %    oneovrei, extload, sngf, trapsum
```

```
43: %------------------------------------------------
44:
45: if nargin < 8, BeamProp=[]; end
46: % Evaluate function value coordinates and 1/EI
47: x=linspace(0,BeamLength,NoSegs)';
48: kk=oneovrei(x,EIdata,BeamProp);
49:
50: % External load contributions to shear and
51: % moment interior to span and at right end
52: [ve,me]=extload(x,Force,ExtRamp);
53: [vv,mm]=extload(BeamLength,Force,ExtRamp);
54:
55: % Deflections and position of interior supports
56: ns=size(IntSup,1);
57: if ns > 0
58:    ysprt=IntSup(:,1); r=IntSup(:,2);
59:    snf=sngf(x,r,1);
60: else
61:    ysprt=[]; r=[]; snf=zeros(NoSegs,0);
62: end
63:
64: % Form matrix governing y''(x)
65: smat=kk(:,ones(1,ns+3)).* ...
66:        [x,ones(NoSegs,1),snf,me];
67:
68: % Integrate twice to get slope and deflection
69: % matrices
70: smat=trapsum(0,BeamLength,smat);
71: ymat=trapsum(0,BeamLength,smat);
72:
73: % External load contributions to
74: % slope/deflection at the right end
75: ss=smat(NoSegs,ns+3); yy=ymat(NoSegs,ns+3);
76:
77: % Equations to solve for left end conditions
78: % and internal reactions
79: ns4=ns+4; j=1:4; a=zeros(ns4,ns4);
80: b=zeros(ns4,1); js=1:ns; js4=js+4;
81:
82: % Account for four independent boundary
83: % conditions.  Usually two conditions will be
84: % imposed at each end.
85: for k=1:4
86:    val=EndCond(k,1); typ=EndCond(k,2);
87:    wchend=EndCond(k,3);
```

```
88:    if wchend==1
89:      b(k)=val; row=zeros(1,4); row(typ)=1;
90:      a(k,j)=row;
91:    else
92:      if typ==1       % Shear
93:        a(k,j)=[1,0,0,0]; b(k)=val-vv;
94:        if ns>0
95:          a(k,js4)=sngf(BeamLength,r,0);
96:        end
97:      elseif typ==2 % Moment
98:        a(k,j)=[BeamLength,1,0,0]; b(k)=val-mm;
99:        if ns>0
100:         a(k,js4)=sngf(BeamLength,r,1);
101:       end
102:     elseif typ==3 % Slope
103:       a(k,j)=[smat(NoSegs,1:2),1,0];
104:       b(k)=val-ss;
105:       if ns>0
106:         a(k,js4)=smat(NoSegs,3:ns+2);
107:       end
108:     else            % Deflection
109:       a(k,j)=[ymat(NoSegs,1:2),BeamLength,1];
110:       b(k)=val-yy;
111:       if ns>0
112:         a(k,js4)=ymat(NoSegs,3:ns+2);
113:       end
114:     end
115:   end
116: end
117:
118: % Interpolate to assess how support deflections
119: % are affected by end conditions, external
120: % loads, and support reactions.
121: if ns>0
122:   a(js4,1)=table1([x,ymat(:,1)],r);
123:   a(js4,2)=table1([x,ymat(:,2)],r);
124:   a(js4,3)=r; a(js4,4)=ones(ns,1);
125:   for j=1:ns-1
126:     a(j+5:ns+4,j+4)= ...
127:       table1([x,ymat(:,j+2)],r(j+1:ns));
128:   end
129: end
130: b(js4)=ysprt-table1([x,ymat(:,ns+3)],r);
131:
132: % Solve for unknown reactions and end conditions
```

```
133: c=a\b; v0=c(1); m0=c(2); s0=c(3); y0=c(4);
134: Reactions=c(5:ns+4);
135:
136: % Compute the shear, moment, slope, deflection
137: % for all x
138: if ns > 0
139:     V=v0+ve+sngf(x,r,0)*Reactions;
140:     M=m0+v0*x+me+sngf(x,r,1)*Reactions;
141:     Theta=s0+smat(:,ns+3)+smat(:,1:ns+2)* ...
142:             [v0;m0;Reactions];
143:     Delta=y0+s0*x+ymat(:,ns+3)+ ...
144:             ymat(:,1:ns+2)*[v0;m0;Reactions];
145: else
146:     Reactions=[]; V=v0+ve; M=m0+v0*x+me;
147:     Theta=s0+smat(:,ns+3)+smat(:,1:2)*[v0;m0];
148:     Delta=y0+s0*x+ymat(:,ns+3)+ ...
149:             ymat(:,1:2)*[v0;m0];
150: end
```

Function extload

```
1:  function [V,M,EITheta,EIDelta]=extload ...
2:              (x,Force,ExtRamp)
3:  % [V,M,EITheta,EIDelta]=extload ...
4:  %                       (x,Force,ExtRamp)
5:  % ~~~~~~~~~~~~~~~~~~~~~~~~~~~~~~~~~~~~~~~~~~~
6:  %
7:  % This function computes the shear, moment,
8:  % slope and deflection in a uniform depth Euler
9:  % beam which is loaded by a series of
10: % concentrated loads and ramp loads. The values
11: % of shear, moment, slope and deflection all
12: % equal zero when x=0.
13: %
14: % x       - location along beam
15: % Force   - concentrated force matrix
16: % ExtRamp - distributed load matrix
17: %
18: % V       - shear
19: % M       - moment
20: % EITheta - slope
21: % EIDelta - deflection
22: %
```

```
23: % User m functions required: sngf
24: %-------------------------------------------------
25:
26: nf=size(Force,1); nr=size(ExtRamp,1);
27: nx=length(x); V=zeros(nx,1); M=V;
28: EITheta=V; EIDelta=V;
29: % Concentrated load contributions
30: if nf > 0
31:   F=Force(:,1); f=Force(:,2); V=V+sngf(x,f,0)*F;
32:   M=M+sngf(x,f,1)*F;
33:   if nargout > 2
34:     EITheta=EITheta+sngf(x,f,2)*(F/2);
35:     EIDelta=EIDelta+sngf(x,f,3)*(F/6);
36:   end
37: end
38: % Ramp load contributions
39: if nr > 0
40:   P=ExtRamp(:,1); Q=ExtRamp(:,2);
41:   p=ExtRamp(:,3); q=ExtRamp(:,4);
42:   S=(Q-P)./(q-p); sp2=sngf(x,p,2);
43:   sq2=sngf(x,q,2); sp3=sngf(x,p,3);
44:   sq3=sngf(x,q,3); sp4=sngf(x,p,4);
45:   sq4=sngf(x,q,4);
46:   V=V+sngf(x,p,1)*P-sngf(x,q,1)* ... % Shear
47:     Q+(sp2-sq2)*(S/2);
48:   M=M+sp2*(P/2)-sq2*(Q/2)+ ...        % Moment
49:     (sp3-sq3)*(S/6);
50:   if nargout > 2
51:     EITheta=EITheta+sp3*(P/6)- ...  % EI*Theta
52:             sq3*(Q/6)+(sp4-sq4)*(S/24);
53:     EIDelta=EIDelta+sp4*(P/24)- ... % EI*Delta
54:             sq4*(Q/24)+(sngf(x,p,5)- ...
55:             sngf(x,q,5))*(S/120);
56:   end
57: end
```

Function oneovrei

```
1: function val=oneovrei(x,EIdata,BeamProp)
2: % [val]=oneovrei(x,EIdata,BeamProp)
3: % ~~~~~~~~~~~~~~~~~~~~~~~~~~~~~~~~~~~
4: %
5: % This function computes 1/EI by piecewise
```

```
6:  % linear interpolation through a set of data
7:  % values.
8:  %
9:  % x         - location along beam
10: % EIdata    - EI or depth values
11: % BeamProp  - null or width values
12: %
13: % val       - computed value for 1/EI
14: %
15: % User m functions required: none
16: %-----------------------------------------------
17:
18: if size(EIdata,1) < 2  % uniform depth case
19:     v=EIdata(1,1);
20:     EIdata=[v,min(x);v,max(x)];
21: end
22: if ( nargin > 2 ) & ( sum(size(BeamProp)) > 0)
23:     % Compute properties assuming the cross
24:     % section is rectangular and EIdata(:,1)
25:     % contains depth values
26:     width=BeamProp(1); E=BeamProp(2);
27:     EIdata(:,1)=E*width/12*EIdata(:,1).^3;
28: end
29: val=1./lintrp(EIdata(:,2),EIdata(:,1),x);
```

Function sngf

```
1:  function y=sngf(x,x0,n)
2:  % y=sngf(x,x0,n)
3:  % ~~~~~~~~~~~~~~
4:  %
5:  % This function computes the singularity
6:  % function defined by
7:  %     y=<x-x0>^n for n=0,1,2,...
8:  %
9:  % User m functions required: none
10: %-----------------------------------------------
11:
12: if nargin < 3, n=0; end
13: x=x(:); nx=length(x); x0=x0(:)'; n0=length(x0);
14: x=x(:,ones(1,n0)); x0=x0(ones(nx,1),:); d=x-x0;
15: s=(d>=zeros(size(d))); v=d.*s;
16: if n==0
```

```
17:    y=s;
18: else
19:    y=v;
20:    for j=1:n-1; y=y.*v; end
21: end
```

Function trapsum

```
 1: function v=trapsum(a,b,y,n)
 2: %
 3: % v=trapsum(a,b,y,n)
 4: % ~~~~~~~~~~~~~~~~~~
 5: %
 6: % This function evaluates:
 7: %
 8: %    integral(a=>x, y(x)*dx) for a<=x<=b
 9: %
10: % by the trapezoidal rule (which assumes linear
11: % function variation between successive function
12: % values).
13: %
14: % a,b - limits of integration
15: % y    - integrand which can be a vector valued
16: %        function returning a matrix such that
17: %        function values vary from row to row.
18: %        It can also be input as a matrix with
19: %        the row size being the number of
20: %        function values and the column size
21: %        being the number of components in the
22: %        in the vector function.
23: % n    - the number of function values used to
24: %        perform the integration.  When y is a
25: %        matrix then n is computed as the number
26: %        of rows in matrix y.
27: %
28: % v    - integral value
29: %
30: % User m functions called:  none
31: %-----------------------------------------------
32:
33: if isstr(y)
34:    % y is an externally defined function
35:    x=linspace(a,b,n)'; h=x(2)-x(1);
```

```
36:    Y=feval(y,x); % Function values must vary in
37:                  % row order rather than column
38:                  % order or computed results
39:                  % will be wrong.
40:    m=size(Y,2);
41: else
42:    % y is column vector or a matrix
43:    Y=y; [n,m]=size(Y); h=(b-a)/(n-1);
44: end
45: v=[zeros(1,m); ...
46:    h/2*cumsum(Y(1:n-1,:)+Y(2:n,:))];
```

Chapter 12

Applications of Analytic Functions

12.1 Properties of Analytic Functions

Complex valued functions of a single complex variable are useful in various disciplines such as physics and numerical approximation theory. The current chapter summarizes a number of attractive properties of analytic functions and presents some applications in which MATLAB is helpful. Excellent textbooks presenting the theory of analytic functions [18, 73, 117] are available which fully develop various theoretical concepts employed in this chapter. Therefore, only the properties which may be helpful in subsequent discussions are included.

12.2 Definition of Analyticity

We consider a complex valued function

$$F(z) = u(x, y) + \imath v(x, y) \qquad z = x + \imath y$$

which depends on the complex variable z. The function $F(z)$ is analytic at point z if it is differentiable in the neighborhood of z. Differentiability requires that the limit

$$\lim_{|\Delta z| \to 0} \left[\frac{F(z + \Delta z) - F(z)}{\Delta z} \right] = F'(z)$$

exists independent of how $|\Delta z|$ approaches zero. Necessary and sufficient conditions for analyticity are continuity of the first partial derivatives of u and v and satisfaction of the Cauchy-Riemann conditions

$$\frac{\partial u}{\partial x} = \frac{\partial v}{\partial y} \qquad \frac{\partial u}{\partial y} = -\frac{\partial v}{\partial x}$$

These conditions can be put in more general form as follows. Let n denote an arbitrary direction in the z-plane and let s be the direction obtained by a 90° counterclockwise rotation from the direction of n. The generalized Cauchy-Riemann conditions are

$$\frac{\partial u}{\partial n} = \frac{\partial v}{\partial s} \qquad \frac{\partial u}{\partial s} = -\frac{\partial v}{\partial n}$$

Satisfaction of the CRC (Cauchy-Riemann conditions) implies that both u and v are solutions of Laplace's equation

$$\nabla^2 u = \frac{\partial^2 u}{\partial x^2} + \frac{\partial^2 u}{\partial y^2} = 0$$

and

$$\frac{\partial^2 v}{\partial x^2} + \frac{\partial^2 v}{\partial y^2} = 0$$

These functions are called harmonic. Functions related by the CRC are also said to be harmonic conjugates. When one function u is known, its harmonic conjugate v can be found within an additive constant by using

$$v = \int dv = \int \frac{\partial v}{\partial x}\, dx + \int \frac{\partial v}{\partial y}\, dy = \int \left(-\frac{\partial u}{\partial y}\, dx + \frac{\partial u}{\partial x}\, dx \right) + \text{constant}$$

Harmonic conjugates also have the properties that curves $u = \text{constant}$ and $v = \text{constant}$ intersect orthogonally. This follows because $u = \text{constant}$ implies du/dn is zero in a direction tangent to the curve. However $du/dn = dv/ds$ so $v = \text{constant}$ along a curve intersecting $u = \text{constant}$ orthogonally.

Sometimes it is helpful to regard a function of x and y as a function of $z = x + \imath y$ and $\bar{z} = x - \imath y$. The inverse is $x = (z + \bar{z})/2$ and $y = (z - \bar{z})/(2\imath)$. Chain rule differentiation applied to a general function ϕ yields

$$\frac{\partial \phi}{\partial x} = \frac{\partial \phi}{\partial z} + \frac{\partial \phi}{\partial \bar{z}} \qquad \frac{\partial \phi}{\partial y} = \imath \frac{\partial \phi}{\partial z} - \imath \frac{\partial \phi}{\partial \bar{z}}$$

$$\left(\frac{\partial}{\partial x} - \imath \frac{\partial}{\partial y} \right) \phi = 2\frac{\partial \phi}{\partial z} \qquad \left(\frac{\partial}{\partial x} + \imath \frac{\partial}{\partial y} \right) \phi = 2\frac{\partial \phi}{\partial \bar{z}}$$

So Laplace's equation becomes

$$\frac{\partial^2 \phi}{\partial x^2} + \frac{\partial^2 \phi}{\partial y^2} = 4\frac{\partial^2 \phi}{\partial z \partial \bar{z}}$$

It is straightforward to show the condition that a function F be an analytic function of z is expressible as

$$\frac{\partial F}{\partial \bar{z}} = 0$$

It is important to note that most of the functions routinely employed with real arguments are analytic in some part of the z-plane. These include, for example,

$$z^n, \ \sqrt{z}, \ \log(z), \ e^z, \ \sin(z), \ \cos(z), \ \arctan(z)$$

to mention a few. The real and imaginary parts of these functions are harmonic and they arise in various physical applications. Especially significant are the integral powers of z. We can write

$$z = re^{i\theta} \qquad r = \sqrt{x^2 + y^2} \qquad \theta = \tan^{-1}\left(\frac{y}{x}\right)$$

and get

$$z^n = u + iv \qquad u = r^n \cos(n\theta) \qquad v = r^n \sin(n\theta)$$

The reader can verify by direct differentiation that both u and v are harmonic.

Points where $F(z)$ is non-differentiable are called singular points and these are categorized as isolated or non-isolated. Isolated singularities are termed either poles or essential singularities. Branch points are the most common type of non-isolated singularity. Singular points and their significance are discussed further below.

12.3 Series Expansions

If $F(z)$ is analytic inside and on the boundary of an annulus defined by $a \leq |z - z_0| \leq b$ then $F(z)$ is representable in a Laurent series of the form

$$F(z) = \sum_{n=-\infty}^{\infty} a_n(z - z_0)^n \qquad a \leq |z - z_0| \leq b$$

where

$$F(z) = \frac{1}{2\pi i} \int_L \frac{F(t)\, dt}{(t - z_0)^{n+1}}$$

and L represents any closed curve encircling z_0 and lying between the inner circle $|z - z_0| = a$ and the outer circle $|z - z_0| = b$. If $F(z)$ is also analytic for $|z - z_0| < a$, the negative powers in the Laurent series drop out to give Taylor's series

$$F(z) = \sum_{n=0}^{\infty} a_n(z - z_0)^n \qquad |z - z_0| \leq b$$

Special cases of the Laurent series lead to classification of isolated singularities as poles or essential singularities. Suppose the inner radius can be made arbitrarily small but nonzero. If the coefficients below some order, say $-m$, vanish but $a_{-m} \neq 0$, we classify z_0 as a pole of order m. Otherwise, we say z_0 is an essential singularity.

Another term of importance in connection with Laurent series is a_{-1}, the coefficient of $(z - z_0)^{-1}$. This coefficient, called the residue at z_0, arises in evaluation of integrals.

12.4 Integral Properties

Analytic functions have many useful integral properties. One of these properties which concerns integrals around closed curves is:

Cauchy-Goursat Theorem: If $F(z)$ is analytic at all points in a simply connected region R then

$$\int_L F(z)\, dz = 0$$

for every closed curve L in the region.

An immediate consequence of this theorem is that the integral of $F(z)$ along any path between two end points z_1 and z_2 is independent of the path (this only applies for simply connected regions).

12.4.1 Cauchy Integral Formula

If $F(z)$ is analytic inside and on a closed curve L bounding a simply connected region R then

$$F(z) \;=\; \frac{1}{2\pi i} \int_L \frac{F(t)\, dt}{t - z} \qquad \text{for } z \text{ inside } L$$

$$F(z) \;=\; 0 \qquad \text{for } z \text{ outside } L$$

The Cauchy integral formula provides a simple means for computing $F(z)$ at interior points when its boundary values are known. We refer to any integral of the form

$$I(z) = \frac{1}{2\pi i} \int_L \frac{F(t)\, dt}{t - z}$$

as a Cauchy integral, regardless of whether $F(t)$ is the boundary value of an analytic function. $I(z)$ defines a function analytic in the complex plane cut along the curve L. When $F(t)$ is the boundary value of a function analytic inside a closed curve L, $I(z)$ is evidently discontinuous across L since $I(z)$ approaches $F(z)$ as z approaches L from the inside but gives zero for an approach from the outside. The theory of Cauchy integrals for both open and closed curves is thoroughly developed in Muskhelishvili's texts [71, 72] and is used to solve many practical problems.

12.4.2   Residue Theorem

If $F(z)$ is analytic inside and on a closed curve L except at isolated singularities $z_1, z_2, \ldots, z_n$ where it has Laurent expansions, then

$$\int_L F(z)\, dz = 2\pi i \sum_{\text{residues}} F(z)$$

where the residue summation means the addition of the coefficient of $(z - z_i)^{-1}$ in the Laurent series at each isolated singularity inside L. In the instance where z_i is a pole of order m, the residue can be computed as

$$a_{-1} = \frac{1}{(m-1)!} \left\{ \frac{d^{m-1}}{dz^{m-1}} [F(z)(z - z_i)^m] \right\}_{\lim z \to z_i}$$

12.5   Physical Problems Leading to Analytic Functions

Several physical phenomena require solutions involving real valued functions satisfying Laplace's equation. Since analytic functions have harmonic real and imaginary parts, it often happens that two-dimensional problems can be formulated concisely in terms of analytic functions. Such useful tools as Taylor series expansions can yield effective computational devices. One of the simplest practical examples involves determining a function u harmonic inside the unit circle $|z| \leq 1$ and having boundary values described by a Fourier series

$$u(r, \theta) = \sum_{n=-\infty}^{\infty} c_n \sigma^n \qquad r = 1,\ \sigma = e^{i\theta}$$

with $c_{-n} = \bar{c}_n$ because u is real. The desired function can be found as

$$u = \mathbf{Real}\,[F(z)]$$

where

$$F(z) = c_0 + 2 \sum_{n=1}^{\infty} c_n z^n \qquad |z| \leq 1$$

This solution is useful because the Fast Fourier Transform (FFT) can be employed to generate Fourier coefficients handling quite general boundary conditions, and the series for $F(z)$ converges rapidly when $|z| < 1$. This series will be employed below to solve both the problem where boundary values are given (the Dirichlet problem) and where normal derivative values are known on the boundary (the Neumann problem). Several applications where analytic functions occur are summarized below.

12.5.1   Steady-State Heat Conduction

The steady-state temperature distribution in a homogeneous two-dimensional body is harmonic. We can take $u = \mathbf{Real}[F(z)]$. Boundary curves where $u = $ constant lead to conditions

$$F(z) + \overline{F(z)} = \text{constant}$$

in the complex variable. Boundary curves which are insulated to prevent heat flow lead to $\partial u / \partial n = 0$ which implies

$$F(z) - \overline{F(z)} = \text{constant}$$

12.5.2   Incompressible Fluid Flow

Some flow problems for incompressible, nonviscous fluids involve velocity components which are obtainable in terms of the first derivative of an analytic function. A complex velocity potential $F(z)$ exists such that

$$u - \imath v = F'(z)$$

At impermeable boundaries the flow normal to the boundary must vanish which implies

$$F(z) - \overline{F(z)} = \text{constant}$$

at a boundary. Furthermore a uniform flow field with $u = U$, $v = V$ is easily described by

$$F(z) = (U - \imath V)z$$

12.5.3   Torsion and Flexure of Elastic Beams

The distribution of stresses in a cylindrical elastic beam subjected to torsion or bending can be computed using analytic functions [88]. For example, in the torsion problem shear stresses τ_{XZ} and τ_{YZ} can be sought as

$$\tau_{XZ} - \imath \tau_{YZ} = \mu \alpha [f'(z) - \imath \bar{z}]$$

and the condition of zero traction on the lateral faces of the beam is described by

$$f(z) - \overline{f(z)} = \imath z \bar{z}$$

If the function $z = \omega(\zeta)$ which maps $|\zeta| \leq 1$ onto the beam cross section is known, then an explicit integral formula solution can be written as

$$f(\zeta) = \frac{1}{2\pi} \int_{|\sigma|=1} \frac{\omega(\sigma)\overline{\omega(\sigma)}d\sigma}{\sigma - \zeta}$$

Consequently, the torsion problem for a beam of simply-connected cross section is represented concisely in terms of the function which maps a circular disk onto the cross section.

12.5.4   Plane Elastostatics

Analyzing the elastic equilibrium of two-dimensional bodies satisfying conditions of plane stress or plane strain can be reduced to determining two analytic functions. The formulas to find three stress components and two displacement components are more involved than the ones just stated. They will be investigated later when stress concentrations in a plate having a circular or elliptic hole are discussed.

12.5.5   Electric Field Intensity

Electromagnetic field theory is concerned with the field intensity ϵ which is described in terms of the electrostatic potential ϕ [90] such that

$$\mathcal{E} = E_x + \imath E_y = -\frac{\partial \phi}{\partial x} - \imath \frac{\partial \phi}{\partial y}$$

where ϕ is a harmonic function at all points not occupied by charge. Consequently a complex electrostatic potential $\Omega(z)$ exists such that

$$\mathcal{E} = -\overline{\Omega'(z)}$$

The electromagnetic problem is closely analogous to fluid flow problems which will be explored later in this chapter. We will also find that harmonic functions remain harmonic under the geometry change of a conformal transformation. This produces interesting situations where solutions for new problems can sometimes be derived by simple geometry changes.

12.6   Branch Points and Multivalued Behavior

Before specific types of maps are examined, we need to consider the concept of branch points. A type of singular point quite different from isolated singularities such as poles arises when a singular point of $F(z)$ cannot be made the interior of a small circle on which $F(z)$ is single valued. Such singularities are called branch points and the related behavior is typified by functions such as $\sqrt{z - z_0}$ and $\log(z-z_0)$. If $p = \log(z-z_0)$, then we should regard as acceptable any value p such that e^p produces the value $z - z_0$. Using polar form we can write

$$(z - z_0) = |z - z_0| e^{\imath(\theta + 2\pi k)} \qquad \theta = \arg(z - z_0)$$

with k being any integer. Taking

$$p = \log|z - z_0| + \imath(\theta + 2\pi k)$$

yields an infinity of values all satisfying $e^p = z - z_0$. Furthermore, if z traverses a counterclockwise circuit around a circle $|z - z_0| = \delta$,

θ increases by 2π and $\log(z - z_0)$ does not return to its initial value, (i.e. $\log(z - z_0)$ is discontinuous on a path containing z_0). A similar behavior is exhibited by $\sqrt{z - z_0}$ which changes sign for a circuit about $|z - z_0| = \delta$.

Functions with branch points have the characteristic behavior that the relevant functions are discontinuous on contours enclosing the branch points. Computing the function involves selection among a multiplicity of possible values. Hence $\sqrt{4}$ can equal $+2$ or -2, and choosing the proper value depends on the functions involved. For sake of definiteness MATLAB uses what are called principal branch definitions such that

$$\sqrt{z} = |z|^{1/2} e^{i\theta/2} \qquad -\pi < \theta = \tan^{-1}\left(\frac{y}{x}\right) \leq \pi$$

and

$$\log(z) = \log|z| + i\theta$$

The functions defined this way have discontinuities across the negative real axis. Futhermore, $\log(z)$ becomes infinite at $z = 0$.

Dealing carelessly with multivalued functions can produce strange results. Consider the function

$$p = \sqrt{z^2 - 1}$$

which will have discontinuities on lines such that $z^2 - 1 = -|h|$. This exhibits trouble when

$$z = \pm\sqrt{1 - |h|}$$

Taking $0 \leq |h| \leq 1$ gives a discontinuity line on the real axis between -1 and $+1$, and taking $|h| > 1$ leads to a discontinuity on the imaginary axis. Figure 12.1 illustrates the odd behavior exhibited by `sqrt(z.^2-1)`. The reader can easily verify that using `sqrt(z-1).*sqrt(z+1)` defines a different function which is continuous in the plane cut along a straight line between -1 and $+1$.

Multivalued functions arise quite naturally in solutions of boundary value problems, and the choices of branch cuts and branch values are usually evident from physical circumstances. For instance, consider a steady-state temperature problem for the region $|z| < 1$ with boundary conditions requiring $u(r, \theta) = 1$, $r = 1$, $0 < \theta < \pi$ and $u(r, \theta) = 0$, $r = 1$, $\pi < \theta < 2\pi$. It can be shown that the desired solution is

$$u = \mathbf{Real}\left\{\frac{1}{\pi i}\left[\log(z + 1) - \log(z - 1)\right]\right\} + \frac{3}{2}$$

where the logarithms must be defined so u is continuous inside the unit circle and u equals $1/2$ at $z = 0$. Appropriate definitions result by taking

$$-\pi < \arg(z + 1) \leq \pi \qquad 0 \leq \arg(z - 1) \leq 2\pi$$

MATLAB does not provide this definition intrinsically, so the user must handle each problem individually when branch points arise.

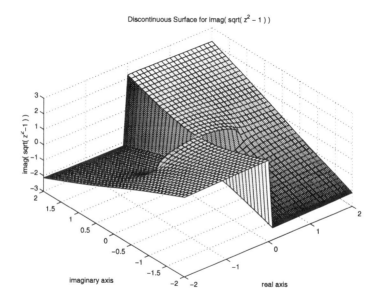

Figure 12.1. Discontinuous Surface for $\mathbf{imag}(\mathbf{sqrt}(z^2 - 1)^{1/2})$

12.7   Conformal Mapping and Harmonic Functions

A transformation of the form

$$x = x(\xi, \eta) \qquad y = y(\xi, \eta)$$

is said to be conformal if the angle between intersecting curves in the (ξ, η) plane remains the same for corresponding mapped curves in the (x, y) plane. Consider the transformation implied by $z = \omega(\zeta)$ where ω is an analytic function of ζ. Since

$$dz = \omega'(\zeta)\, d\zeta$$

it follows that

$$|dz| = |\omega'(\zeta)|\, |d\zeta| \qquad \arg(dz) = \arg[\omega'(\zeta)] + \arg(d\zeta)$$

This implies that the element of length $|d\zeta|$ is stretched by a factor of $|\omega'(\zeta)|$ and the line element $d\zeta$ is rotated by an angle $\arg[\omega'(\zeta)]$. The transformation is conformal at all points where $\omega'(\zeta)$ exists and is nonzero.

Much of the interest in conformal mapping results from the fact that harmonic functions remain harmonic under a conformal transformation. To see why this is true, examine Laplace's equation written in the form

$$\nabla^2_{xy} u = 4 \frac{\partial^2 u}{\partial z \partial \bar{z}}$$

For a conformal map we have

$$z = \omega(\zeta) \qquad \bar{z} = \overline{\omega(\zeta)}$$

$$\frac{\partial u}{\partial z} = \frac{1}{\omega'(\zeta)} \frac{\partial u}{\partial \zeta} \qquad \frac{\partial u}{\partial \bar{z}} = \frac{1}{\overline{\omega'(\zeta)}} \frac{\partial u}{\partial \bar{\zeta}}$$

Since z depends only on ζ and $\bar{z}$ depends only on $\bar{\zeta}$ we find that

$$\nabla^2_{xy} u = 4 \frac{1}{\omega'(\zeta)\overline{\omega'(\zeta)}} \frac{\partial^2 u}{\partial \zeta \partial \bar{\zeta}} = \frac{1}{|\omega'(\zeta)|^2} \nabla^2_{\xi\eta} u$$

It follows that $\nabla^2_{xy} u = 0$ implies $\nabla^2_{\xi\eta} u = 0$ wherever the transformation is conformal, or $\omega'(\zeta) \neq 0$. This is a remarkable and highly useful property. Normally, changing coordinates in a differential equation produces a new differential equation of very different structure. For instance, the polar coordinate transformation $x = r\cos(\theta)$, $y = r\sin(\theta)$ leads to

$$\nabla^2 u = \frac{\partial^2 u}{\partial r^2} + \frac{1}{r}\frac{\partial u}{\partial r} + \frac{1}{r^2}\frac{\partial^2 u}{\partial \theta^2}$$

The differential equation has a very different form because $x + iy$ is not an analytic function of $r + i\theta$. On the other hand, a transformation $z = \log(\zeta)$ produces

$$\nabla^2_{xy} u = (\zeta\bar{\zeta})\nabla^2_{\xi\eta} u$$

and $\nabla^2_{xy} u = 0$ implies $\nabla^2_{\xi\eta} u = 0$ at points other than $\zeta = 0$ or $\zeta = \infty$.

Because solutions to Laplace's equation are very important in physical applications, and such functions remain harmonic under a conformal map, an analogy between problems in two regions often can be useful. This is particularly attractive for problems where the harmonic function has constant values or zero normal gradient on critical boundaries. An instance pertaining to fluid flow about an elliptic cylinder will be used later to illustrate the harmonic function analogy. In the subsequent sections we will discuss several transformations and their relevant geometrical interpretation.

12.8 Mapping onto the Exterior or the Interior of an Ellipse

We will examine in some detail the transformation

$$z = \left(\frac{a+b}{2}\right)\zeta + \left(\frac{a-b}{2}\right)\zeta^{-1} = R(\zeta + m\zeta^{-1}) \qquad \zeta \geq 1$$

where $R = (a+b)/2$ and $m = (a-b)/(a+b)$. The derivative

$$z'(\zeta) = R(1 - m\zeta^{-2})$$

becomes nonconformal when $z'(\zeta) = 0$ or $\zeta = \pm\sqrt{m}$. For sake of discussion, we temporarily assume $a \geq b$ to make $\sqrt{m}$ real rather than purely imaginary. A circle $\zeta = \rho_0 e^{i\theta}$ transforms into

$$x + iy = R(\rho_0 + m\rho_0^{-1})\cos(\theta) + iR(\rho_0 - m\rho_0^{-1})\sin(\theta)$$

yielding an ellipse. When $\rho_0 = 1$ we get $x = a\cos(\theta)$, $y = b\sin(\theta)$. This mapping function is interesting because it is useful in problems such as flow around an elliptic cylinder or stress concentration around an elliptic hole in a plate. Furthermore, the mapping function can be inverted by solving a quadratic equation to give

$$\zeta = \frac{z + \sqrt{(z-\alpha)(z+\alpha)}}{a+b} \qquad \alpha = \sqrt{a^2 - b^2}$$

The radical should be defined to have a branch cut on the x-axis from $-\alpha$ to α and to behave like $+z$ for large $|z|$. Computing the radical in MATLAB as `sqrt(z-alpha).*sqrt(z+alpha)` works fine when α is real because MATLAB uses

$$-\pi < \arg(z \pm \alpha) \leq \pi$$

and the sign change discontinuities experienced by both factors on the negative real axis cancel to make the product of radicals continuous. However, when $a < b$ the branch points occur at $\pm z_0$ where $z_0 = \imath\sqrt{b^2 - a^2}$, and a branch cut is needed along the imaginary axis. We can give a satisfactory definition by requiring

$$-\frac{\pi}{2} < \arg(z \pm z_0) \le \frac{3\pi}{2}$$

The function **elipinvr** provided below handles general a and b.

Before leaving the problem of ellipse mapping we mention the fact that mapping the interior of a circle onto the interior of an ellipse is rather complicated but can be formulated by use of elliptic functions [73]. However, a simple solution to compute boundary point correspondence between points on the circle and points on the ellipse appears in [51]. This can be used to obtain mapping functions in rational form which are quite accurate. The function **elipdplt** produces the mapping. Results showing how a polar coordinate grid in the ζ-plane maps onto a two to one ellipse appears in Figure 12.2. Note that the distortion of line elements is large.

Often it is desirable to see how a rectangular or polar coordinate grid distorts under a mapping transformation. To accomplish this, a surface of constant height in the mapped plane can be drawn. Then, viewing the surface from above shows the curvilinear coordinate lines. A function called **topview** implements this idea for general input arrays x, y. If the input data are vectors instead of arrays, then the routine draws a single curve instead of a surface. When **topview** is executed with no input, it generates the plot, appearing in Figure 12.3, showing how a polar coordinate grid in the ζ-plane maps into conformal ellipses under the transformation

$$z = R\left(\zeta + \frac{m}{\zeta}\right)$$

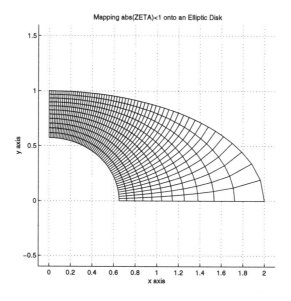

Figure 12.2. Mapping $|z| < 1$ onto an Elliptic Disk

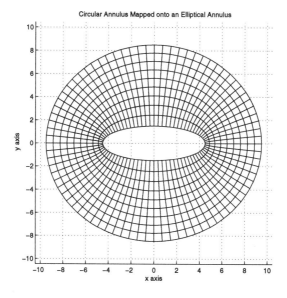

Figure 12.3. Circular Annulus Mapped onto an Elliptic Annulus

12.8.1   Program Output and Code

Function sqrtsurf

```
1: function sqrtsurf
2: %
3: % sqrtsurf
4: % ~~~~~~~~
5: %
6: % This function illustrates the discontinuity
7: % in the function w=sqrt(z*z-1).
8: %
9: % User m functions called: genprint
10: %-------------------------------------------------
11:
12: xx=linspace(-2,2,41); [x,y]=meshgrid(xx,xx);
13: z=x+i*y; w=sqrt(z.*z-1);
14: %colormap('gray'); brighten(0.75);
15: surf(x,y,imag(w)); view(-40,50);
16: xlabel('real axis'); ylabel('imaginary axis');
17: zlabel('imag( sqrt( z^2-1 ) )');
18: title(['Discontinuous Surface for imag( sqrt', ...
19:         '( z^2 - 1 ) )']);
20: grid on; figure(gcf);
21: %genprint('sqrtsurf');
```

Function elipinvr

```
1: function zeta=elipinvr(a,b,z)
2: %
3: % zeta=elipinvr(a,b,z)
4: % ~~~~~~~~~~~~~~~~~~~~
5: %
6: % This function inverts the transformation
7: % z=(a+b)/2*zeta+(a-b)/2/zeta which maps
8: % abs(zeta)>=1 onto (x/a).^2+(y/b).^2 >= 1
9: %
10: % a    - semi-diameter on x-axis
11: % b    - semi-diameter on y-axis
12: % z    - array of complex values
13: %
14: % zeta - array of complex values for the
15: %          inverse mapping function
```

```
16: %
17: % User m functions called:  none
18: %-----------------------------------------------
19:
20: z0=sqrt(a^2-b^2); ab=a+b;
21: if a==b
22:   zeta=z/a;
23: elseif a>b  % branch cut along the real axis
24:   zeta=(z+sqrt(z-z0).*sqrt(z+z0))/ab;
25: else        % branch cut along the imaginary axis
26:   ap=angle(z+z0); ap=ap+2*pi*(ap<=-pi/2);
27:   am=angle(z-z0); am=am+2*pi*(am<=-pi/2);
28:   zeta=(z+sqrt(abs(z.^2-z0.^2)).*exp(...
29:       i/2*(ap+am)))/ab;
30: end
```

Function elipdisk

```
1: function [z,a,b]=elipdisk(zeta,rx,ry)
2: %
3: % [z,a,b]=elipdisk(zeta,rx,ry)
4: % ~~~~~~~~~~~~~~~~~~~~~~~~~~~~~
5: %
6: % This function computes a rational function
7: % mapping abs(zeta)<=1 onto an elliptical disk
8: % defined by (x/rx)^2+(y/ry)^2<=1. Boundary
9: % points are computed using theory from
10: %    P. Henrici, Applied Complex Analysis,
11: %    Vol 3, p391.
12: % The rational function approximation has the
13: % form:
14: %         z=sum(a(j)*zeta^(2*j-1)) /
15: %           (1+sum(b(j)*zeta^(2*j)));
16: %
17: %   zeta  - matrix of points with abs(zeta)<=1
18: %   rx,ry - ellipse semidiameters on x and y
19: %           axes
20: %
21: %   z     - points into which zeta maps
22: %   a,b   - coefficients in the rational
23: %           function defining the map
24: %
25: % User m functions called: ratcof
```

```
26: %--------------------------------------------------
27:
28: ntrms=100; ntheta=251;
29: tau=(0:2*pi/ntheta:2*pi)';
30: ep=(rx-ry)/(rx+ry);
31: z=exp(i*tau); z=z+ep*conj(z);
32: j=1:ntrms;  ep=ep.^j; ep=ep./(j.*(1+ep.*ep));
33: theta=tau+2*( sin((2*tau+pi)*j)*ep');
34: zta=exp(i*theta); z=rx/max(real(z))*z;
35: [a,b]=ratcof(zta.^2,z./zta,8);
36: a=fix(real(1e8*a))/1e8; b=fix(real(1e8*b))/1e8;
37: af=flipud(a(:)); bf=flipud([1;b(:)]);
38: zta2=zeta.^2;
39: z=zeta.*polyval(af,zta2)./polyval(bf,zta2);
40:
```

Function topview

```
1: function topview(x,y,xlabl,ylabl,titl)
2: %
3: % topview(x,y,xlabl,ylabl,titl)
4: % ~~~~~~~~~~~~~~~~~~~~~~~~~~~~~~~
5: %
6: % This function views a surface from the top
7: % to show the coordinate lines of the surface.
8: % It is useful for illustrating how coordinate
9: % lines distort under a conformal transformation.
10: % Calling topview with no arguments depicts the
11: % mapping of a polar coordinate grid map under
12: % a transformation of the form
13: % z=R*(zeta+m/zeta).
14: %
15: %  x,y         - real matrices defining a
16: %                curvilinear coordinate system
17: %  xlabl,ylabl - labels for x and y axes
18: %  titl        - title for the graph
19: %
20: % User m functions called:  cubrange, genprint
21: %-------------------------------------------------
22:
23: close
24: if nargin<5
25:    xlabl='x axis'; ylabl=' y axis'; titl='';
```

```
26: end
27:
28: % Default example using z=R*(zeta+m/zeta)
29: if nargin==0
30:    zeta=linspace(1,3,10)'* ...
31:        exp(i*linspace(0,2*pi,81));
32:    z=3*(zeta+1/2./zeta); x=real(z); y=imag(z);
33:    titl=['Circular Annulus Mapped onto an ', ...
34:          'Elliptical Annulus'];
35: end
36:
37: range=cubrange([x(:),y(:)],1.1);
38:
39: % The data defines a curve
40: if size(x,1)==1 | size(x,2)==1
41:    plot(x,y); xlabel(xlabl); ylabel(ylabl);
42:    title(titl); axis('equal'); axis(range);
43:    grid on; figure(gcf);
44:    if nargin==0
45:      %genprint('topviewl');
46:    end
47: % The data defines a surface
48: else
49:    mesh(x,y,zeros(size(x)));
50:    xlabel(xlabl); ylabel(ylabl); title(titl);
51:    axis('equal'); axis(range); grid on;
52:    view([0,90]); figure(gcf);
53:    if nargin==0
54:      %genprint('topview');
55:    end
56: end
```

Function elipdplt

```
1: function [z,a,b]=elipdplt(rx,ry)
2: %
3: % [z,a,b]=elipdplt(rx,ry)
4: % ~~~~~~~~~~~~~~~~~~~~~~~~
5: % This function plots contour lines showing
6: % how a circular disk maps onto an elliptic
7: % disk.
8: %
9: % User m functions called: elipdisk, genprint
```

```
10: %------------------------------------------------
11:
12: if nargin==0, rx=2; ry=1; end
13: zeta=linspace(.5,1,12)'* ...
14:      exp(i*linspace(0,pi/2,61));
15: [z,a,b]=elipdisk(zeta,rx,ry);
16: x=real(z); y=imag(z);
17: topview(x,y,'x axis','y axis',...
18:    'Mapping abs(ZETA)<1 onto an Elliptic Disk');
19: %genprint('elipdisk');
```

Function ratcof

```
 1: function [a,b]=ratcof(xdata,ydata,ntop,nbot)
 2: %
 3: % [a,b]=ratcof(xdata,ydata,ntop,nbot)
 4: % ~~~~~~~~~~~~~~~~~~~~~~~~~~~~~~~~~~~~~
 5: % Determine a and b to approximate ydata as
 6: % a rational function of the variable xdata.
 7: % The function has the form:
 8: %
 9: %     y(x) = sum(1=>ntop) ( a(j)*x^(j-1) ) /
10: %            ( 1 + sum(1=>nbot) ( b(j)*x^(j)) )
11: %
12: % xdata,ydata - input data vectors (real or
13: %                 complex)
14: % ntop,nbot   - number of series terms used in
15: %                 the numerator and the
16: %                 denominator.
17: %
18: % User m functions called: none
19: %------------------------------------------------
20:
21: ydata=ydata(:); xdata=xdata(:);
22: m=length(ydata);
23: if nargin==3, nbot=ntop; end;
24: x=ones(m,ntop+nbot); x(:,ntop+1)=-ydata.*xdata;
25: for i=2:ntop, x(:,i)=xdata.*x(:,i-1); end
26: for i=2:nbot
27:   x(:,i+ntop)=xdata.*x(:,i+ntop-1);
28: end
29: ab=x\ydata;
30: a=ab(1:ntop); b=ab(ntop+1:ntop+nbot);
```

12.9   Linear Fractional Transformations

The mapping function defined by

$$w = \frac{az + b}{cz + d}$$

is called a linear fractional, or bilinear, transformation where a, b, c, and d are constants. It can be inverted to yield

$$z = \frac{-dw + b}{cw - a}$$

If c is zero the transformation is linear. Otherwise, we can divide out c to get

$$w = \frac{Az + B}{z + D}$$

The three remaining constants can be found by making three points in the z-plane map to three given points in the w-plane. Note that $z = \infty$ maps to $w = A$ and $z = -D$ maps to $w = \infty$.

The transformation has the attractive property that circles or straight lines map into circles or straight lines. An equation defining a circle or straight line in the z-plane has the form

$$Pz\bar{z} + Qz + \bar{Q}\bar{z} + S = 0$$

where P and S are real. A straight line is obtained when P is zero. Putting $z = (-dw + b)/(cw - a)$ into the previous equation and clearing fractions leads to an equation of the form

$$P_0 w\bar{w} + Q_0 w + \bar{Q}_0 \bar{w}_0 + S_0 = 0$$

which defines a circle in the w-plane when P_0 is nonzero. Otherwise, a straight line in the w-plane results.

Determining the bilinear transformation to take three z-points to three w-points is straightforward except for special cases. Let

$$\texttt{Z=[z1;z2;z3]} \qquad \text{and} \qquad \texttt{W=[w1;w2;w3]}$$

If $\texttt{det([Z,W,ones(3,1)])}$ vanishes then a linear transformation with $c = 0$ and $d = 1$ applies. If $z = \infty$ maps to w_1 we take $a = w_1$, $c = 1$. If $z = z_1$ maps to $w = \infty$ we take $c = 1$, $d = -z_1$. In the usual situation we simply write $w(z + D) = Az + B$ and solve the system

$$\texttt{[Z,ones(3,1),-W]*[A;B;D]=W.*Z}$$

Function **linfrac**, used to compute the coefficients in the transformation, is provided at the end of this section. Points at infinity are handled by including ∞ (represented in MATLAB by **inf**) as a legitimate value in the components of z or w. For example, the transformation $w = (2z + 3)/(z - 1)$ takes $z = \infty$ to $w = 2$, $z = 1$ to $w = \infty$, and $z = 1 + \imath$ to $w = 2 - 5\imath$. The expression

```
cz=linfrac([inf,1,1+i],[2,inf,2-5i]);
```

produces the coefficients in the transformation. Similarly, the transformation is inverted by

```
cw=linfrac([2,inf,2-5i],[inf,1,1+i]);
```

or equivalently by

```
cw=linfrac([0,1,2i],[-1.5,-4,-0.25-1.25i]);
```

Another type of problem of interest in connection with a known bilinear transformation is to find the circle or straight line into which a given circle or straight line maps. Function **crc2crc** performs this task. The coefficients c are given along with three points lying on a circle or a straight line. Then parameters w_0, r_0 pertaining to the w-plane are computed. If parameter **type** equals 1, then w_0 and r_0 specify the center and radius of a circle. Otherwise, w_0 and r_0 are two points defining a straight line.

The linear fractional transformation can be used to map an eccentric annulus such as that in Figure 12.4 onto a concentric annulus. Suppose a region $1 \leq |z| \leq R$ is to be mapped onto the region defined by

$$|w| \geq R_1 \qquad |w - w_0| \leq R_0$$

The radius R and mapping coefficients c can be obtained by solving a system of nonlinear simultaneous equations. Function **ecentric** accomplishes the task. A function call of

```
[c,r]=ecentric(0.25,-0.25,1);
```

produces

$$w = \frac{3.4821z + 0.25}{z + 13.9282} \qquad R = 3.7321$$

and the plot shown in Figure 12.4.

To demonstrate the utility of the transformation just discussed, consider the problem of determining the steady state temperature field in an eccentric annulus with the inner and outer boundaries held at u_1 and u_0, respectively. The temperature field will be a harmonic function which remains harmonic under a conformal transformation. The related problem for the concentric annulus has the simple form

$$u = u_1 + \frac{(u_0 - u_1)\ln(r)}{\ln(R)} \qquad 1 \le r \le R$$

By analogy, expressing $r = |z|$ in terms of w gives the temperature distribution at points in the w-plane.

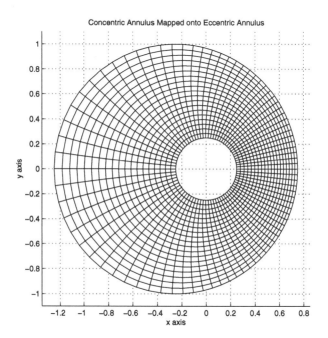

Figure 12.4. Concentric Annulus Mapped onto Eccentric Annulus

12.9.1   Program Output and Code

Function linfrac

```
 1: function c=linfrac(z,w)
 2: %
 3: % c=linfrac(z,w)
 4: % ~~~~~~~~~~~~~~~
 5: %
 6: % This function determines the linear
 7: % fractional transformation to map any three
 8: % points in the z-plane into any three points
 9: % in the w plane. Not more than one point in
10: % either the z or w plane may be located at
11: % infinity.
12: %
13: % z  - vector of complex values [z1,z2,z3]
14: % w  - vector of complex values [w1,w2,w3]
15: %
16: % c  - vector defining the bilinear
17: %        transformation
18: %          w=(c(1)*z + c(2))/(c(3)*z + c(4))
19: %
20: % User m functions called:  none
21: %-------------------------------------------------
22:
23: z=z(:); w=w(:); c=ones(4,1);
24: k=find(z==inf); j=find(w==inf); kj=[k;j];
25:
26: % z and w both contain points at infinity
27: if length(kj)==2
28:   c(1)=w(k); c(4)=-z(j); w(kj)=[]; z(kj)=[];
29:   c(2)=(w-c(1))*z+w*c(4);
30:   return
31: end
32:
33: % z=infinity maps to a finite w point
34: if ~isempty(k) & isempty(j)
35:   c(1)=w(k); z(k)=[]; w(k)=[];
36:   c([2 4])=[[1;1],-w]\[(w-c(1)).*z];
37:   return
38: end
39:
40: % a finite z point maps to w = infinity
```

```
41: if ~isempty(j) & isempty(k)
42:   c(4)=-z(j); z(j)=[]; w(j)=[];
43:   c([1 2])=[z,[1;1]]\[w.*(z+c(4))];
44:   return
45: end
46:
47: % case where all  points are finite
48: mat=[z,ones(3,1),-w];
49:
50: % case for a general transformation
51: if det(mat)~=0
52:   c([1 2 4])=mat\[w.*z];
53: % case where transformation is linear
54: else
55:   c(3)=0; c([1 2])=[z,ones(3,1)]\w;
56: end
```

Function crc2crc

```
1: function [w0,r0,type]=crc2crc(c,z)
2: %
3: % [w0,r0,type]=crc2crc(c,z)
4: % ~~~~~~~~~~~~~~~~~~~~~~~~~
5: %
6: % This function determines the circle or
7: % straight line into which a circle or straight
8: % line maps under a linear fractional
9: % transformation.
10: %
11: % c     - coefficients defining a linear
12: %           fractional transformation
13: %             w=(c(1)*z+c(2))/(c(3)*z*c(4))
14: %           where c(2)*c(3)-c(1)*c(4) is nonzero
15: % z     - a vector of three complex values
16: %           lying on a circle or a straight line
17: %
18: % w0    - center of a circle in the w plane
19: %           if type=1, or a point on a straight
20: %           line if type=2
21: % r0    - radius of a circle in the w plane
22: %           if type=1, or a point on a straight
23: %           line if type=2
24: % type  - equals 1 to denote a circle or 2 to
```

```
25: %          denote a straight line in the w plane
26: %
27: % User m functions called:  none
28: %----------------------------------------------
29:
30: % check for degenerate transformation
31: if c(2)*c(3)==c(1)*c(4)
32:   disp(['Degenerate transformation in ', ...
33:         'function crc2crc']);
34:   w0=[]; r0=[]; type=[]; return;
35: end
36:
37: % evaluate the mapping of the z points
38: w=(c(1)*z(:)+c(2))./(c(3)*z(:)+c(4));
39:
40: % check whether a point passes to infinity or
41: % the three z points defines a straight line
42: k=find(w==inf);
43: dt=det([real(w),imag(w),ones(3,1)]);
44: if ~isempty(k); w(k)=[]; end
45:
46: % case for a straight line in the w plane
47: % defined by two point on the line
48: if dt==0 | ~isempty(k)
49:   type=2; w0=w(1); r0=w(2);
50: % case for a circle in the w plane defined by
51: % a center point and the circle radius
52: else
53:   type =1;
54:   v=[2*real(w),2*imag(w),ones(3,1)]\abs(w).^2;
55:   w0=v(1)+i*v(2); r0=sqrt(v(3)+abs(w0)^2);
56: end
```

Function ecentric

```
1: function [c,r]=ecentric(ri,wo,ro)
2: %
3: % [c,r]=ecentric(ri,wo,ro)
4: % ~~~~~~~~~~~~~~~~~~~~~~~~~
5: %
6: % This function determines the bilinear
7: % transformation which maps the region
8: % 1<=abs(z)<=r onto an eccentric annulus
```

```
 9: % defined by
10: %       abs(w)>=ri & abs(w-wo)<=ro
11: %
12: % The coefficients c in the transformation
13: %       w=(c(1)*z+c(2))/(c(3)*z+c(4))
14: % must be found as well as the outer radius r
15: % of the annulus in the z plane.
16: %
17: % ri  - radius of inner circle abs(w)=ri
18: % wo  - center of outer circle abs(w-wo)=ro
19: % ro  - radius of outer circle
20: %
21: % c   - coefficients in the mapping function
22: % r   - radius of outer circle abs(z)=r
23: %
24: % User m functions called: topview, genprint
25: %------------------------------------------------
26:
27: c1=(wo+ro)/ri; c2=(wo-ro)/ri; c3=2/(c1+c2);
28: c4=(c2-c1)/(c1+c2); c5=c3-c1-c1*c4; c6=1-c1*c3;
29: rt=sqrt(c5^2-4*c4*c6);
30: r1=(-c5+rt)/(2*c4); r2=(-c5-rt)/(2*c4);
31: r=max([r1,r2]); d=c3+c4*r; c=[ri*d;ri;1;d];
32:
33: % Show the region onto which a polar coordinate
34: % grid in the z-plane maps.
35: z=linspace(1,r,20)'*exp(i*linspace(0,2*pi,81));
36: w=(c(1)*z+c(2))./(c(3)*z+c(4));
37: titl=['Concentric Annulus Mapped onto ', ...
38:       'Eccentric Annulus'];
39: topview(real(w),imag(w),'x axis','y axis',titl);
40: %genprint('ecentric');
```

12.10   Schwarz-Christoffel Mapping onto a Square

The Schwarz-Christoffel transformation [73] provides integral formulas defining transformations to map the interior of a circle onto the interior or exterior of a polygon. Special cases obtained by allowing selected vertices to pass to infinity lead to a variety of results [57]. In general situations, evaluating the parameters and integrals in the Schwarz-Christoffel transformation is difficult and requires use of special software [34]. We will examine only two cases: a) where the interior of a circle is mapped onto the interior of a square, and b) where the exterior of a circle is mapped onto the exterior of a square. The function

$$z = \text{constant} \int_0^\zeta (1 + t^4)^{-1/2} \, dt$$

maps $|\zeta| \le 1$ inside the square defined by $|x| \le 1$, $|y| \le 1$. Expanding this radical by the binominal expansion gives

$$z = c \sum_{n=0}^\infty (-1)^n \left[\frac{\Gamma(n + \frac{1}{2})}{n!(4n + 1)} \right] \zeta^{1+4n} \qquad |\zeta| \le 1$$

A reasonably good approximation to the mapping function can be obtained by taking several hundred terms in the mapping function and adjusting the constant c to make $\zeta = 1$ match $z = 1$. This series expansion converges slowly and rounds the corners of the square because the derivative of the mapping function behaves like $(\zeta - \zeta_o)^{-1/2}$ at $\zeta_o = \pm e^{\pm i\pi/4}$.

The transformation to map $|\zeta| \ge 1$ onto the exterior of the square $|x| \le 1$, $|y| \le 1$ has the form

$$z = \text{constant} \int_1^\zeta (1 + t^{-4})^{1/2} \, dt + c_o$$

Using the binomial expansion again leads to

$$z = c \sum_{n=0}^\infty (-1)^n \left[\frac{\Gamma(n - \frac{1}{2})}{n!(4n - 1)} \right] \zeta^{1-4n} \qquad |\zeta| \ge 1$$

The function **swcsqmap** provides both interior and exterior polynomial maps. Once again, truncating the series after a specified number of terms and making $\zeta = 1$ map to $z = 1$ gives an approximate mapping function which converges much more rapidly than the series for the interior problem. Rounding of the square corners is greatly reduced because the mapping function derivative behaves like $(\zeta - \zeta_o)^{1/2}$ at $\zeta_o = \pm e^{\pm i\pi/4}$. Figures 12.5 and 12.6 also illustrate results produced by the ten term series for both interior and exterior regions. Using rational functions to produce better results than polynomials was discussed earlier in Chapter 3. The function **squarat**, which provides both interior and exterior maps, appears below.

It should be noted that inverting a mapping function $z = \omega(\zeta)$ to get $\zeta = g(z)$ explicitly is often difficult, if not impossible. For example, consider the form

$$z = \frac{\zeta(a + b\zeta^4 + c\zeta^8)}{1 + d\zeta^4 + e\zeta^8} \qquad |\zeta| \leq 1$$

which requires solving the polynomial

$$c\zeta^9 - ez\zeta^8 + b\zeta^5 - dz\zeta^4 + a\zeta - z = 0$$

and picking the root inside the unit circle. Even though MATLAB has an efficient root finder to factor polynomials with complex coefficients, inverting the mapping function for hundreds or thousands of values can be time consuming.

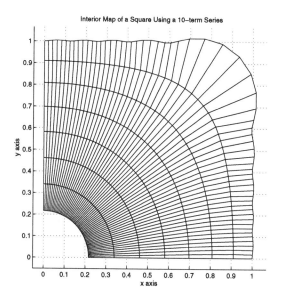

Figure 12.5. Interior Map Using a 10-term Series

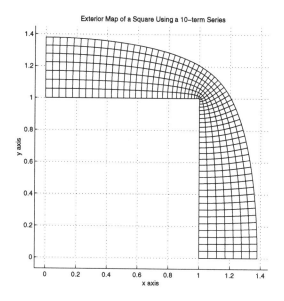

Figure 12.6. Exterior Map Using a 10-term Series

12.10.1   Program Output and Code

Script File swcsq10

```
 1: % Example: swcsq10
 2: % ~~~~~~~~~~~~~~~~~~
 3: %
 4: % This example demonstrates square map
 5: % approximations pertaining to truncated
 6: % Schwarz-Christoffel transformations.
 7: %
 8: % User m functions called: pauz, genprint
 9: %-----------------------------------------------
10:
11: zeta=linspace(0.2,1,8)'* ...
12:      exp(i*linspace(0,pi/2,61));
13: [z,a]=swcsqmap(zeta,10);
14: topview(real(z),imag(z),'x axis','y axis', ...
15:         ['Interior Map of a Square Using', ...
16:         ' a 10-term Series']);
17: disp('Press [Enter] to continue'); pauz;
18: %genprint('insqr10');
19:
20: zeta=linspace(1,1.25,8)'* ...
21:      exp(i*linspace(0,pi/2,61));
22: [z0,a]=swcsqmap(zeta,10,1);
23: topview(real(z0),imag(z0),'x axis','y axis', ...
24:         ['Exterior Map of a Square Using ', ...
25:         'a 10-term Series']);
26: %genprint('outsqr10');
```

Function swcsqmap

```
 1: function [z,a]=swcsqmap(zeta,ntrms,ifout)
 2: %
 3: % [z,a]=swcsqmap(zeta,ntrms,ifout)
 4: % ~~~~~~~~~~~~~~~~~~~~~~~~~~~~~~~~~~
 5: %
 6: % This function evaluates power series
 7: % approximations for mapping either the inside
 8: % of a circle onto the inside of a square, or
 9: % mapping the outside of a circle onto the
10: % outside of a square. The Schwarz-Christoffel
11: % integrals defining the mapping functions are
12: % expanded in Taylor series and are truncated
13: % to produce approximations in the following
14: % polynomial forms:
15: %
16: %    For the interior problem:
17: %       z=sum(a(n)*zeta^(4*n-3),n=1:ntrms)
18: %
19: %    For the exterior problem:
20: %       z=sum(a(n)*zeta^(-4*n+5),n=1:ntrms)
21: %
22: % The side length of the square is adjusted
23: % to equal 2.
24: %
25: % zeta  - complex values where the mapping
26: %          function is evaluated
27: % ntrms - number of terms used in the
28: %          truncated series
29: % ifout - a parameter omitted if an interior
30: %          map applies. ifout can have any
31: %          value (such as 1) to show that an
32: %          exterior map is to be performed.
33: %
34: % z    - values of the mapping function
35: % a    - coefficients in the mapping series
36: %
37: % User m functions called:  none
38: %------------------------------------------------
39:
40: n=0:ntrms-2;
41: if nargin==2    % recursion formula for mapping
42:                 % interior on interior
```

```
43:    p1=(n+1/2)./(n+1); p2=(n+1/4)./(n+5/4);
44: else              % recursion formula for mapping
45:                   %exterior on exterior
46:    p1=(n-1/2)./(n+1); p2=(n-1/4)./(n+3/4);
47: end
48: a=[1,cumprod(-p1.*p2)]; a=a(:)/sum(a);
49: z4=zeta.^4;
50: if nargin ==3, z4=1./z4; end;
51: z=zeta.*polyval(flipud(a(:)),z4);
```

Function squarat

```
1: function [z,a,b]=squarat(zeta,ifout)
2: %
3: % [z,a,b]=squarat(zeta,ifout)
4: % ~~~~~~~~~~~~~~~~~~~~~~~~~~~~~
5: %
6: % This function maps either the interior of a
7: % circle onto the interior of a square, or maps
8: % the exterior of a circle onto the exterior of
9: % a square using a rational function having the
10: % approximate form:
11: %
12: %   z(zeta) = zeta *
13: %
14: %           Sum(a(j)*zeta4^j)/(1+Sum(b(j)*zeta4^j)
15: %
16: % where zeta4=zeta^4 for an interior problem,
17: % or zeta4=zeta^(-4) for an exterior problem.
18: %
19: % zeta  - matrix of complex values such that
20: %           abs(zeta)<=1 for an interior map,
21: %           or abs(zeta)>=1 for an exterior map
22: % ifout - parameter present in the call list
23: %           only when an exterior mapping is
24: %           required
25: %
26: % z     - matrix of values of the mapping
27: %           function
28: % a,b   - coefficients of the polynomials
29: %           defining the rational mapping
30: %           function
31: %
```

```
32: % User m functions called: none
33: %-------------------------------------------------
34:
35: zeta4=zeta.^4;
36:
37: if nargin==1 % map interior on interior
38:   a=[ 1.07835, 1.37751,-0.02642, -0.09129, ...
39:       0.13460,-0.15763, 0.07430,  0.14858, ...
40:       0.01878,-0.00354 ]';
41:   b=[ 1.37743, 0.07157,-0.11085,  0.12778, ...
42:      -0.13750, 0.05313, 0.14931,  0.02683, ...
43:      -0.00350,-0.000120 ]';
44: else           % map exterior on exterior
45:   a = [1.18038, 1.10892, 0.13365, -0.02910]';
46:   b = [1.10612, 0.27972, 0.00788]';
47:   zeta4=1./zeta4;
48: end
49:
50: % Evaluate the mapping function
51: af=flipud(a); bf=flipud([1;b]);
52: z=zeta.*polyval(af,zeta4)./polyval(bf,zeta4);
```

12.11   Determining Harmonic Functions in a Circular Disk

The problem of determining a function harmonic for $|z| < 1$ and satisfying certain boundary conditions can be analyzed effectively using complex series methods. Three basic problems will be considered.

I) Dirichlet Problem

$$\nabla^2 u = 0 \qquad |z| < 1$$

$$u(r, \theta) = f(\theta) \qquad r = 1 \qquad 0 \leq \theta \leq 2\pi$$

We assume $f(\theta)$ is a real piecewise continuous function representable in a Fourier series as

$$f(\theta) = \sum_{n=-\infty}^{\infty} f_n \sigma^n \qquad \sigma = e^{i\theta} \qquad f_{-n} = \overline{f_n}$$

Then u is given by the series

$$u = f_0 + 2\,\textbf{Real}\left[\sum_{n=1}^{\infty} f_n z^n\right] \qquad z = r\sigma \qquad 0 \leq r \leq 1$$

II) Neumann Problem

$$\nabla^2 u = 0 \qquad |z| < 1$$

$$\frac{\partial u}{\partial r} = g(\theta) \qquad r = 1 \qquad 0 \leq \theta \leq 2\pi$$

We assume that the gradient function g is expandable in a Fourier series as

$$g(\theta) = \sum_{n=-\infty}^{\infty} g_n \sigma^n \qquad \sigma = e^{i\theta} \qquad g_{-n} = \overline{g_n}$$

where

$$g_0 = \frac{1}{2\pi} \int_0^{2\pi} g(\theta)\, d\theta = 0$$

Vanishing of g_0 is a fundamental existence requirement for the Neumann problem to be solvable. The series solution is

$$u = 2\,\textbf{Real}\left[\sum_{n=1}^{\infty} \left(\frac{g_n}{n}\right) z^n\right] + c \qquad z = r\sigma \qquad 0 \leq r \leq 1$$

where c is an arbitrary real constant.

III) **Mixed Problem**

In the third type of problem the function value is specified on one part of the boundary and the normal gradient is specified on the remainder. In the general situation a solution can be constructed by methods using Cauchy integrals [72]. Only a simple case will be examined here. We require

$$\nabla^2 u = 0 \qquad |z| < 1$$

$$u(r, \theta) = f(\theta) \qquad r = 1 \qquad \theta_1 < \theta < \theta_2$$

$$\frac{\partial u}{\partial r} = g(\theta) \qquad r = 1 \qquad \theta_2 < \theta < (2\pi + \theta_1)$$

For convenience use the notation

$$a = e^{i\theta_1} \qquad b = e^{i\theta_2}$$
$$L: \qquad z = e^{i\theta} \qquad \theta_1 < \theta < \theta_2$$
$$L': \qquad z = e^{i\theta} \qquad \theta_2 < \theta < (2\pi + \theta_1)$$

The mixed problem can be reduced to a case where g is zero by first solving a Neumann problem for a harmonic function v such that

$$\frac{\partial v}{\partial r} = g(\theta) \qquad \text{on} \qquad L'$$

$$\frac{\partial v}{\partial r} = -\frac{\int_{\theta_2}^{2\pi+\theta_1} g(\theta)\, d\theta}{\theta_2 - \theta_1} \qquad \text{on} \qquad L$$

Then we replace $f(\theta)$ by $f(\theta) - v(1, \theta)$ to get a problem where

$$u = f(\theta) - v(1, \theta) \qquad \text{on} \qquad L$$

$$\frac{\partial u}{\partial r} = 0 \qquad \text{on} \qquad L'$$

The complete solution then equals the sum of u and v. Consequently, no loss of generality results in dealing with the problem

$$u = f \qquad \text{on} \qquad L$$

$$\frac{\partial u}{\partial r} = 0 \qquad \text{on} \qquad L'$$

Consider the function

$$R(z) = \sqrt{(z - a)(z - b)}$$

defined in the complex plane cut along L. We choose the branch of R satisfying

$$R(0) = e^{i(\theta_1 + \theta_2)/2}$$

The solution to the mixed boundary value problem can be expressed as

$$u = \textbf{Real} \left[\frac{R(z)}{\pi \imath} \int_L \frac{f(t)\, dt}{R^+(t)(t-z)} \right] \qquad t = e^{\imath\theta} \qquad \theta_1 < \theta < \theta_2$$

where $R^+(t)$ means the boundary value of $R(z)$ on the inside of the arc. As an example take

$$\theta_1 = +\frac{\pi}{2} \qquad \theta_2 = -\frac{\pi}{2}$$

$$a = -\imath \qquad b = +\imath$$

$$R(z) = \sqrt{z^2 + 1} \qquad R(0) = 1$$

$$u = \cos(\theta) \qquad -\frac{\pi}{2} \le \theta \le \frac{\pi}{2}$$

Carrying out the integration gives

$$u = \textbf{Real}\,[F(z)]$$

$$F(z) = \frac{z + z^{-1} + (1 - z^{-1})\sqrt{z^2 + 1}}{2} \qquad |z| \le 1$$

This function is employed as a test case in subsequent calculations. The exact solution is evaluated in function **mbvtest**.

12.11.1   Numerical Results

The function **lapcrcl** solves either Dirichlet or Neumann problems for the unit disk. The boundary values are specified as piecewise linear functions of the polar angle. Then function **lintrp** is used to obtain a dense set of boundary values which are transformed by the FFT to produce coefficients in the series solution. When **lapcrcl** is executed with no input data, a Dirichlet problem is solved having the boundary condition

$$u(1, \theta) = 1 + \frac{\cos(16\theta)}{10} \qquad -\frac{\pi}{2} < \theta < \frac{\pi}{2}$$

$$u(1, \theta) = \frac{\cos(16\theta)}{10} \qquad \frac{\pi}{2} < \theta < \frac{3\pi}{2}$$

This boundary condition was selected to produce the interesting surface plot shown in Figure 12.7

The mixed boundary value problem is more difficult to handle than the Dirichlet or Neumann problems because numerical evaluation of the Cauchy integral must be performed cautiously. As z approaches a point on L, the integrand becomes singular. Theoretical developments involving Cauchy principal value integrals and the Plemelj formulas are needed to handle this situation thoroughly [72]. Even when z is close to the boundary, large integrand magnitude may cause inaccurate numerical integration. Furthermore, the integrand will have square root type singularities at the ends of L unless $f(a) = f(b) = 0$. Regularization procedures which can cope fully with these difficulties [25] will not be investigated in this text. Instead a simplified approach is presented.

The function **cauchint** was written to evaluate a contour integral of the form

$$F(z) = \frac{1}{2\pi i} \int_L \frac{f(\zeta)\, d\zeta}{\zeta - z}$$

with the density function f, as well as the shape of L, being defined using cubic spline interpolation. A set of points

$$[\zeta_1, \zeta_2, \ldots, \zeta_m] \qquad \zeta = \xi + i\eta$$

lying on L, along with boundary values

$$[f(\zeta_1), f(\zeta_2), \ldots, f(\zeta_m)] = [f_1, f_2, \ldots, f_m]$$

are given. Spline functions $\zeta(t)$, $f(t)$ are defined for $1 \le t \le m$ such that

$$\zeta(j) = \zeta_j \qquad \text{and} \qquad f(j) = f_j \qquad j = 1, 2, \ldots, n$$

The integrand in parametric form becomes

$$F(z) = \frac{1}{2\pi i} \int_1^m \frac{f(t)\,[\xi'(t) + i\eta'(t)]\, dt}{\zeta(t) - z}$$

and this integral is evaluated using function **gcbpwf** which produces Gaussian base points and weight factors. This routine uses eigenvalue methods and computes integration formulas of order twenty or so very rapidly.

Function **cauchtst** was employed to produce an approximate solution of the problem cited above. A surface plot of the exact solution appears in Figure 12.8. A plot of the difference between the exact and approximate solutions for $0 \le r \le 0.95$ is shown in Figure 12.9. This error is about three orders-of-magnitude smaller than the maximum function values in the solution. The reader can verify that using $r = 0.999$ and $-\pi/2 < \theta < \pi/2$ will lead to much larger errors. Nevertheless, the authors have found function **cauchint** to be helpful provided proper caution is exercised when it is used.

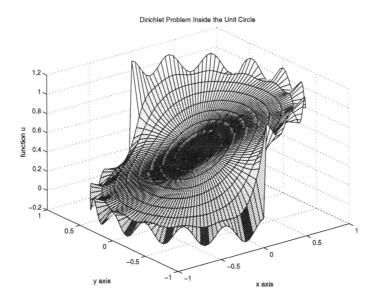

Figure 12.7. Dirichlet Problem Inside the Unit Circle

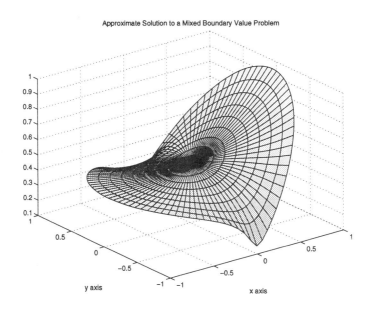

Figure 12.8. Approximate Solution to a Mixed Boundary Value Problem

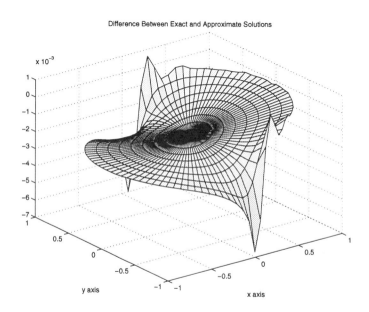

Figure 12.9. Difference Between Exact and Approximate Solutions

12.11.2   Program Output and Code

Function lapcrcl


```
 1: function [u,r,th]=lapcrcl ...
 2:                     (bvtyp,bvdat,rvec,thvec,nsum)
 3: %
 4: % [u,r,th]=lapcrcl(bvtyp,bvdat,rvec,thvec,nsum)
 5: % ~~~~~~~~~~~~~~~~~~~~~~~~~~~~~~~~~~~~~~~~~~~~~~~
 6: %
 7: % This function solves Laplace's equation
 8: % inside a circle of unit radius. Either a
 9: % Dirichlet problem or a Neumann problem can be
10: % analyzed using boundary values defined by
11: % piecewise linear interpolation of data
12: % specified in terms of the polar angle.
13: %
14: % bvtyp       - parameter determining what type
15: %               of boundary value problem is
16: %               solved. If bvtyp equals one,
17: %               boundary data specifies function
18: %               values and a Dirichlet problem
19: %               is solved. Otherwise, the
20: %               boundary data specifies values
21: %               of normal gradient, and a Neumann
22: %               problem is solved if, in accord
23: %               with the existence conditions for
24: %               this problem, the average value
25: %               of gradient on the boundary is
26: %               zero (negligibly small in an
27: %               approximate solution).
28: % bvdat       - a matrix of boundary data. Each
29: %               bvdat(j,:) gives a function value
30: %               and polar angle (in degrees) of
31: %               a data point used by function
32: %               lintrp to linearly interpolate
33: %               for all other boundary values
34: %               needed to generate the solution.
35: % rvec,thvec  - vectors of radii and polar
36: %               coordinate values used to form a
37: %               polar coordinate grid of points
38: %               inside the unit circle. No values
39: %               of r exceeding unity are allowed.
40: % nsum        - the number of terms summed in the
```

```
41: %                    series expansion of the analytic
42: %                    function which has u as its real
43: %                    part. Typically, no more than one
44: %                    hundred terms are needed to
45: %                    produce a good solution.
46: %
47: % u              - values of the harmonic function
48: %                    evaluated at a set of points on
49: %                    a polar coordinate grid inside
50: %                    the unit circle.
51: % r,th           - the grid of polar coordinate
52: %                    values in which the function is
53: %                    evaluated
54: %
55: % User m functions called:  lintrp, genprint
56: %-------------------------------------------------
57:
58: % Default test case solves a Dirichlet problem
59: % for a function having the following exact
60: % solution:
61: %
62: % -1/2+imag(log((z-i)/(z+i))/pi)+real(z^16))/10
63: %
64: if nargin ==0
65:   bvtyp=1; th=linspace(0,2*pi,201)';
66:   bv=1-(th>pi/2)+(th>3*pi/2)+cos(16*th)/10;
67:   bvdat=[bv,180/pi*th];
68:   rvec=linspace(1,0,10);
69:   thvec=linspace(0,360,161); nsum=80;
70: end
71:
72: nft=512;
73: thfft=linspace(0,2*pi*(nft-1)/nft,nft);
74: if nargin<5, nsum=80; end;
75: nsum=min(nsum,nft/2-1);
76: fbv=bvdat(:,1); thbv=pi/180*bvdat(:,2);
77: nev=size(bvdat,1); nr=length(rvec);
78: nth=length(thvec); neval=nr*nth;
79: [R,Th]=meshgrid(rvec,pi/180*thvec);
80: r=R(:); th=Th(:);
81:
82: % Check for any erroneous points outside the
83: % unit circle
84: rvec=rvec(:);
85: kout=find(rvec>1); nout=length(kout);
```

```
86: if length(kout)>0
87:    print('Input data is incorrect. The ');
88:    print('following r values lie outside the ');
89:    print('unit circle:'); disp(rvec(kout)');
90:    return
91: end
92:
93: if bvtyp==1 % Solve a Dirichlet problem
94:    % Check for points on the boundary where
95:    % function values are known. Interpolate
96:    % these directly
97:    konbd=find(r==1); onbndry=length(konbd);
98:    if onbndry > 0
99:       u(konbd)=lintrp(thbv,fbv,th(konbd));
100:   end
101:
102:   % Evaluate the series solution
103:   kinsid=find(r<1); inside=length(kinsid);
104:
105:   if inside > 0
106:      a=fft(lintrp(thbv,fbv,thfft));
107:      a=a(1:nsum)/(nft/2);
108:      a(1)=a(1)/2; Z=r(kinsid).*exp(i*th(kinsid));
109:      u(kinsid)=real(polyval(flipud(a(:)),Z));
110:   end
111:
112:   titl= ...
113:   'Dirichlet Problem Inside the Unit Circle';
114:
115: else % Solve a Neumann problem
116:   gbv=lintrp(thbv,fbv,thfft);
117:   a=fft(gbv)/(nft/2);
118:   erchek=abs(a(1))/sum(abs(gbv));
119:   if erchek>1e-3
120:      disp(' ');
121:      disp('ERROR DUE TO NONZERO AVERAGE VALUE');
122:      disp('OF NORMAL GRADIENT ON THE BOUNDARY.');
123:      disp('CORRECT THE INPUT DATA AND RERUN.');
124:      disp(' '); u=[]; r=[]; th=[]; return;
125:   end
126:   a=a(2:nsum)./(1:nsum-1)'; z=r.*exp(i*th);
127:   u=real(polyval(flipud([0;a(:)]),z));
128:   titl='Neumann Problem Inside the Unit Circle';
129: end
130:
```

```
131: u=reshape(u,nth,nr); r=R; th=Th;
132: surf(r.*cos(th),r.*sin(th),u);
133: xlabel('x axis'); ylabel('y axis');
134: zlabel('function u'); title(titl);
135: %colormap('gray'); brighten(0.95);
136: colormap('default');
137: grid on; figure(gcf);
138: %genprint('dirich');
```

Function cauchint

```
 1: function F=cauchint(fb,zb,z,nquad)
 2: %
 3: % F=cauchint(fb,zb,z,nquad)
 4: % ~~~~~~~~~~~~~~~~~~~~~~~~~~
 5: %
 6: % This function numerically evaluates a Cauchy
 7: % integral of the form:
 8: %
 9: %   F(z)=1/(2*pi*i)*Integral(f(t)/(t-z)*dt)
10: %
11: % where t denotes points on a curve in the
12: % complex plane. The boundary curve is defined
13: % by spline interpolation through data points
14: % zb lying on the curve. The values of f(t)
15: % are also specified by spline interpolation
16: % through values fb corresponding to the
17: % points zb. Numerical evaluation of the
18: % integral is performed using a composite
19: % Gauss formula of arbitrary order.
20: %
21: % fb    - values of density function f
22: %           at point on the curve
23: % zb    - points where fb is given. The
24: %           number of values of zb must be
25: %           adequate to define the curve
26: %           accuratley.
27: % z     - a matrix of values at which the
28: %           Cauchy integral is to be evaluated.
29: %           If any of the z-values lie on path
30: %           of integration or too close to the
31: %           path of integration, incorrect
32: %           results will be obtained.
```

```
33: % nquad - the order of Gauss quadrature
34: %          formula used to perform numerical
35: %          integration
36: %
37: % F      - The value of the Cauchy integral
38: %          corresponding to matrix argument z
39: %
40: % User m functions called: gcbpwf, spc, spltrp
41: %------------------------------------------------
42:
43: n=length(fb); [nr,nc]=size(z); z=z(:).';
44: nz=length(z); t=1:n;
45: [bp,wf]=gcbpwf(1,n,nquad,n-1);
46: matf=spc(t,fb,[3,3],[0,0]);
47: matz=spc(t,zb,[3,3],[0,0]);
48: fq=spltrp(bp,matf,0);
49: zq=spltrp(bp,matz,0); zq=zq(:);
50: zqd=spltrp(bp,matz,1); fq=fq(:).*zqd(:);
51: nq=length(fq);
52: bdrylen=sum(abs(zq(2:nq)-zq(1:nq-1)));
53: closnes=zq(:,ones(1,nz))-z(ones(nq,1),:);
54: closnes=min(abs(closnes(:))); bigz=max(abs(z));
55:
56: if closnes/bdrylen<.01 | closnes/bigz<.01
57:   disp(['WARNING! SOME DATA VALUES ARE ', ...
58:          'EITHER NEAR OR ON']);
59:   disp(['THE BOUNDARY. COMPUTED RESULTS ', ...
60:          'MAY BE INACCURATE']);
61: end
62: F=wf(:)'*(fq(:,ones(1,nz))./(zq(:,ones(1,nz))...
63:                             -z(ones(nq,1),:)));
64: F=reshape(F,nr,nc)/(2*pi*i);
```

Function gcbpwf

```
1: function [cbp,cwf]=gcbpwf ...
2:                     (xlow,xhigh,nquad,mparts)
3: %
4: % [cbp,cwf]=gcbpwf(xlow,xhigh,nquad,mparts)
5: % ~~~~~~~~~~~~~~~~~~~~~~~~~~~~~~~~~~~~~~~~~~
6: %
7: % This function computes base points, cbp, and
8: % weight factors, cwf, in a composite
```

```
 9: % quadrature formula which integrates an
10: % arbitrary function from xlow to xhigh by
11: % dividing the interval of integration into
12: % mparts equal parts and integrating over each
13: % part using a Gauss formula requiring nquad
14: % function values.  Suppose a function f(x) is
15: % being integrated by use of the composite
16: % formula. Denote the base points and weight
17: % factors of the standard Gauss formula as bp
18: % and wf. Then the integration can be
19: % accomplished with the following algorithm:
20: %
21: %    x=b
22: %    integral( f(x)*dx ) =
23: %    x=a
24: %
25: %            j=n k=m
26: %        d1*sum sum( wf(j)*fun(a1+d*k+d1*bp(j)) )
27: %            j=1 k=1
28: %
29: % where
30: %    bp are base points
31: %    wf are weight factors
32: %    n = nquad,  m = mparts
33: %    d = (b-a)/m, d1 = d/2, a1 = a-d1
34: %
35: % The base points bp and weight factors wf are
36: % first generated by eigenvalue methods.
37: %
38: % The same calculation can be performed with a
39: % single summation using the composite base
40: % points and weight factors as follows:
41: %
42: %    x=b
43: %    integral( f(x)*dx ) =
44: %    x=a
45: %
46: %                 j=nquad*mparts
47: %                 sum( cwf(j)*fun( cbp(j) )
48: %                 j=1
49: %
50: % User m functions called:  none
51: %-------------------------------------------------
52:
53: % Compute base points and weight factors for a
```

```
54: % single interval
55: u=(1:nquad-1)./sqrt((2*(1:nquad-1)).^2-1);
56: [vc,bp]=eig(diag(u,-1)+diag(u,1));
57: [bp,k]=sort(diag(bp)); wf=2*vc(1,k)'.^2;
58:
59: % Evaluate the composite base points and weight
60: % factors in terms of bp and wf
61: d=(xhigh-xlow)/mparts;  d1=d/2;
62: dbp=d1*bp(:); dwf=d1*wf(:);  dr=d*(1:mparts);
63: cbp=dbp(:,ones(1,mparts))+ ...
64:     dr(ones(nquad,1),:)+(xlow-d1);
65: cwf=dwf(:,ones(1,mparts)); cwf=cwf(:);cbp=cbp(:);
```

Function cauchtst

```
 1: function u=cauchtst(z,nquad)
 2: %
 3: % u=cauchtst(z,nquad)
 4: % ~~~~~~~~~~~~~~~~~~~
 5: %
 6: % This function solves a mixed boundary
 7: % value problem for the interior of a circle
 8: % by numerical evaluation of a Cauchy integral.
 9: %
10: %  z     - matrix of complex coordinates where
11: %          function values are computed
12: %  nquad - order of Gauss quadrature used to
13: %          perform numerical integration
14: %
15: %  u     - computed values of the approximate
16: %          solution
17: %
18: % User m functions called: cauchint, mbvtest,
19: %                          pauz, genprint
20: %-------------------------------------------------
21:
22: if nargin<2, nquad=8; end; nbdat=20;
23: if nargin==0
24:   z=linspace(0,.95,10)'* ...
25:     exp(i*linspace(0,2*pi,91));
26: end
27: th=linspace(-pi/2,pi/2,nbdat); zb=exp(i*th);
28: fb=sqrt(zb-i).*sqrt(zb+i); fb(1)=1; fb(nbdat)=1;
```

```
29: fb=cos(th)./fb; F=cauchint(fb,zb,z,nquad);
30: F=F.*sqrt(z-i).*sqrt(z+i); u=2*real(F);
31:
32: surf(real(z),imag(z),u); xlabel('x axis');
33: ylabel('y axis');
34: title(['Approximate Solution to ', ...
35:        'a Mixed Boundary Value Problem']);
36: %colormap('gray'); brighten(0.95);
37: grid on; figure(gcf);
38: fprintf('\nPress [Enter] to continue\n'); pauz;
39: %genprint('caucher1');
40:
41: uexact=mbvtest(z,1); udif=u-uexact;
42: clf; surf(real(z),imag(z),udif);
43: title(['Difference Between Exact and ', ...
44:        'Approximate Solutions']);
45: xlabel('x axis'); ylabel('y axis');
46: %colormap('gray'); brighten(0.95);
47: grid on; figure(gcf);
48: %genprint('caucher2');
```

Function mbvtest

```
1: function u=mbvtest(z,noplot)
2: %
3: % u=mbvtest(z,noplot)
4: % ~~~~~~~~~~~~~~~~~~~
5: %
6: % This function determines a function which is
7: % harmonic for abs(z)<1 and satisfies at r=1,
8: %   u=cos(theta),  -pi/2<theta<pi/2
9: %   du/dr=0,        pi/2<theta<3*pi/2
10: % The solution only applies for points inside
11: % or on the unit circle.
12: %
13: % z       - matrix of complex values where the
14: %           solution is computed.
15: % noplot  - option set to one if no plot is
16: %           requested, otherwise option is not
17: %           required.
18: %
19: % u       - values of the harmonic function
20: %           defined inside the unit circle
```

```
21: %
22: % User m functions called:  genprint
23: %------------------------------------------------
24:
25: if nargin==0
26:   noplot=0;
27:   z=linspace(0,1,10)'* ...
28:     exp(i*linspace(0,2*pi,81));
29: end
30: [n,m]=size(z); z=z(:); u=1/2*ones(size(z));
31: k=find(abs(z)>0); Z=z(k);
32: U=(Z+1./Z+(1-1./Z).*sqrt(Z-i).*sqrt(Z+i))/2;
33: u(k)=real(U); u=reshape(u,n,m);
34: if nargin==1 | noplot==0
35:   z=reshape(z,n,m);
36:   surf(real(z),imag(z),u); xlabel('x axis');
37:   ylabel('y axis');
38:   title(['Mixed Boundary Value Problem ', ...
39:          'for a Circular Disk']);
40:   grid; figure(gcf);
41:   %genprint('mbvtest');
42: end
```

12.12   Fluid Flow about an Elliptic Cylinder

This section demonstrates how analytic functions can be used to analyze potential flow about a circular cylinder in an infinite flow field. Then the function which maps the exterior of a circle onto the exterior of an ellipse is used, in conjunction with the invariance of harmonic functions under a conformal transformation, to deduce the solution for flow about an elliptic cylinder.

First, let us analyze the flow around a circular cylinder in the region $|\zeta| \geq 1$, $\zeta = \xi + i\eta$ with the requirement that the velocity components at infinity have constant values

$$u = U \qquad v = V$$

The velocity components are derivable from a potential function ϕ such that

$$u = \frac{\partial \phi}{\partial \xi} \qquad v = \frac{\partial \phi}{\partial \eta}$$

where ϕ is a harmonic function. The velocity normal to the cylinder boundary must be zero. This requires that the function ψ, conjugate to ϕ, be constant on the boundary. The constant can be taken as zero without loss of generality. In terms of the complex velocity potential

$$f(\zeta) = \phi + i\psi$$

we need

$$f(\zeta) - \overline{f(\zeta)} = 0 \quad \text{for} \quad |\zeta| = 1$$

The velocity field is related to the complex velocity potential by

$$u - iv = f'(\zeta)$$

so the flow condition at infinity is satisfied by

$$f(\zeta) = p\zeta + O(1) \qquad p = U - iV$$

A Laurent series can be used to represent $f(\zeta)$ in the form

$$f(\zeta) = p\zeta + a_0 + \sum_{n=1}^{\infty} a_n \zeta^{-n}$$

Imposition of the boundary condition on the cylinder surface requiring

$$f(\sigma) - \overline{f(\sigma)} = 0 \qquad \sigma = e^{i\theta}$$

leads to

$$p\sigma + a_0 + \sum_{n=1}^{\infty} a_n \sigma^{-n} - \bar{p}\sigma^{-1} - \overline{a_0} - \sum_{n=1}^{\infty} \overline{a_n}\sigma^n = 0$$

Taking $a_0 = 0$, $a_1 = \bar{p}$, and $a_n = 0, n \geq 2$ satisfies all conditions of the problem and yields

$$f(\zeta) = p\zeta + \bar{p}\zeta^{-1}$$

as the desired complex potential function giving the velocity field as

$$u - iv = f'(\zeta) = p - \bar{p}\zeta^{-2} \qquad |\zeta| \geq 1$$

Now consider flow about an elliptic cylinder lying in the z-plane. If the velocity at infinity has components (U, V) then we need a velocity potential $F(z)$ such that $F'(\infty) = U - iV$ and

$$F(z) - \overline{F(z)} = 0 \quad \text{for} \quad \left(\frac{x}{a}\right)^2 + \left(\frac{y}{b}\right)^2 = 1$$

This is nearly the same problem as was already solved in the ζ-plane except that

$$\frac{dF}{dz} = \frac{d\zeta}{dz}\frac{dF}{d\zeta} = \frac{1}{\omega'(\zeta)}\frac{dF}{d\zeta}$$

where $\omega(\zeta)$ is the mapping function

$$z = \omega(\zeta) = R(\zeta + m\zeta^{-1}) \qquad R = \frac{a+b}{2} \qquad m = \frac{a-b}{a+b}$$

In terms of ζ we would need

$$\frac{dF}{d\zeta} = \omega'(\infty)[U - iV] = R(U - iV) \qquad \text{at} \qquad \zeta = \infty$$

Consequently, the velocity potential for the elliptic cylinder problem expressed in terms of ζ is

$$F = p\zeta + \bar{p}\zeta^{-1} \qquad p = R(U - iV)$$

and the velocity components in the z-plane are given by

$$u - iv = \frac{1}{\omega'(\zeta)}\left[p - \bar{p}\zeta^{-2}\right]$$

or

$$u - iv = \frac{(U - iV) - (U - iV)\zeta^{-2}}{1 - m\zeta^{-2}}$$

To get values for a particular choice of z we can use the inverse mapping function

$$\zeta = \frac{z + \sqrt{z^2 - 4mR^2}}{2R}$$

to eliminate ζ completely or we can compute results in terms of ζ.

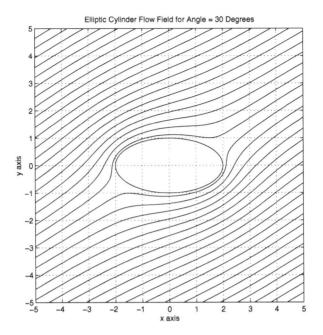

Figure 12.10. Elliptic Cylinder Flow Field for Angle $= 30°$

To complete our discussion of this flow problem we will graph the lines characterizing the directions of flow. The velocity potential $F = \phi + \imath\psi$ satisfies

$$u = \frac{\partial \phi}{\partial x} = \frac{\partial \psi}{\partial y} \qquad v = \frac{\partial \phi}{\partial y} = -\frac{\partial \psi}{\partial x}$$

so a curve tangent to the velocity field obeys

$$\frac{dy}{dx} = \frac{v}{u} = -\frac{\partial \psi / \partial x}{\partial \psi / \partial y}$$

or

$$\frac{\partial \psi}{\partial x} \, dx + \frac{\partial \psi}{\partial y} \, dy = 0 \qquad \psi = \text{constant}$$

Consequently, the flow lines are the contours of the imaginary part of the complex velocity potential. The function we want to contour does not exist inside the ellipse, but we can circumvent this problem by computing ψ in the ellipse exterior and then setting ψ to zero inside the ellipse. The function **elipcyl** analyzes the cylinder flow and produces the accompanying contour plot shown in Figure 12.10.

12.12.1   Program Output and Code

Function elipcyl

```
 1: function [x,y,F]=elipcyl(a,n,rx,ry,ang)
 2: %
 3: % [x,y,F]=elipcyl(a,n,rx,ry,ang)
 4: % ~~~~~~~~~~~~~~~~~~~~~~~~~~~~~~~~
 5: %
 6: % This function computes the flow field around
 7: % an elliptic cylinder. The velocity direction
 8: % at infinity is also produced.
 9: %
10: % a      - defines the region -a<x<a, -a<y<a
11: %           within which the flow field is
12: %           computed
13: % n      - this determines the grid size which
14: %           uses n by n points
15: % rx,ry - major and minor semi-diameters af the
16: %           ellipse lying on the x and y axes,
17: %           respectively
18: % ang -    the angle in degrees which the
19: %           velocity at infinity makes with the
20: %           x axis
21: %
22: % x,y    - matrices of points where the velocity
23: %           potential is computed
24: % F      - matrix of complex velocity potential
25: %           values. This function is set to zero
26: %           inside the ellipse, where the
27: %           potential is actually not defined
28: %
29: % User m functions called:  none
30: %-----------------------------------------------
31:
32: % default data for a 2 by 1 ellipse
33: if nargin==0
34:   a=5; n=81; rx=2; ry=1; ang=30;
35: end
36:
37: % Compute a square grid in the z plane.
38: ar=pi/180*ang; p=(rx+ry)/2*exp(-i*ar);
39: cp=conj(p); d=linspace(-a,a,n);
40: [x,y]=meshgrid(d,d); m=sqrt(rx^2-ry^2);
```

```
41:
42: % Obtain points in the zeta plane outside
43: % the ellipse
44: z=x(:)+i*y(:); k=find((x/rx).^2+(y/ry).^2>=1);
45: Z=z(k); zeta=(Z+sqrt(Z-m).*sqrt(Z+m))/(rx+ry);
46: F=zeros(n*n,1);
47:
48: % Evaluate the potential for a circular
49: % cylinder
50: F(k)=p*zeta+cp./zeta; F=reshape(F,n,n);
51:
52: % Contour the stream function to show the
53: % direction of flow
54: contour(x,y,imag(F),40); axis('square');
55: zb=exp(i*linspace(0,2*pi,101));
56: xb=rx*real(zb); yb=ry*imag(zb);
57: hold on; plot(xb,yb);
58: xlabel('x axis'); ylabel('y axis');
59: title(['Elliptic Cylinder Flow Field for ', ...
60:         'Angle = ',num2str(ang),' Degrees']);
61: figure(gcf); hold off;
62: %genprint('elipcyl');
```

12.13 Torsional Stresses in a Beam Mapped onto a Unit Disk

Torsional stresses in a cylindrical beam can be computed from an integral formula when the function $z = \omega(\zeta)$ mapping $|\zeta| \leq 1$ onto the beam cross section is known [88]. The complex stress function

$$f(\zeta) = \frac{1}{2\pi} \int_\gamma \frac{\omega(\sigma)\overline{\omega(\sigma)}\,d\sigma}{\sigma - \zeta} + \text{constant}$$

can be evaluated exactly by contour integration in some cases. However, an approach employing series methods is easy to implement and gives satisfactory results if enough series terms are taken. When $\omega(\zeta)$ is a polynomial, $f(\zeta)$ is a polynomial of the same order as $\omega(\zeta)$. Furthermore, when $\omega(\zeta)$ is a rational function, residue calculus can be employed to compute $f(\zeta)$ exactly provided the poles of $\overline{\omega}(1/\zeta)$ can be found. A much simpler approach is to use the FFT to expand $\omega(\sigma)\overline{\omega(\sigma)}$ in a complex Fourier series and write

$$\omega(\sigma)\overline{\omega(\sigma)} = \sum_{n=-\infty}^{\infty} c_n \sigma^n \qquad \sigma = e^{i\theta}$$

Then the complex stress function is

$$f(\zeta) = i \sum_{n=1}^{\infty} c_n \zeta^n + \text{constant}$$

where the constant has no influence on the stress state. The shear stresses relative to the curvilinear coordinate system are obtainable from the formula

$$\frac{\tau_{\rho z} - i\tau_{\alpha z}}{\mu \mathcal{X}} = \frac{\left[f'(\zeta) - i\overline{\omega(\zeta)}\omega'(\zeta) \right]\zeta}{|\zeta\omega'(\zeta)|}$$

where μ is the shear modulus and $\mathcal{X}$ is the angle of twist per unit length. The capital Z subscript on shear stresses refers to the direction of the beam axis normal to the xy plane rather than the complex variable $z = x + iy$. The series expansion gives

$$f'(\zeta) = i \sum_{n=1}^{\infty} nc_n \zeta^{n-1}$$

and this can be used to compute stresses. Differentiated series expansions often converge slowly or may even be divergent. To test the series expansion solution, a rational function mapping $|\zeta| < 1$ onto a square defined by $|x| \leq 1$ and $|y| \leq 1$ was employed. Function **mapsqr** which computes $z(\zeta)$ and $z'(\zeta)$ is used by function **torstres** to evaluate stresses

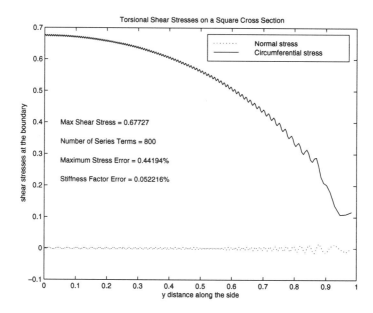

Figure 12.11. Torsional Shear Stresses on a Square Cross Section

in terms of ζ. A short driver program **runtors** evaluates stresses on the boundary for $x = 1$, $0 \le y \le 1$. Stresses divided by the side length of 2 are plotted and results produced from a highly accurate solution [88] are compared with values produced using 800 terms in $f(\zeta)$. Results depicted in Figure 12.11 show that the error in maximum shear stress was only 0.44% and the torsional stiffness was accurate within 0.05%. Even though the differentiated series converges slowly, computation time is still small. The reader can verify that using 1500 terms reduces the boundary stress oscillations to negligible magnitude and produces a maximum stress error of 0.03%. Although taking 1500 terms to achieve accurate results seems excessive, less than 400 nonzero terms are actually involved because geometrical symmetry implies a series increasing in powers of four. For simplicity and generality, no attempt was made to account for geometrical symmetry exhibited by a particular mapping function. It appears that a series solution employing a mapping function is a viable computational tool to deal with torsion problems.

12.13.1 Program Output and Code

Script File runtors

```
 1: % Example: runtors
 2: % ~~~~~~~~~~~~~~~~~
 3: %
 4: % Example showing torsional stress computation
 5: % for a beam of square cross section using
 6: % conformal mapping and a complex stress
 7: % function.
 8: %
 9: % User m functions called: torstres, genprint
10: %-------------------------------------------------
11:
12: % Generate zeta values defining half of a side
13: theta=linspace(0,pi/4,501); zeta=exp(i*theta);
14: ntrms=800;
15:
16: % Compute stresses using an approximate rational
17: % function mapping function for the square
18: [tr,ta,z,c,C]= ...
19:    torstres('mapsqr',zeta,ntrms,4*1024);
20:
21: % Results from the exact solution
22: n=1:2:13;
23: tmexact=1-8/pi^2*sum(1./(n.^2.*cosh(n*pi/2)));
24: err=abs(ta(1)/2-tmexact)*100/tmexact;
25: stfexct=16/3-1024/pi^5*sum(tanh(pi/2*n)./n.^5);
26: stfaprx=8/3-pi*sum((1:ntrms)'.* ...
27:          abs(C(2:ntrms+1)).^2);
28: ster=100*abs(stfaprx-stfexct)/stfexct;
29:
30: % Plot circumferential and normal stresses at
31: % the boundary
32: th=180/pi*theta;
33: clf; plot(imag(z),tr/2,':',imag(z),ta/2,'-')
34: xlabel('y distance along the side');
35: ylabel('shear stresses at the boundary');
36: title(['Torsional Shear Stresses on a ', ...
37:         'Square Cross Section']);
38: text(.05,.40, ...
39:    ['Max Shear Stress = ',num2str(max(ta)/2)]);
40: text(.05,.34, ...
```

```
41:    ['Number of Series Terms = ',num2str(ntrms)]);
42: text(.05,.28, ...
43:    ['Maximum Stress Error = ',num2str(err),'%']);
44: text(.05,.22,['Stiffness Factor Error = ', ...
45:    num2str(ster),'%']);
46: legend('Normal stress','Circumferential stress');
47: figure(gcf);
48: %disp('Use mouse to locate legend block');
49: %disp('Press [Enter] when finished'); pauz;
50: %genprint('torsion');
```

Function torstres

```
 1: function [trho,talpha,z,c,C]= ...
 2:                   torstres(mapfun,zeta,ntrms,nft)
 3: %
 4: % [trho,talpha,z,c,C]= ...
 5: %                 torstres(mapfun,zeta,ntrms,nft)
 6: % ~~~~~~~~~~~~~~~~~~~~~~~~~~~~~~~~~~~~~~~~~~~~~~~~~~
 7: %
 8: % This function computes torsional stresses in
 9: % a beam such that abs(zeta)<=1 is mapped onto
10: % the beam cross section by a function named
11: % mapfun.
12: %
13: % mapfun - a character string giving the name
14: %          of the mapping function
15: % zeta   - values in the zeta plane
16: %          corresponding to which torsional
17: %          stresses are computed
18: % ntrms  - the number of terms used in the
19: %          series expansion of the mapping
20: %          function
21: % nft    - the number of function values
22: %          employed to compute Fourier
23: %          coefficients of the complex stress
24: %          function
25: %
26: % trho   - torsional stresses in directions
27: %          normal to the lines into which
28: %          abs(zeta)=const map. These values
29: %          should be zero at the boundary
30: %          corresponding to abs(zeta)=1.
31: % talpha - torsional stresses in directions
32: %          tangent to the curves into which
33: %          abs(zeta)=const map. The maximum
34: %          value of shear stress always occurs
35: %          at some point on the boundary defined
36: %          by abs(zeta)=1.
37: % z      - values of z where stresses are
38: %          computed
39: % c      - coefficients in the series expansion
40: %          of the complex stress funtion
41: % C      - complex Fourier coefficients of
42: %          z.*conj(z) on the boundary of the
```

```
43: %              beam cross section
44: %
45: % User m functions called:  none
46: %-------------------------------------------------
47:
48: if nargin<4, nft=4096; end;
49: if nargin<3, ntrms=800; end
50:
51: % Compute boundary values of the mapping
52: % function needed to construct the complex
53: % stress function
54: zetab=exp(i*linspace(0,2*pi*(nft-1)/nft,nft));
55: zb=feval(mapfun,zetab); zb=zb(:);
56:
57: % Evaluate z and z'(zeta) at other
58: % desired points
59: [z,zp]=feval(mapfun,zeta);
60:
61: % Compute Fourier coefficients for the complex
62: % stress function and its derivative
63: C=fft(zb.*conj(zb))/nft;
64: c=i*C(2:ntrms+1).*(1:ntrms)';
65: fp=polyval(flipud(c),zeta);
66:
67: % Evaluate stresses relative to the curvilinear
68: % coordinate system
69: tcplx=zeta./abs(zeta.*zp).*(fp-i*conj(z).*zp);
70:
71: % trho is the radial shear stress which should
72: % vanish at the boundary
73: trho=real(tcplx);
74:
75: % talpha is the circumferential stress which
76: % gives the maximum stress of interest at the
77: % boundary
78: talpha=-imag(tcplx);
```

Function mapsqr

```
1: function [z,zp]=mapsqr(zeta);
2: %
3: % [z,zp]=mapsqr(zeta)
4: % ~~~~~~~~~~~~~~~~~~~~
```

```
 5: %
 6: % This function maps the interior of a circle
 7: % onto the interior of a square using a rational
 8: % function of the approximate form:
 9: %
10: % z(zeta)=zeta*Sum(a(j)* ...
11: %          zeta4^(j-1)/(1+Sum(b(j)*zeta4^(j-1))
12: %
13: % where zeta4=zeta^4
14: %
15: % zeta - matrix of complex values such that
16: %          abs(zeta)<=1
17: % z,zp - matrices of values of the mapping
18: %          function and its first derivative
19: %
20: % User m functions called:  none
21: %-------------------------------------------------

23: a=[ 1.07835,  1.37751, -0.02642, -0.09129, ...
24:     0.13460, -0.15763,  0.07430,  0.14858, ...
25:     0.01878, -0.00354 ]';
26: b=[ 1.37743,  0.07157, -0.11085,  0.12778, ...
27:    -0.13750,  0.05313,  0.14931,  0.02683, ...
28:    -0.00350, -0.000120 ]';

30: % Evaluate the mapping function
31: zeta4=zeta.^4; p=zeta.*polyval(flipud(a),zeta4);
32: q=polyval(flipud([1;b]),zeta4); z=p./q;

34: % Exit if the derivative of z is not needed
35: if nargout==1, return, end

37: % evaluate z'(zeta)
38: na=length(a); nb=length(b);
39: pp=polyval(flipud((4*(1:na)'-3).*a),zeta4);
40: qp=4*zeta.^3.*polyval(flipud((1:nb)'.*b),zeta4);
41: zp=(q.*pp-p.*qp)./q.^2;
```

12.14 Stress Analysis by the Kolosov-Muskhelishvili Method

Two-dimensional problems in linear elastostatics of homogeneous bodies can be analyzed with the use of analytic functions. The primary quantities of interest are cartesian stress components τ_{xx}, τ_{yy}, and τ_{xy} and displacement components u and v. These can be expressed as

$$\tau_{xx} + \tau_{yy} = 2[\Phi(z) + \overline{\Phi(z)}]$$

$$-\tau_{xx} + \tau_{yy} + 2\imath\tau_{xy} = 2[\bar{z}\Phi'(z) + \Psi(z)]$$

$$2\mu(u + \imath v) = \kappa\phi(z) - z\overline{\Phi(z)} - \overline{\psi(z)}$$

$$\phi(z) = \int \Phi(z)\,dz \qquad \psi(z) = \int \Psi(z)\,dz$$

where μ is the shear modulus and κ depends on Poisson's ratio ν according to $\kappa = 3 - 4\nu$ for plane strain or $\kappa = (3 - \nu)/(1 + \nu)$ for plane stress. The above relations are known as the Kolosov-Muskhelishvili formulas [72] and they have been used to solve many practical problems employing series or integral methods. Bodies such as the circular disk, a plate with a circular hole, and a circular annulus can be handled for quite general boundary conditions. Solutions can also be developed for geometries such that a rational function is known to map the interior of a circle onto the desired geometry. Futhermore, complex variable methods provide the most general techniques available for solving a meaningful class of mixed boundary value problems such as contact problems typified by pressing a rigid punch into a half plane.

Fully understanding all of the analyses presented in [71, 72] requires familiarity with contour integration, conformal mapping, and multivalued functions. However, some of the closed form solutions given in these texts can be used without the need for extensive background in complex variable methods or the physical concepts of elasticity theory. With that perspective let us examine the problem of computing stresses in an infinite plate uniformly stressed at infinity and having a general normal stress $N(\theta)$ and tangential shear $T(\theta)$ applied to the hole. We will use the general solution of Muskhelishvili[1] [71] to evaluate stresses anywhere in the plate with particular interest on stress concentrations occurring around the hole. The stress functions Ψ and Φ can be represented as follows

$$\Phi(z) = -\frac{1}{2\pi\imath}\int_\gamma \frac{(N + \imath T)d\sigma}{\sigma - z} + \alpha + \beta z^{-1} + \delta z^{-2} \qquad \sigma = e^{\imath\theta}$$

---

[1] Chapter 20.

where γ denotes counterclockwise contour integration around the boundary of the hole and the other constants are given by

$$\alpha = \frac{T_{xx}^{\infty} + T_{yy}^{\infty}}{4} \qquad \delta = \frac{-T_{xx}^{\infty} + T_{yy}^{\infty} + 2\imath T_{xy}^{\infty}}{2}$$

$$\beta = -\frac{\kappa}{1+\kappa} \frac{1}{2\pi} \int_0^{2\pi} (N + \imath T) e^{\imath \theta} \, d\theta$$

Parameters α and δ depend only on the components of stress at infinity, while β is determined by the force resultant on the hole caused by the applied loading. The quantity $N + \imath T$ is the boundary value of radial stress τ_{rr} and shear stress $\tau_{r\theta}$ in polar coordinates. Hence

$$N + \imath T = \tau_{rr} + \imath \tau_{r\theta} \qquad |z| = 1$$

The transformation formulas relating cartesian stresses τ_{xx}, τ_{yy}, τ_{xy} and polar coordinate stresses τ_{rr}, $\tau_{\theta\theta}$, $\tau_{r\theta}$ are

$$\tau_{rr} + \tau_{\theta\theta} = \tau_{xx} + \tau_{yy} \qquad -\tau_{rr} + \tau_{\theta\theta} + 2\imath\tau_{r\theta} = (-\tau_{xx} + \tau_{yy} + 2\imath\tau_{xy})e^{2\imath\theta}$$

Let us assume that $N + \imath T$ is expandable in a Fourier series of the form

$$N + \imath T = \sum_{n=-\infty}^{\infty} c_n \sigma^n \qquad \sigma = e^{\imath\theta}$$

where c_n can be obtained by integration as

$$c_n = \frac{1}{2\pi} \int_0^{2\pi} (N + \imath T)\sigma^{-n} \, d\theta$$

or we can compute the approximate coefficients more readily by use of the FFT.

The stress function $\Psi(z)$ is related to $\Phi(z)$ according to

$$\Psi = \frac{1}{z^2}\overline{\Phi\left(\frac{1}{\bar{z}}\right)} - \frac{d}{dz}\left[\frac{1}{z}\Phi(z)\right] \qquad |z| \geq 1$$

Substituting the complex Fourier series into the integral formula for Φ gives

$$\Phi = -\sum_{n=0}^{\infty} c_n z^n + \alpha + \beta z^{-1} + \delta z^{-2} \qquad |z| \leq 1$$

$$\Phi = \sum_{n=1}^{\infty} c_{-n} z^{-n} + \alpha + \beta z^{-1} + \delta z^{-2} \qquad |z| \geq 1$$

$$\Phi = \sum_{n=0}^{\infty} a_n z^{-n} \qquad |z| \geq 1$$

These two relations then determine Ψ in the form

$$\Psi = \bar{\delta} + \bar{\beta}z^{-1} + (\alpha + a_0 - \overline{c_0})z^{-2} + \sum_{n=3}^{\infty} \left[(n-1)a_{n-2} - \overline{c_{n-2}}\right] z^{-n}$$

$$\Psi = \sum_{n=0}^{\infty} b_n z^{-n} \qquad |z| \geq 1$$

Hence the series expansions of functions $\Phi(z)$ and $\Psi(z)$ can be generated in terms of the coefficients c_n and the stress components at infinity. The stresses can be evaluated by using the stress functions. Displacements can also be obtained by integrating Φ and Ψ, but this straightforward calculation is not discussed here.

The program **runplate** was written to evaluate the above formulas by expanding $N+iT$ using the FFT. Truncating the series for harmonics above some specified order, say np, gives approximations for $\Phi(z)$ and $\Psi(z)$ which exactly represent the solution corresponding to the boundary loading defined by the truncated Fourier series. Using the same approach employed in Chapter 6 we can define N and T as piecewise linear functions of the polar angle θ.

The program utilizes several routines described in the table below.

runplate	define N, T, stresses at infinity, z-points where results are requested, and the number of series terms used.
platecrc	computes series coefficients defining the stress functions.
strfun	evaluates Φ, Ψ, and Φ'.
cartstrs	evaluates cartesian stresses for given values of z and the stress functions.
rec2polr	transforms from cartesian stresses to polar coordinate stresses.
polflip	simplified interface to function **polyval**.

The sample problem analyzes a plate having $tau_{yy}^{\infty} = 1$, $tau_{xx}^{\infty} = tau_{xy}^{\infty} = 0$. As indicated by the surface plot in Figure 12.12, the circumferential stress on the hole varies between -1 and 3, producing a stress concentration factor of three due to the presence of the hole. Readers should consider investigating how stresses around the hole are affected by different normal stress distributions on the hole. For example, taking

```
T=0;  ti=[0,0,0];
th=linspace(0,2*pi,81);
N=[cos(4*th), 180/pi*th];
```

gives the interesting results depicted in Figure 12.13.

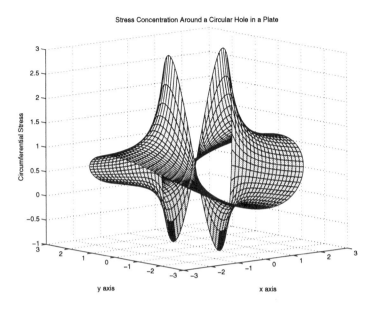

Figure 12.12. Stress Concentration around a Circular Hole in a Plate

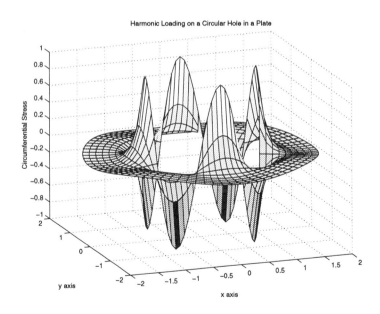

Figure 12.13. Harmonic Loading on a Circular Hole in a Plate

12.14.1   Program Output and Code

Script File runplate

```
 1: % Example: runplate
 2: % ~~~~~~~~~~~~~~~~~
 3: %
 4: % Example to compute stresses around a
 5: % hole in a plate by use of the complex
 6: % variable method of Kolosov-Muskhelishvili.
 7: %
 8: % User m functions required:
 9: %    runplate, platecrc, strfun, cartstrs,
10: %    rec2polr, polflip, pauz, genprint
11: %-----------------------------------------------
12:
13: Problem=1;
14: if Problem == 1
15:    titl=['Stress Concentration Around a ', ...
16:          'Circular Hole in a Plate'];
17:    N=0; T=0; ti=[0,1,0]; kapa=2; np=50;
18:    rz=linspace(1,3,20)'; tz=linspace(0,2*pi,81);
19:    z=rz*exp(i*tz); x=real(z); y=imag(z);
20:    plotfile='strconc1'; viewpnt=[-40,10];
21: else
22:    titl=['Harmonic Loading on a Circular', ...
23:          ' Hole in a Plate'];
24:    th=linspace(0,2*pi,81)';
25:    N=[cos(4*th),180/pi*th];
26:    T=0; ti=[0,0,0]; kapa=2; np=10;
27:    rz=linspace(1,2,10)'; tz=linspace(0,2*pi,81);
28:    z=rz*exp(i*tz); x=real(z); y=imag(z);
29:    plotfile='strconc2'; viewpnt=[-20,20];
30: end
31:
32: fprintf('\nSTRESSES IN A PLATE WITH A ');
33: fprintf('CIRCULAR HOLE');
34: fprintf('\n\nStress components at infinity ');
35: fprintf('are: '); fprintf('%g ',ti);
36: fprintf('\nNormal stresses on the hole are ');
37: fprintf('defined by N =');
38: fprintf('\n   %g',N);
39: fprintf('\nTangential stresses on the hole ');
40: fprintf('are defined by T = %g',T);
```

```
41: fprintf('\nElastic constant kapa equals: ');
42: fprintf('%s',num2str(kapa));
43: fprintf('\nHighest harmonic order used is: ');
44: fprintf('%s',num2str(np));
45:
46: [a,b,c]=platecrc(N,T,ti,kapa,np);
47:
48: fprintf('\n');
49: fprintf('\nThe Kolosov-Muskhelishvili stress ');
50: fprintf('functions have\nthe series forms:');
51: fprintf('\nPhi=sum(a(k)*z^(-k+1), k=1:np+1)');
52: fprintf('\nPsi=sum(b(k)*z^(-k+1), k=1:np+3)');
53: fprintf('\n');
54: fprintf('\nCoefficients defining stress ');
55: fprintf('function Phi are:\n');
56: disp(a(:).');
57: fprintf('Coefficients defining stress ');
58: fprintf('function Psi are:\n');
59: disp(b(:).');
60:
61: % Evaluate the stress functions
62: [Phi,Psi,Phip]=strfun(a,b,z);
63:
64: % Compute the cartesian stresses and the
65: % principal stresses
66: [tx,ty,txy,pt1,pt2]=cartstrs(z,Phi,Psi,Phip);
67: theta=angle(z./abs(z)); x=real(z); y=imag(z);
68: [tr,tt,trt]=rec2polr(tx,ty,txy,theta);
69: pmin=num2str(min([pt1(:);pt2(:)]));
70: pmax=num2str(max([pt1(:);pt2(:)]));
71:
72: disp(...
73: ['Minimum Principal Stress = ',num2str(pmin)]);
74: disp(...
75: ['Maximum Principal Stress = ',num2str(pmax)]);
76: fprintf('\nPress [Enter] for a surface ');
77: fprintf('plot of the\n  circumferential stress ');
78: fprintf('in the plate\n'); pauz; clf;
79: colormap('hsv');
80: %colormap('gray'); brighten(0.95);
81: surf(x,y,tt); xlabel('x axis'); ylabel('y axis');
82: zlabel('Circumferential Stress');
83: title(titl); grid on; view(viewpnt); figure(gcf);
84: %genprint(plotfile);
85: fprintf('\nAll Done\n');
```

Function platecrc

```
 1: function [a,b,c]=platecrc(N,T,ti,kapa,np)
 2: %
 3: % [a,b,c]=platecrc(N,T,ti,kapa,np)
 4: % ~~~~~~~~~~~~~~~~~~~~~~~~~~~~~~~~~
 5: %
 6: % This function computes coefficients in the
 7: % series expansions which define the Kolosov-
 8: % Muskhelishvili stress functions for a plate
 9: % having a circular hole of unit radius. The
10: % plate is uniformly stressed at infinity. On
11: % the surface of the hole, normal and tangential
12: % stress distributions N and T defined as
13: % piecewise linear functions are applied.
14: %
15: % N      - a two column matrix with each row
16: %          containing a value of normal stress
17: %          and polar angle in degrees used to
18: %          specify N as a piecewise linear
19: %          function of the polar angle. Step
20: %          discontinuities can be included by
21: %          using successive values of N with the
22: %          same polar angle values.  The data
23: %          should cover the range of theta from
24: %          0 to 360.  N represents boundary values
25: %          of the polar coordinate radial stress.
26: %          A single constant value can be input
27: %          when N is constant (including zero
28: %          if desired).
29: % T      - a two column matrix defining values of
30: %          the polar coordinate shear stress on
31: %          the hole defined as a piecewise linear
32: %          function. The points where function
33: %          values of T are specified do not need
34: %          not need be the same as as those used
35: %          to specify N. Input a single constant
36: %          when T is constant on the boundary.
37: % ti     - vector of cartesian stress components
38: %          [tx,ty,txy] at infinity.
39: % kapa - a constant depending on Poisson's ratio
40: %          nu.
41: %              kapa=3-4*nu for plane strain
42: %              kapa=(3-nu)/(1+nu) for plane stress
```

```
43: %           When the resultant force on the hole
44: %           is zero, then kapa has no effect on
45: %           the solution.
46: % np   - the highest power of exp(i*theta) used
47: %           in the series expansion of N+i*T. This
48: %           should not exceed 255.
49: %
50: % a    - coefficients in the series expansion
51: %           defining the stress function
52: %             Phi=sum(a(k)*z^(-k+1), k=1:np+1)
53: % b    - coefficients in the series expansion
54: %           defining the stress function
55: %             Psi=sum(b(k)*z^(-k+1), k=1:np+3)
56: %
57: % User m functions called:  lintrp
58: %-----------------------------------------------
59:
60: % Handle case of constant boundary stresses
61: if length(N(:))==1; N=[N,0;N,360]; end
62: if length(T(:))==1; T=[T,0;T,360]; end
63:
64: % Expand the boundary stresses in a Fourier
65: % series
66: f=pi/180; nft=512; np=min(np,nft/2-1);
67: thta=linspace(0,2*pi*(nft-1)/nft,nft);
68:
69: % Interpolate linearly for values at the
70: % Fourier points
71: Nft=lintrp(f*N(:,2),N(:,1),thta);
72: Tft=lintrp(f*T(:,2),T(:,1),thta);
73: c=fft(Nft(:)+i*Tft(:))/nft;
74:
75: % Evaluate auxiliary parameters in the
76: % series solutions
77: alp=(ti(1)+ti(2))/4; bet=-kapa*c(nft)/(1+kapa);
78: sig=(-ti(1)+ti(2)-2*i*ti(3))/2;
79:
80: % Generate a and b coefficients using the
81: % Fourier coefficients of N+i*T.
82: a=zeros(np+1,1); b=zeros(np+3,1); j=(1:np)';
83: a(j+1)=c(nft+1-j); a(1)=alp;
84: a(2)=bet+c(nft); a(3)=sig+c(nft-1);
85: j=(3:np+2)'; b(j+1)=(j-1).*a(j-1)-conj(c(j-1));
86: b(1)=conj(sig); b(2)=conj(bet);
87: b(3)=alp+a(1)-conj(c(1));
```

```
88:
89: % Discard any negligibly small high order
90: % coefficients.
91: tol=max(abs([N(:);T(:);ti(:)]))/1e4;
92: ka=max(find(abs(a)>tol));
93: if isempty(ka), a=0; else, a(ka+1:np+1)=[]; end
94: kb=max(find(abs(b)>tol));
95: if isempty(kb), b=0; else, b(kb+1:np+3)=[]; end
```

Function strfun

```
1: function [Phi,Psi,Phip]=strfun(a,b,z)
2: %
3: % [Phi,Psi,Phip]=strfun(a,b,z)
4: % ~~~~~~~~~~~~~~~~~~~~~~~~~~~~~
5: %
6: % This function evaluates the complex
7: % stress functions Phi(z), and Psi(z)
8: % as well as the derivative function Phi'(z)
9: % using series coefficients determined from
10: % function platecrc. The calculation also
11: % uses a function polflip defined such that
12: % polflip(a,z)=polyval(flipud(a(:)),z).
13: %
14: % a,b      - series coefficients defining Phi
15: %            and Psi
16: % z        - matrix of complex values
17: %
18: % Phi,Psi - complex stress function values
19: % Phip    - derivative Phi'(z)
20: %
21: % User m functions called: polflip
22: %-----------------------------------------------
23:
24: zi=1./z; np=length(a); a=a(:);
25: Phi=polflip(a,zi); Psi=polflip(b,zi);
26: Phip=-polflip((1:np-1)'.*a(2:np),zi)./z.^2;
```

Function rec2polr

```
 1: function [tr,tt,trt]=rec2polr(tx,ty,txy,theta)
 2: %
 3: % [tr,tt,trt]=rec2polr(tx,ty,txy,theta)
 4: % ~~~~~~~~~~~~~~~~~~~~~~~~~~~~~~~~~~~~~~~~
 5: %
 6: % This function transforms cartesian stress
 7: % components tx,ty,txy to polar coordinate
 8: % stresses tr,tt,trt.
 9: %
10: % tx,ty,txy - matrices of cartesian stress
11: %               components
12: % theta     - a matrix of polar coordinate
13: %               values.  This can also be a
14: %               single value if all stress
15: %               components are rotated by the
16: %               same angle.
17: %
18: % tr,tt,trt - matrices of polar coordinate
19: %               stresses
20: %
21: % User m functions called:  none
22: %-------------------------------------------------
23:
24: if length(theta(:))==1
25:   theta=theta*ones(size(tx)); end
26: a=(tx+ty)/2;
27: b=((tx-ty)/2-i*txy).*exp(2*i*theta);
28: c=a+b; tr=real(c); tt=2*a-tr; trt=-imag(c);
```

Function polflip

```
1: function y=polflip(a,x)
2: %
3: % y=polflip(a,x)
4: % ~~~~~~~~~~~~~~
5: %
6: % This function evaluates polyval(a,x) with
7: % the order of the elements reversed.
8: %
9: % User m functions called:  none
10: %---------------------------------------------------
11:
12: y=polyval(a(length(a):-1:1),x);
```

Function cartstrs

```
1: function [tx,ty,txy,tp1,tp2]= ...
2:                         cartstrs(z,Phi,Psi,Phip)
3: %
4: % [tx,ty,txy,tp1,tp2]=cartstrs(z,Phi,Psi,Phip)
5: % ~~~~~~~~~~~~~~~~~~~~~~~~~~~~~~~~~~~~~~~~~~~~~~~
6: %
7: % This function uses values of the complex
8: % stress functions to evaluate cartesian stress
9: % components relative to the x,y axes.
10: %
11: % z         - matrix of complex values where
12: %             stresses are required
13: % Phi,Psi   - matrices containing complex stress
14: %             function values
15: % Phip      - values of  Phi'(z)
16: %
17: % tx,ty,txy - values of the cartesian stress
18: %             components for the x,y axes
19: % tp1,tp2   - values of maximum and minimum
20: %             principal stresses
21: %
22: % User m functions called:  none
23: %---------------------------------------------------
24:
25: A=2*real(Phi); B=conj(z).*Phip+Psi;
```

```
26: C=A-B; R=abs(B);
27: tx=real(C); ty=2*A-tx; txy=-imag(C);
28: tp1=A+R; tp2=A-R;
```

12.14.2   Stressed Plate with an Elliptic Hole

We end this section with an example using conformal mapping in elasticity theory. Plane elastostatic problems lead to the biharmonic differential equation

$$\nabla^4 U = 0$$

where U is the Airy stress function [88]. Even though harmonic functions remain harmonic under a conformal transformation, biharmonic functions do not except for a linear map. Consequently, elasticity solutions for one geometry do not map onto elasticity solutions for another geometry through conformal transformation. This does not prevent the use of conformal mapping in elasticity, but we do have to deal with a problem having a different structure in the mapped variables. We will examine that problem enough to show the kind of differences involved. Let a mapping function $z = \omega(\zeta)$ define curvilinear coordinate lines in the z-plane. A polar coordinate grid corresponding to $\arg(\zeta) = $ constant and $|\zeta| = $ constant maps into curves we term ρ lines and α lines, respectively. Plotting of such lines was demonstrated previously with function **topview** (mapping the exterior of a circle onto the exterior of an ellipse). It can be shown that curvilinear coordinate stresses $\tau_{\rho\rho}, \tau_{\alpha\alpha}, \tau_{\rho\alpha}$ are related to cartesian stresses according to

$$\tau_{\rho\rho} + \tau_{\alpha\alpha} = \tau_{xx} + \tau_{yy} \qquad -\tau_{\rho\rho} + \tau_{\alpha\alpha} + 2\imath\tau_{\rho\alpha} = h(-\tau_{xx} + \tau_{yy} + 2\imath\tau_{xy})$$

where

$$h = \frac{\zeta\omega'(\zeta)}{\overline{\zeta\omega'(\zeta)}}$$

Muskhelishvili [71] has developed a general solution for a plate with an elliptic hole allowing general boundary tractions. Here we use one solution from his text which employs the mapping function

$$z = \omega(\zeta) = R\left(\zeta + \frac{m}{\zeta}\right)$$

and the stress functions

$$\phi(z) = \int \Phi(z)\,dz \qquad \psi(z) = \int \Psi(z)\,dz$$

When ζ is selected as the primary reference variable, we have to perform chain rule differentiation and write

$$\Phi(z) = \frac{\phi'(\zeta)}{\omega'(\zeta)} \qquad \Psi(z) = \frac{\psi'(\zeta)}{\omega'(\zeta)}$$

$$\Phi'(z) = \frac{\omega'(\zeta)\phi''(\zeta) - \omega''(\zeta)\phi'(\zeta)}{\omega'(\zeta)^3}$$

in order to compute stresses in terms of the ζ-variable. Readers unaccustomed to using conformal mapping in this context should remember that there is no stress state in the ζ-plane comparable to the analogous velocity components which can be envisioned in a potential flow problem, for instance. We are simply using ζ as a convenient reference variable to analyze physical stress and displacement quantities existing only in the z-plane.

Suppose the infinite plate has an elliptic hole defined by

$$\left(\frac{x}{r_x}\right)^2 + \left(\frac{y}{r_y}\right)^2 = 1$$

and the hole is free of applied tractions. The stress state at infinity consists of a tension p inclined at angle λ with the x-axis. The stress functions relating to that problem are found to be ([71], page 338)

$$\phi(\zeta) = b\zeta + \frac{c}{\zeta}$$

$$\psi = d\zeta + \frac{e}{\zeta} + \frac{f\zeta}{\zeta^2 - m} \qquad |\zeta| \geq 1$$

$$a = e^{2i\lambda} \qquad b = \frac{pr}{4} \qquad c = b(2a - m)$$

$$d = -\frac{pr\bar{a}}{2} \qquad e = -\frac{pra}{2m} \qquad f = \frac{pr(m + \frac{1}{m})}{2}$$

Clearly these functions have no obvious relation to the simpler results shown earlier for a plate with a circular hole. The function **eliphole** computes curvilinear coordinate stresses in the z-plane expressed in terms of the ζ-variable. When $\lambda = \pi/2$, the plate tension acts along the y-axis and the maximum circumferential stress occurs at $z = r_x$ corresponding to $\zeta = 1$. A surface plot produced by **eliphole** for the default data case using $r_x = 2$ and $r_y = 1$ is shown in Figure 12.14. It is also interesting to graph $\tau_{\alpha\alpha}^{\max}/\tau_{yy}^{\infty}$ as a function of r_x/r_y. The program **elpmaxst** produces the plot in Figure 12.15 showing that the circumferential stress concentration increases linearly according to

$$\frac{\tau_{\alpha\alpha}^{\max}}{p} = 1 + 2\left(\frac{r_x}{r_y}\right)$$

which can also be verified directly from the stress functions.

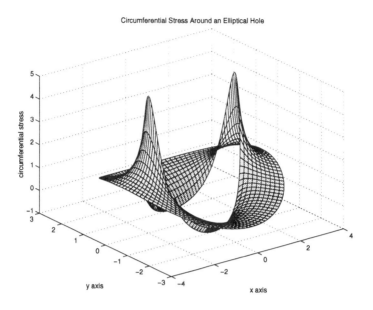

Figure 12.14. Circumferential Stress Around an Elliptical Hole

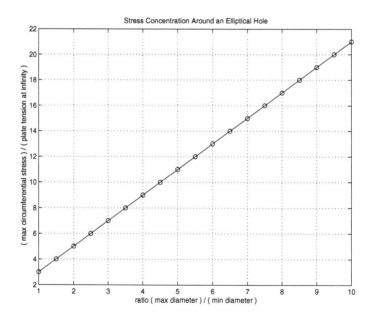

Figure 12.15. Stress Concentration Around an Elliptical Hole

12.14.3 Program Output and Code

Script File elpmaxst

```
 1: % Example: elpmaxst
 2: % ~~~~~~~~~~~~~~~~~~
 3: %
 4: % MATLAB example to plot the stress
 5: % concentration around an elliptic hole
 6: % as a function of the semi-diameter ratio.
 7: %
 8: % User m functions required:
 9: %    eliphole, genprint
10: %-------------------------------------------------
11:
12: r=linspace(1.001,10,19); tamax=zeros(size(r));
13: for j=1:19
14:    [tr,tamax(j)]=eliphole(r(j),1,1,90,1);
15: end
16: plot(r,tamax,'-',r,tamax,'o');
17: title(['Stress Concentration Around an ', ...
18:        'Elliptical Hole']);
19: xlabel(['ratio ( max diameter ) / ', ...
20:         '( min diameter )']);
21: ylabel(['( max circumferential stress ) / ',...
22:         '( plate tension at infinity )']);
23: grid on; figure(gcf);
24: %genprint('elpmaxst');
```

Function eliphole

```
 1: function [tr,ta,tra,z]=eliphole ...
 2:                        (rx,ry,p,ang,zeta)
 3: %
 4: % [tr,ta,tra,z]=eliphole(rx,ry,p,ang,zeta)
 5: % ~~~~~~~~~~~~~~~~~~~~~~~~~~~~~~~~~~~~~~~~~~
 6: %
 7: % This function determines curvilinear
 8: % coordinate stresses around an elliptic hole
 9: % in a plate uniformly stressed at infinity.
10: %
11: % rx,ry    - ellipse semidiameters on the x and
12: %            y axes
```

```
13: % p            - values of uniaxial tension at
14: %                infinity
15: % ang          - angle of inclination in degrees
16: %                of the tensile stress at infinity
17: % zeta         - curvilinear coordinate values for
18: %                which stresses are evaluated
19: %
20: % tr           - tensile stress normal to an
21: %                elliptical coordinate line
22: % ta           - tensile stress in a direction
23: %                tangential to the elliptical
24: %                coordinate line
25: % tra          - shear stress complementary to the
26: %                normal stresses
27: % z            - points in the z plane where
28: %                stresses are computed
29: %
30: % User m functions called: genprint
31: %-----------------------------------------------
32:
33: if nargin==0
34:    rx=2; ry=1; p=1; ang=90;
35:    zeta=linspace(1,2,11)'* ...
36:        exp(i*linspace(0,2*pi,121));
37: end
38:
39: % The complex stress functions and mapping
40: % function have the form
41: %    phi(zeta)=b*zeta+c/zeta
42: %    psi(zeta)=d*zeta+e/zeta+f*zeta/(zeta^2-m)
43: %    z=w(zeta)=r(zeta+m/zeta)
44: %    Phi(zeta)=phi'(zeta)/w'(zeta)
45: %    Psi(zeta)=psi'(zeta)/w'(zeta)
46: %    d(Phi)/dz=(w'(zeta)*phi''(zeta)-...
47: %              w''(zeta)*phi'(zeta))/w'(zeta)^3
48:
49: r=(rx+ry)/2; m=(rx-ry)/(rx+ry);
50: z=r*(zeta+m./zeta); zeta2=zeta.^2;
51: zeta3=zeta.^3; wp=r*(1-m./zeta2);
52: wpp=2*r*m./zeta3; a=exp(2*i*pi/180*ang);
53: b=p*r/4; c=b*(2*a-m); d=-p*r/2*conj(a);
54: e=-p*r/2*a/m; f=p*r/2*(m+1/m)*(a-m);
55: phip=b-c./zeta2; phipp=2*c./zeta3;
56: h=wp.*zeta; h=h./conj(h);
57: Phi=phip./wp; Phipz=(wp.*phipp-wpp.*phip)./wp.^3;
```

```
58: Psi=(d-e./zeta2-f*(zeta2+m)./(zeta2-m).^2)./wp;
59: A=2*real(Phi); B=(conj(z).*Phipz+Psi).*h;
60: C=A-B; tr=real(C); ta=2*A-tr; tra=imag(B);
61: if nargin==0
62:   %colormap('gray'); brighten(.95);
63:   surf(real(z),imag(z),ta);
64:   xlabel('x axis'); ylabel('y axis');
65:   zlabel('circumferential stress');
66:   title(['Circumferential Stress Around ', ...
67:          'an Elliptical Hole']);
68:   grid on; figure(gcf);
69:   %genprint('eliphole');
70: end
```

Chapter 13

Nonlinear Optimization Applications

13.1 Basic Concepts

Optimization problems occur for a diverse range of topics. Perhaps the simplest type of optimization problem involves a scalar function of several variables. For example, the cost of a product having several ingredients may need to be minimized. This problem can be represented by a function $F(x)$ which depends on the vector $x = [x_1; x_2; \ldots; x_n]$ in n-dimensional space. Function F is called the *objective function* and cases where the independent variables x_i can vary arbitrarily are considered unconstrained. Most problems have constraints requiring x_i to remain within given bounds or satisfy other functional equations. Different analysis procedures exist for solving problems depending on whether they are linear or nonlinear, constrained or unconstrained. General solutions are available to handle linear objective functions with linear equality and inequality constraints. The discipline devoted to such problems is known as *linear programming* [40] and applications involving thousands of independent variables can be analyzed.[1] Although this class of linear problems is important, it does not offer the versatility of methods used to address nonlinear problems (which are more compute intensive for problems of similar dimensionality).[2] The material in this chapter addresses nonlinear problems with a few independent variables which are either constrained or restricted to lie within bounds of the form

$$a_i \leq x_i \leq b_i$$

This type of constraint can be satisfied by taking

$$x_i = a_i + (b_i - a_i) \sin^2(z_i)$$

---

[1] High dimensionality linear problems should always be solved using the appropriate specialized software.

[2] The MathWorks markets an "Optimization Toolbox" intended to satisfy a number of specialized optimization needs.

and letting z_i vary arbitrarily. The MATLAB intrinsic functions **fmin** and **fmins** are employed for solving this class of problem. The following five problems are presented to illustrate the nature of nonlinear optimization methods:

1. computing the inclination angle necessary to cause a projectile to strike a stationary distant object,

2. determining the point on a surface closest to a point external to the surface,

3. finding parameters of a nonlinear equation to closely fit a set of data values,

4. determining components of end force on a statically loaded cable necessary to make the endpoint assume a desired position, and

5. computing the shape of a curve between two points such that a smooth particle slides from one end to the other in the minimum time.

Before addressing specific problems, some of the general concepts of optimization will be discussed.

The minimum of an unconstrained differentiable function

$$F(x_1, x_2, \ldots, x_n)$$

will occur at a point where the function has a zero gradient. Thus the condition

$$\frac{\partial F}{\partial x_i} = 0 \qquad 1 \leq i \leq n$$

leads to n nonlinear simultaneous equations. Such systems often have multiple solutions, and a zero gradient occurs at either maxima or minima. No reliable general methods currently exist to obtain all solutions to a general system of nonlinear equations. However, practical situations exist where one unique point providing a relative minimum is expected to exist. In such cases $F(x)$ is called *unimodal* and we seek x_0 which makes

$$F(x_0) < F(x_0 + \Delta) \quad \text{for} \quad |\Delta| > 0$$

Most unconstrained nonlinear programming software starts from an initial point and searches iteratively for a point where the gradient vanishes. *Multimodal*, or non-unimodal, functions can sometimes be solved by initiating searches from multiple starting points and using the best result obtained among all the searches. Since situations such as false convergence are fairly common with nonlinear optimization methods, results obtained warrant greater scrutiny than might be necessary for linear problems.

The intrinsic MATLAB functions **fmin** and **fmins** are adequate to address many optimization problems. Readers should study the documentation available for **fmin**, which performs a one-dimensional search within specified limits, and **fmins**, which performs an unconstrained multi-dimensional search starting from a user selected point. Both functions require objective functions of acceptable syntactical form. Various options controlling convergence tolerances and function evaluation counts should be studied to insure that the parameter choices are appropriately defined.

13.2 Initial Angle for a Projectile

In Chapter 8, equations of motion for motion of a projectile with atmospheric drag were formulated and a function **traject** producing a solution $y(x)$ passing through $(x, y) = (0, 0)$ with arbitrary inclination was developed. The solution is generated for $0 \leq x \leq x_f$ assuming the initial velocity is large enough for the projectile to reach x_f. Therefore, program execution terminates if dx/dt goes to zero. In order to hit a target at position (x_f, y_f), the starting angle of the trajectory must be selected iteratively because the equations of motion cannot be solved exactly (except for the undamped case). With the aid of an optimization method we calculate $|y(x_f) - y_f)|$ and minimize this quantity (described in function **missdis** which has the firing angle as its argument). Function **fmin** seeks the angle to minimize the "miss" distance. Program **runtraj** illustrates the solution to the problem described and Figure 13.1 shows the trajectory required for the projectile to strike the object.

Depending on the starting conditions, zero, one, or two solutions exist to cause the "miss" distance to approach zero. Function **fmin** terminates at either a local minimum or at one of the search limits. The reader will need to examine how the initial data correlates to the final answers. For example, if the projectile misses the target by a significant amount, the initial projectile velocity was not large enough to reach the target.

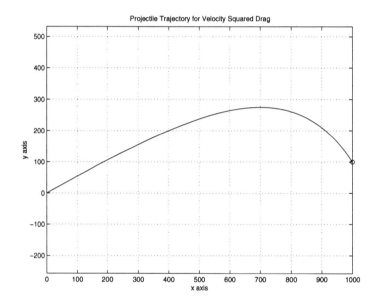

Figure 13.1. Projectile Trajectory for v^2 Drag Condition

13.2.1   Program Output and Code

Script File runtraj

```
 1: % Example: runtraj
 2: % ~~~~~~~~~~~~~~~~~
 3: %
 4: % This program integrates the differential
 5: % equations governing two-dimensional motion
 6: % of a projectile subjected to gravity loading
 7: % and atmospheric drag proportional to the
 8: % velocity squared. The initial inclination
 9: % angle needed to hit a distant target is
10: % computed repeatedly and function fmin is
11: % employed to minimize the square of the
12: % distance by which the target is missed. The
13: % optimal value of the miss distance is zero
14: % and the optimum angle will typically be found
15: % unless the initial velocity is too small
16: % and the horizontal velocity becomes zero
17: % before the target is passed. The initial
18: % velocity of the projectile must be large
19: % enough to make the problem well posed.
20: % Otherwise, the program will terminate with
21: % an error message.
22: %
23: % User m functions called: missdis, traject,
24: %                                 projcteq
25: %-------------------------------------------------
26:
27: clear all;
28: global Vinit Gravty Cdrag Xfinl Yfinl
29:
30: vinit=600; gravty=32.2; cdrag=0.002;
31: xfinl=1000; yfinl=100;
32:
33: disp(' ');
34: disp('SEARCH FOR INITIAL INCLINATION ANGLE ');
35: disp('TO MAKE A PROJECTILE STRIKE A DISTANT');
36: disp('OBJECT'); disp(' ');
37: disp(['Initial velocity = ',num2str(vinit)]);
38: disp(['Gravity constant = ',num2str(gravty)]);
39: disp(['Drag coefficient = ',num2str(cdrag)]);
40: disp(['Coordinates of target = (', ...
```

```
41:      num2str(xfinl),',',...
42:      num2str(yfinl),')']); disp(' ');
43:
44: % Replicate input data as global variables
45: Vinit=vinit; Gravty=gravty; Cdrag=cdrag;
46: Xfinl=xfinl; Yfinl=yfinl;
47:
48: % Perform the minimization search
49: fstart=180/pi*atan(yfinl/xfinl); fend=75;
50: fprintf('Please wait for completion\n')
51: fprintf('of the search calculation\n'); fp=flops;
52: bestang=fmin('missdis',fstart,fend);
53: fp=num2str((flops-fp)/1e6);
54: % Display final results
55: [y,x,t]=traject ...
56:         (bestang,vinit,gravty,cdrag,xfinl);
57: fprintf(...
58: ['\nThe search computation used ',fp,'\n']);
59: fprintf(...
60: 'million floating point operations\n\n');
61:
62: dmiss=abs(yfinl-y(length(y)));
63: disp(['Final miss distance is ', ...
64:     num2str(dmiss),' when the']);
65: disp(['initial inclination angle is ', ...
66:     num2str(bestang),...
67:      ' degrees']);
```

Function missdis

```
1: function [dsq,x,y]=missdis(angle)
2: %
3: % [dsq,x,y]=missdis(angle)
4: % ~~~~~~~~~~~~~~~~~~~~~~~~~
5: %
6: % This function is used by fmin. It returns an
7: % error measure indicating how much the target
8: % is missed for a particular initial inclination
9: % angle of the projectile.
10: %
11: % angle - the initial inclination angle of
12: %           the projectile in degrees
13: %
14: % dsq    - the square of the difference between
15: %           Yfinal and the final value of y found
16: %           using function traject.
17: % x,y    - points on the trajectory.
18: %
19: % Several global parameters (Vinit, Gravty,
20: % Cdrag, Xfinl) are passed to missdis by the
21: % driver program runtraj.
22: %
23: % User m functions called: traject
24: %-----------------------------------------------
25:
26: global Vinit Gravty Cdrag Xfinl Yfinl
27: [y,x,t]=traject ...
28:          (angle,Vinit,Gravty,Cdrag,Xfinl,1);
29: dsq=(y(length(y))-Yfinl)^2;
```

13.3   Closest Point on a Surface

A common challenge in disciplines such as robotics is to design a system
in which two objects must avoid contact with one another. A related
problem involves finding the shortest distance from an arbitrary point
R_0 in space to any point $R(u, v)$ on a surface described in terms of scalar
parameters u and v. Minimizing $|R(u, v) - R_0|$ entails a two-dimensional
search which will be referred to as the closest point problem. Many
familiar shapes such as spheres, cylinders, volumes of revolution, and
conical frusta are readily parameterized by equations where

$$0 \le u \le A \qquad 0 \le v \le B$$

with A and B being constants. This allows us to generate a discrete set
of points

$$0 \le u_i \le A \qquad 1 \le i \le n$$
$$0 \le v_j \le B \qquad 1 \le j \le m$$

leading to a matrix D with elements

$$D_{ij} = |R(u_i, v_j) - R_0|$$

for which we need to identify the smallest element. The matrix indices
for the smallest element can be found using

```
[dmin,J]=min(min(D)); [dmin,I]=min(min(D'));
```

and the closest point on the surface is approximated as $R(u_I, v_J)$ with
the minimal distance being D_{IJ}. An optimization routine such as **fmins**
can be used in a two-dimensional search to solve this problem and that
strategy will be utilized in subsequent examples. However, in the current
example we employ a simple approximate strategy of evaluating the
objective function at a dense set of points and choosing the best answer
obtained.

Consider the closest point problem for a conical frustum (a cone with
the tip cut off). The middle of the base is located at an arbitrary point
R_e and the longitudinal axis is in the direction of a vector R_a. The
shape of the frustum is then completely specified by giving the base
radius r_b, the top radius r_t, and the height h. Imagine a planar section
passing through the longitudinal axis to cut the frustum in two parts.
The length of the lateral surface dimension is

$$r_s = \sqrt{h^2 + (r_b - r_t)^2}$$

with points on the surface being described by an axial coordinate u pertaining to movement along the edge of the planar section and a circumferential coordinate which rotates the plane through $360°$. This implies

that
$$0 \leq u \leq (r_b + r_s + r_r) \qquad 0 \leq v \leq 2\pi$$

Let $\hat{e}_1$ and $\hat{e}_2$ be unit orthogonal base vectors tangent to the base and let $\hat{e}_3$ be a unit vector along the longitudinal axis. Any point on the surface can be described as

$$R = R_e + x\hat{e}_1 + y\hat{e}_2 + z\hat{e}_3$$

where x, y, and z are defined by the following three sets of equations:

I)
$$x = u\cos(v) \qquad y = u\sin(v) \qquad z = 0 \qquad 0 \leq u \leq r_b$$

II)
$$r = r_b + \frac{(r_t - r_b)(u - r_b)}{r_s} \qquad r_b \leq u \leq (r_b + r_s)$$

$$x = r\cos(v) \qquad y = r\sin(v) \qquad z = \frac{h(u - r_b)}{r_s}$$

III)
$$r = r_b + r_s + r_t - u \qquad (r_b + r_s) \leq u \leq (r_b + r_s + r_t)$$

$$x = r\cos(v) \qquad y = r\sin(v) \qquad z = h$$

The function **frusdist** solves the shortest distance problem and produces a plot of the surface. The plot includes a line between the minimum distance point on the surface and the remote point as depicted in Figure 13.2. Figure 13.3 provides a surface plot and shows the negative of the distance matrix as a function of the unitized longitudinal and circumferential parameters. The peak on the surface identifies the point closest to the exterior point. The example summarized in the figures involves a frustum with its axis along a line passing through the origin and $(1, 1, 1)$. The base radius is 1, the top radius is 2, and the height is $4\sqrt{3}$. The remote point r_0 is at $(4, 0, 8)$. Other parameters of the problem can be observed from the default data generated by **frusdist** when no input arguments are provided. Notice that the surface describing d is fairly irregular and has slope discontinuities associated with the base and top edges of the frustum. It can be seen that the constant height surface lines parallel to the circumferential coordinate axis pertain to points at the center of the base and the top. Therefore, changing the circumferential angle does not actually move the points. Depending on where point r_0 is selected, the surface can change shape dramatically. Readers may find it interesting to see what $r_0 = (0, 0, 0)$ produces.

Function **frusdist** calls function **frustum**, which generates points on a surface having the coordinate origin at the center of the base. The arguments resemble those of **frusdist** and are documented within the computer code.

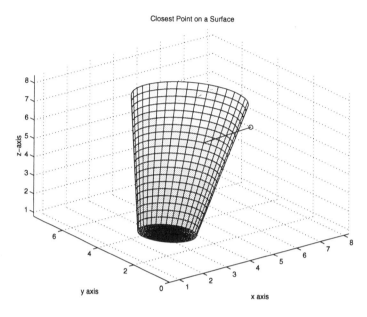

Figure 13.2. Closest Point on a Surface

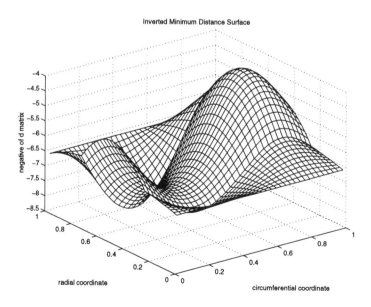

Figure 13.3. Inverted Minimum Distance Surface

13.3.1 Program Output and Code

Function frusdist

```
 1: function [x,y,z,rmin,dmin,d]=frusdist ...
 2:         (rb,rt,h,nb,nt,ns,nc,re,vecs,r0)
 3: %
 4: % [x,y,z,rmin,dmin,d]=
 5: %     frusdist(rb,rt,h,nb,nt,ns,nc,re,vecs,r0)
 6: % ~~~~~~~~~~~~~~~~~~~~~~~~~~~~~~~~~~~~~~~~~~~~~
 7: %
 8: % This function determines the point on the
 9: % surface of a conical frustum which is closest
10: % to another point outside the surface. A grid
11: % of points on the surface is formed along with
12: % a matrix of distances from the remote point to
13: % the surface points. The closest point is
14: % determined by sorting elements of the
15: % distance matrix. The surface and the line from
16: % the nearest point to the closest surface point
17: % are drawn. A surface is also plotted to depict
18: % the distance matrix as a function of the axial
19: % and circumferencial variables used to
20: % construct the surface. For clarity the
21: % negative of the distance matrix is plotted so
22: % the peak on the surface identifies the point
23: % having minimum distance.
24: %
25: % rb,rt - the base and top radii of the frustum.
26: %         The base radius can be larger or
27: %         smaller than the top radius.
28: % h     - the height of the frustum
29: % nb,nt - the number of plot increments taken
30: %         on the base and on the top
31: % ns,nc - the number of increments taken on
32: %         the side and around the circumference.
33: % re    - a vector to the center of the base
34: % vecs  - a matrix having two columns which
35: %         define the orientation of the conical
36: %         frustum. vecs(:,1) gives the direction
37: %         of the axis of the frustum. The cross
38: %         product of vecs(:,1) and vecs(:,2)
39: %         gives the direction of the third base
40: %         vector defining a triad of local base
```

```
41: %           vectors centered at re on the base.
42: %           (See function frustum for more detail.)
43: % r0    - coordinate vector of the remote point
44: %           for which the closest point on the
45: %           frustum is sought
46: %
47: % x,y,z - matrices of coordinate points on the
48: %           surface of the frustum
49: % rmin  - the vector for the point closest to
50: %           the remote point r0
51: % dmin  - the shortest distance
52: % d     - the matrix containing distances from
53: %           the outside point to surface points.
54: %
55: % User m functions called:
56: %    frustum, cubrange, surfesy, pauz, genprint
57: %-----------------------------------------------
58:
59: % Default data case
60: if nargin==0
61:   rb=1; rt=2; Rb=2*[1,1,1]; Rt=2*[3,3,3];
62:   va=Rt-Rb; h=norm(va); re=Rb;
63:   vecs=[va(:),[1;-1;0]]; r0=[4,0,8];
64:   nb=10; nt=10; ns=20; nc=35;
65: end
66:
67: % Call function frustum to generate points on
68: % the surface
69: r=[rb;rt]; n=[nb;ns;nt;nc];
70: [x,y,z]=frustum(r,h,n,vecs);
71: x=x+re(1); y=y+re(2); z=z+re(3);
72:
73: % Form the matrix containing distances from
74: % the outside point to surface points.
75: d=sqrt((x-r0(1)).^2+(y-r0(2)).^2+(z-r0(3)).^2);
76:
77: % Compute the minimum and the related
78: % surface point
79: [dmin,J]=min(min(d)); [dmin,I]=min(min(d'));
80: rmin=[x(I,J);y(I,J);z(I,J)]; R=[rmin,r0(:)];
81:
82: % Generate points to plot the line from the
83: % outside point to the nearest surface point.
84: R=[linspace(rmin(1),r0(1),50);
85:     linspace(rmin(2),r0(2),50);
```

```
86:     linspace(rmin(3),r0(3),50)];
87:
88: % Create a window range for undistorted
89: % plotting
90: v=cubrange([[x(:),y(:),z(:)];r0(:)']);
91:
92: % Draw the line and then the conical frustum
93: hold off, close, colormap('default');
94: [M,N]=size(R);
95: plot3(R(1,:),R(2,:),R(3,:),'-', ...
96:       R(1,N),R(2,N),R(3,N),'o');
97: hold on;
98: surfesy(x,y,z,'x axis','y axis','z-axis',...
99:         'Closest Point on a Surface',v);
100: axis(v); figure(gcf); hold off;
101: %genprint('closept');
102: pauz('Press [Enter] to continue');
103:
104: % Draw a surface showing -d on which the
105: % highest point identifies the minimum distance
106: % point
107: [naxial,mcircum]=size(d); N=(1:naxial)/naxial;
108: M=(1:mcircum)/mcircum; surf(M,N,-d);
109: title('Inverted Minimum Distance Surface');
110: xlabel('circumferential coordinate');
111: ylabel('radial coordinate');
112: zlabel('negative of d matrix'); figure(gcf);
113: %genprint('minsurf');
```

Function frustum

```
1:  function [x,y,z]=frustum(r,h,n,vecs)
2:  %
3:  % [x,y,z]=frustum(r,h,n,vecs)
4:  % ~~~~~~~~~~~~~~~~~~~~~~~~~~~~
5:  %
6:  % This function creates points defining a
7:  % frustum of a cone having its axis in
8:  % the z direction oriented relative to different
9:  % axes depending on parameters specified in
10: % argument vecs.
11: %
12: % r     - a vector containing [rb,rt] where rb
```

```
13: %           is the base radius and rt is the top
14: %           radius. If only one number is input
15: %           then rt is set equal to rb.
16: % h    - height of the frustum
17: % n    - a vector defining the number of radial
18: %           increments taken on the base, the
19: %           side, the end and the circumference.
20: %           n has the form [nb,ns,nt,nc].
21: %           Using [1,1,1,4] generates a cube.
22: % vecs - a matrix having three rows and either
23: %           one or two columns which determines
24: %           the axis orientation of the frustum.
25: %           When vecs is not input the surface has
26: %           its base on the the xy plane and its
27: %           longitudinal axis along the z
28: %           direction. If vecs is present, the
29: %           longitudinal axis is in the direction
30: %           of vecs(:,1) and the shifted y axis
31: %           is in the direction of vecs(:,1)
32: %           crossed into vecs(:,2).
33: %
34: % x,y,z- matrices of coordinate points on the
35: %           surface of the frustum
36: %
37: % User m functions called:  none
38: %------------------------------------------------
39:
40: if nargin<4, vecs=[]; end
41:
42: % The default data creates a cube of unit
43: % side length
44: if nargin==0
45:    r=1/sqrt(2)*[1,1]; h=1; n=[1,1,1,4];
46: end
47:
48:
49: rb=r(1); nb=n(1); ns=n(2); rt=rb;
50: if length(r)==1, rt=rb; else rt=r(2); end
51: if length(n)>2, nt=n(3); else, nt=nb; end
52: if length(n)>3, nc=n(4); else, nc=36; end
53:
54: % Generate radius values for rotation about
55: % the z axis
56: R=[linspace(0,rb*(nb-1)/nb,nb), ...
57:    linspace(rb,rt,ns+1),...
```

```
58:      linspace(rt*(nt-1)/nt,0,nt)]';
59: z=[zeros(1,nb),linspace(0,h,ns+1),h*ones(1,nt)]';
60: z=z(:,ones(1,nc+1));
61:
62: % Make a surface of revolution by rotation
63: % about the z axis
64: th=linspace(pi/nc,pi/nc+2*pi,nc+1);
65: x=R*cos(th); y=R*sin(th);
66: if nargin<4 | isempty(vecs), return, end
67:
68: % If vecs is present shift the axes of the
69: % frustum appropriately.
70: [N,M]=size(x); e3=vecs(:,1); e3=e3/norm(e3);
71: if size(vecs,2)==1
72:   u=null(e3'); u=[u(:,1),cross(e3,u(:,1)),e3];
73: else
74:   e2=cross(e3,vecs(:,2)); e2=e2/norm(e2);
75:   u=[cross(e2,e3),e2,e3];
76: end
77: w=[x(:),y(:),z(:)]*u'; x=reshape(w(:,1),N,M);
78: y=reshape(w(:,2),N,M); z=reshape(w(:,3),N,M);
```

Function surfesy

```
1: function surfesy(x,y,z,xlab,ylab,zlab,titl,v)
2: %
3: % surfesy(x,y,z,xlab,ylab,zlab,titl,v)
4: % ~~~~~~~~~~~~~~~~~~~~~~~~~~~~~~~~~~~~~~
5: %
6: % This function provides an easy input
7: % interface to function surf.
8: %
9: % x,y,z          - data for surface plotting
10: % xlab,ylab,zlab - labels for the coordinate
11: %                  axes
12: % titl           - a title for the plot
13: % v              - a vector to set the axis
14: %                  range. If no value is input,
15: %                  a range is found to make an
16: %                  undistorted plot. If a
17: %                  single number is input, the
18: %                  default scaling is used
19: %
```

```
20: % User m functions called:  cubrange
21: %------------------------------------------------
22:
23: if nargin<8, v=cubrange(x,y,z); end
24: if nargin<7, titl=[]; end
25: if nargin<6, xlab=''; ylab=''; zlab=''; end
26: surf(x,y,z); xlabel(xlab); ylabel(ylab);
27: zlabel(zlab); title(titl);
28: if length(v)>2, axis(v); end
29: grid; figure(gcf);
```

Function cubrange

```
 1: function range=cubrange(xyz,ovrsiz)
 2: %
 3: % range=cubrange(xyz,ovrsiz)
 4: % ~~~~~~~~~~~~~~~~~~~~~~~~~~
 5: % This function determines limits for a square
 6: % or cube shaped region for plotting data values
 7: % in the columns of array xyz to an undistorted
 8: % scale
 9: %
10: % xyz     - a matrix of the form [x,y] or [x,y,z]
11: %           where x,y,z are vectors of coordinate
12: %           points
13: % ovrsiz  - a scale factor for increasing the
14: %           window size. This parameter is set to
15: %           one if only one input is given.
16: %
17: % range   - a vector used by function axis to set
18: %           window limits to plot x,y,z points
19: %           undistorted. This vector has the form
20: %           [xmin,xmax,ymin,ymax] when xyz has
21: %           only two columns or the form
22: %           [xmin,xmax,ymin,ymax,zmin,zmax]
23: %           when xyz has three columns.
24: %
25: % User m functions called:  none
26: %------------------------------------------------
27:
28: if nargin==1, ovrsiz=1; end
29: pmin=min(xyz); pmax=max(xyz); pm=(pmin+pmax)/2;
30: pd=max(ovrsiz/2*(pmax-pmin));
```

```
31: if length(pmin)==2
32:    range=pm([1,1,2,2])+pd*[-1,1,-1,1];
33: else
34:    range=pm([1 1 2 2 3 3])+pd*[-1,1,-1,1,-1,1];
35: end
```

13.4   Fitting Equations to Data

Often an equation of known form is needed to approximately fit some given data values. An equation $y(t)$ to fit m data values (t_i, y_i) might be sought from an equation expressible as

$$y = f(a_1, a_2, \ldots, a_n, t)$$

where n parameters $a_1, a_2, \ldots, a_n$ are needed to minimize the least square error

$$\epsilon(a_1, a_2, \ldots, a_n) = \sum_{j=1}^{n} [y_j - f(a_1, a_2, \ldots, a_n, t_j)]^2$$

The smallest possible error would be zero when the equation passes exactly through all the data values. Function ϵ can be minimized with an optimizer such as **fmins**, or conditions seeking a zero gradient of ϵ which require

$$\frac{\partial \epsilon}{\partial a_i} = 0$$

$$= 2\sum_{j=1}^{n} [f(a_1, a_2, \ldots, a_n, t_j) - y_j] \left(\frac{\partial f}{\partial a_i}\right)$$

can be written. Note that the problem of minimizing a function and the problem of solving a set of nonlinear simultaneous equations are closely related. Solving large systems of nonlinear equations is difficult. Therefore, data fitting by use of function minimization procedures is typically more effective.

The formulation assuming y depends on a single independent variable could just as easily have involved several independent variables $x_1, x_2, \ldots, x_N$ which would yield an equation of the form

$$y = f(a_1, a_2, \ldots, a_n, x_1, x_2, \ldots, x_N)$$

For instance, we might choose the simplest useful equation depending linearly on the independent variables

$$y = \sum_{k=0}^{N} x_k a_k$$

where x_0 means unity. The least square error can be expressed as

$$\epsilon(a_0, a_1, \ldots, a_n) = \sum_{j=1}^{n} \left[y_j - \sum_{k=0}^{N} X_{jk} a_k \right]^2$$

where X_{jk} means the value of the k'th independent variable at the j'th data point. The condition that ϵ have a zero gradient gives

$$\sum_{k=0}^{N}\left[\sum_{j=1}^{n}X_{ji}X_{jk}\right]a_k = \sum_{j=1}^{n}X_{ji}y_j \qquad 1 \leq i \leq N$$

This linear system can be solved using traditional methods. Since the multiple indices in the equation are slightly cryptic, expressing the relationship in matrix notation is helpful. We get

$$Y \approx XA$$

where

$$Y = \begin{bmatrix} y_1 \\ y_2 \\ \vdots \\ y_n \end{bmatrix} \qquad X = [1, X_1, X_2, \ldots, X_N] \qquad A = \begin{bmatrix} a_0 \\ a_1 \\ \vdots \\ a_N \end{bmatrix}$$

with X_i being the column matrix $[x_{i1}, x_{i2}, \ldots, x_{in}]$ and the first column of X contains all ones. The requirement to minimize ϵ is simply

$$(X^T X)A = X^T Y$$

and MATLAB produces the desired solution using

```
A=X\Y;
```

Although taking y as a linear function of parameters $a_0, a_1, \ldots, a_N$ produces solvable linear equations, the general situation yields nonlinear equations, and a minimization search procedure has greater appeal. We conclude this section with an example employing a minimization search.

Consider an experiment where data values (t_i, y_i) are expected to conform to the transient response of a linear harmonic oscillator governed by the differential equation

$$m_0 \ddot{y} + c_0 \dot{y} + k_0 y = 0$$

This equation has a solution representable as

$$y = a_1 e^{-|a_2|t}\cos(|a_3|t + a_4)$$

where $|a_2|$ makes the response decay exponentially and $|a_3|$ assures that the damped natural frequency is positive. Minimizing the error function

$$\epsilon(a_1, a_2, a_3, a_4) = \sum_{j=1}^{n}\left[y_j - a_1 e^{-1|a_2|t_j}\cos(|a_3|t_j + a_4)\right]^2$$

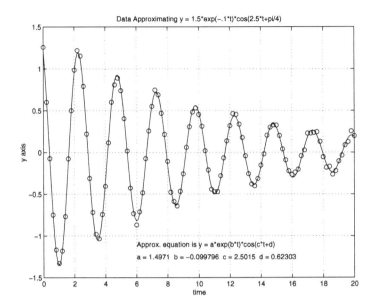

Figure 13.4. Data Approximating $y = 1.5 \exp(-0.1t) \cos(2.5t + \pi/4)$

requires a four-dimensional search.

The program **vibfit** tests data deviating slightly from an equation employing specific values of a_1, a_2, a_3, a_4. Then function **fmins** is used to verify whether the coefficients can be recovered from the data points. Figure 13.4 shows the data values and the equation resulting from the nonlinear least square fit. The results produced are quite acceptable.

13.4.1　Program Output and Code

Script File vibfit

```
 1: % Example: vibfit
 2: % ~~~~~~~~~~~~~~~
 3: %
 4: % This program illustrates use of the Nelder
 5: % and Mead multi-dimensional function
 6: % minimization method to determine an equation
 7: % for y(t) which depends nonlinearly on several
 8: % parameters chosen to fit closely known data
 9: % values. The program minimizes the sum of the
10: % squares of error deviations between the data
11: % values and results produced by the chosen
12: % equation. The example pertains to the time
13: % response curve characterizing free vibrations
14: % of a damped linear harmonic oscillator.
15: %
16: % User m functions called:
17: %    vibfun, pauz, genprint
18: %-----------------------------------------------
19:
20: % Make the data vectors global to allow
21: % access from function vibfun
22: global timdat ydat
23:
24: echo off;
25: disp(' ');
26: disp('        CHOOSING PARAMETERS');
27: disp('   IN THE THE NONLINEAR EQUATION');
28: disp('     Y = A*EXP(B*T)*COS(C*T+D)');
29: disp('TO OBTAIN THE BEST FIT TO GIVEN DATA');
30: fprintf('\nPress [Enter] to see function\n');
31: fprintf('  vibfun which is to be minimized\n');
32: pause;
33:
34: % Generate a set of data to be fitted by a
35: % chosen equation.
36: a=1.5; b=-.1; c=2.5; d=pi/5;
37: timdat=0:.2:20;
38: ydat=a*exp(b*timdat).*cos(c*timdat+d);
39:
40: % Add some random noise to the data
```

```
41: ydat=ydat+.1*(-.5+rand(size(ydat)));
42:
43: % Function vibfun defines the quantity to be
44: % minimized by a search using function fmins.
45: disp(' ');
46: disp('The function to be minimized is:');
47: type vibfun.m; disp(' ');
48: disp('The input data will be plotted next.');
49: disp('Press [Enter] to continue'); pause;
50: plot(timdat,ydat,'o');
51: title('Input Data'); xlabel('time');
52: ylabel('y axis'); grid on; figure(gcf);
53:
54: % Initiate the four-dimensional search
55: x=fmins('vibfun',[1 1 1 1]);
56:
57: % Check how well the computed parameters
58: % fit the data.
59: aa=x(1); bb=-abs(x(2)); cc=abs(x(3)); dd=x(4);
60: as=num2str(aa); bs=num2str(bb);
61: cs=num2str(cc); ds=num2str(dd);
62: ttrp=0:.1:20;
63: ytrp=aa*exp(bb*ttrp).*cos(cc*ttrp+dd);
64: disp(' ');
65: disp('Press [Enter] to see how well the');
66: disp('equation fits the data'); pause;
67: plot(ttrp,ytrp,timdat,ydat,'o');
68: str1=['Approx. equation is y = ', ...
69:        'a*exp(b*t)*cos(c*t+d)'];
70: str2=['a = ',as,'  b = ',bs,'  c = ', ...
71:        cs,'  d = ',ds];
72: text(6,-1.1,str1); text(6,-1.25,str2);
73: xlabel('time'); ylabel('y axis');
74: title(['Data Approximating ', ...
75:        'y = 1.5*exp(-.1*t)*cos(2.5*t+pi/4)']);
76: grid on; figure(gcf);
77: %genprint('apprxdat');
```

Function vibfun

```
 1: function z=vibfun(x)
 2: %
 3: % z=vibfun(x)
 4: % ~~~~~~~~~~~~
 5: %
 6: % This function evaluates the least square
 7: % error for a set of vibration data. The data
 8: % vectors timdat and ydat are passed as global
 9: % variables. The function to be fitted is:
10: %
11: %    y=a*exp(b*t)*cos(c*t+d)
12: %
13: % x - a vector defining a,b,c and d
14: %
15: % z - the square of the norm for the vector
16: %       of error deviations between the data and
17: %       results the equation gives for current
18: %       parameter values
19: %
20: % User m functions called:  none
21: %-------------------------------------------------
22:
23: global timdat ydat
24: a=x(1); b=-abs(x(2)); c=abs(x(3)); d=x(4);
25: z=a*exp(b*timdat).*cos(c*timdat+d);
26: z=norm(z-ydat)^2;
```

13.5   Nonlinear Deflections of a Cable

We will now present an optimization procedure to determine the static equilibrium position of a perfectly flexible inextensible cable having given end positions and a known distributed load per unit length. If $R(s)$ is the position of any point on the cable as a function of arc length $0 \le s \le L$, then the internal tension at position s is

$$T(s) = F_e + \int_s^L q(s)\, ds$$

with $q(s)$ being the applied force per unit length and F_e being the support force at $s = L$. The end force to produce a desired end deflection has to be determined in the analysis. However, the end deflection resulting from any particular choice of end force can be computed by observing that the tangent to the deflection curve will point along the direction of the cable tension. This means

$$\frac{dR}{ds} = \frac{T(s)}{|T(s)|}$$

and

$$R(s) = \int_0^s \frac{T(s)ds}{|T(s)|} = \int_0^s \frac{\left(F_e + \int_s^L q\, ds \right) ds}{|F_e + \int_s^L q\, ds|}$$

where $R(0) = 0$ is taken as the position at the starting end. The deflection at $s = L$ will have some specified position R_e so that requiring $R(L) = R_e$ gives a vector equation depending parametrically on F_e. Thus, we need to solve three nonlinear simultaneous equations in the cartesian components of force F_e. A reasonable analytical approach is to employ an optimization search to minimize $|R(L) - R_e|$ in terms of the components of F_e.

The procedure described for a cable with continuous loading extends easily to a cable having several rigid links connected at frictionless joints where arbitrary concentrated forces are applied. Function **cabldefl** evaluates the position of each joint when the joint forces and outer end force are given. With the end force on the last link treated as a parameter, function **endfl** computes an error measure $|F(L) - R_E|^2$ to be minimized using function **fmins**. The optimization search seeks the components of F_e needed to reduce the error measure to zero. Specifying a sensible problem obviously requires that $|R_e|$ must not exceed the total length of all members in the chain. Initiating the search with a randomly generated starting force leads to a final force produced by **fmins**, which is then employed in another call to **cabldefl** to determine and plot the final deflection position as shown in Figure 13.5. Using a random initial guess for the end force was done to show that choosing bad starting

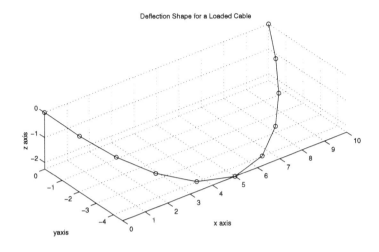

Figure 13.5. Deflected Shape for a Loaded Cable

data, insufficiently stringent convergence tolerances, or too few allow-
able function iterations can sometimes produce erroneous results. This
emphasizes the need to always examine the results from nonlinear search
procedures to assure that satisfactory answers are obtained.

13.5.1 Program Output and Code

Script File cablsolv

```
 1: function [r,t,pends]=cablsolv(Len,P,Rend)
 2: %
 3: % [r,t,pends]=cablsolv(Len,P,Rend)
 4: % ~~~~~~~~~~~~~~~~~~~~~~~~~~~~~~~~~
 5: %
 6: % This function computes the equilibrium
 7: % position for a cable composed of rigid
 8: % weightless links with loads applied at the
 9: % frictionless joints. The ends of the cable
10: % are assumed to have a known position.
11: %
12: % Len    - a vector containing the lengths
13: %          Len(1), ..., Len(n)
14: % P      - matrix of force components applied
15: %          at the interior joints. P(:,i)
16: %          contains the cartesian components of
17: %          the force at joint i.
18: % Rend   - the coordinate vector giving the
19: %          position of the outer end of the last
20: %          link, assuming the outer end of the
21: %          first link is at [0,0,0].
22: %
23: % r      - a matrix with rows giving the
24: %          computed equilibrium positions of all
25: %          ends
26: % t      - a vector of tension values in the
27: %          links
28: % pends  - a matrix having two rows which
29: %          contain the force components acting
30: %          at both ends of the chain to maintain
31: %          equilibrium
32: %
33: % User m functions called:
34: %    endfl, cabldefl, genprint
35: %-------------------------------------------------
36:
37: if nargin < 3
38:    % Example for a ten link cable with vertical
39:    % and lateral loads
40:    Len=1.5*ones(10,1); Rend=[10,0,0];
```

```
41:    P=ones(9,1)*[0,-2,-1];
42: end
43:
44: global len p rend
45: len=Len; rend=Rend; p=P; tol=sum(Len)/1e8;
46:
47: % Start the search with a random force applied
48: % at the far end
49:
50: p0=10*max(abs(p(:)))*rand(size(p,2),1);
51:
52: % Perform a search to minimize the length of
53: % the vector from the ?? of the computed
54: % end position and the desired position Rend.
55: % The final end force will reduce this
56: % deflection error to zero if the search
57: % converges.
58:
59: opts=zeros(1,14); opts(2:3)=[tol,tol];
60: opts(14)=1000;
61: pend=fmins('endfl',p0,opts); pend=pend(:);
62:
63: % Use the computed end force to compute the
64: % final deflection. Also return the
65: % support forces.
66: [r,t,pstart]=cabldefl(len,[p;pend']);
67: x=r(:,1); y=r(:,2); z=r(:,3);
68: pends=[pstart(:)';pend(:)'];
69:
70: % Plot the deflection curve of the cable
71: plot3(x,y,z,x,y,z,'o'); xlabel('x axis');
72: ylabel('yaxis'); zlabel('z axis');
73: title('Deflection Shape for a Loaded Cable');
74: axis('equal'); grid on; figure(gcf);
75: %genprint('defcable');
```

Function cabldefl

```
 1: function [r,t,pbegin]=cabldefl(len,p)
 2: %
 3: % [r,t,pbegin]=cabldefl(len,p)
 4: % ~~~~~~~~~~~~~~~~~~~~~~~~~~~~~~~
 5: %
 6: % This function computes the static equilibrium
 7: % position for a cable of rigid weightless
 8: % links having concentrated loads applied at
 9: % the joints and the outside of the last link.
10: % The outside of the first link is positioned
11: % at the origin.
12: %
13: % len     - a vector of link lengths
14: %           len(1), ..., len(n)
15: % p       - a matrix with rows giving the
16: %           force components acting at the
17: %           interior joints and at the outer
18: %           end of the last link
19: %
20: % r       - matrix having rows which give the
21: %           final positions of each node
22: % t       - vector of member tensions
23: % pbegin  - force acting at the outer end of
24: %           the first link to achieve
25: %           equilibrium
26: %
27: % User m functions called:  none
28: %-------------------------------------------------
29:
30: n=length(len); len=len(:); nd=size(p,2);
31:
32: % Compute the forces in the links
33: T=flipud(cumsum(flipud(p)));
34: t=sqrt(sum((T.^2)')');
35: % Obtain the deflections of the outer ends
36: % and the interior joints
37: r=cumsum(T./t(:,ones(1,nd)).*len(:,ones(1,nd)));
```

```
38: r=[zeros(1,nd);r]; pbegin=-t(1)*r(2,:)/len(1);
```

Function endfl

```
 1: function enderr=endfl(pend)
 2: %
 3: % enderr=endfl(pend)
 4: % ~~~~~~~~~~~~~~~~~~~
 5: %
 6: % This function computes how much the
 7: % position of the outer end of the last link
 8: % deviates from the desired position when an
 9: % arbitrary force pend acts at the cable end.
10: %
11: % pend    - vector of force components applied
12: %             at the outer end of the last link
13: %
14: % enderr - the deflection error defined by the
15: %             square of the norm of the vector
16: %             from the computed end position and
17: %             the desired end position. This error
18: %             should be zero for the final
19: %             equilibrium position
20: %
21: % User m functions called: cabldefl
22: %-------------------------------------------------
23:
24: % Pass the lengths, the interior forces and the
25: % desired position of the outer end of the last
26: % link as global variables.
27: global len p rend
28:
29: % use function cabldefl to compute the
30: % desired error
31: r=cabldefl(len,[p;pend(:)']);
32: rlast=r(size(r,1),:);
33: d=rlast(:)-rend(:); enderr=d'*d;
```

13.6 Quickest Time Descent Curve (the Brachistochrone)

The subject of variational calculus addresses methods to find a function producing the minimum value for an integral depending parametrically on the function. Typically, we have a relationship of the form

$$I(y) = \int_{x_1}^{x_2} G(x, y, y'(x)) \, dx$$

where values of y at $x = x_1$ and $x = x_2$ are known, and $y(x)$ for $x_1 < x < x_2$ is sought to minimize I. A classical example in this subject is determining a curve starting at $(0,0)$ and ending at (a, b) so that a smooth particle will slide from one end to the other in the shortest possible time. Let X and Y be measured positive to the right and downward. Then the descent time for frictionless movement along the curve will be

$$t = \frac{1}{\sqrt{2g}} \int_0^a \sqrt{\frac{1 + Y'(X)^2}{Y(X)}} \, dX \qquad Y(0) = 0 \quad Y(a) = b$$

This problem is solved in various advanced calculus books.[3] The curve is a cycloid expressed in parametric form as

$$X = k[\theta - \sin(\theta)] \qquad Y = k[1 - \cos(\theta)]$$

where $0 < \theta < \theta_f$. Values of θ_f and k are found to make $x(\theta_f) = a$ and $Y(\theta_f) = b$. The exact descent time is

$$t_{\text{best}} = \theta_f \sqrt{\frac{k}{g}}$$

which is significantly smaller than the descent time for a straight line which is

$$t_{\text{line}} = \sqrt{\frac{2(a^2 + b^2)}{gb}}$$

Two functions, **brfaltim** and **bracifun**, are used to compute points on the brachistochrone curve and evaluate the descent time.

The main purpose of this section is to illustrate how optimization search can be used to minimize an integral depending parametrically on a function. The method used chooses a set of base points through which an interpolation curve is constructed to specify the function. Using numerical integration gives a value for the integral. Holding the x values

---

[3] Weinstock [103] provides an excellent discussion of the brachistochrone problem using calculus of variation methods.

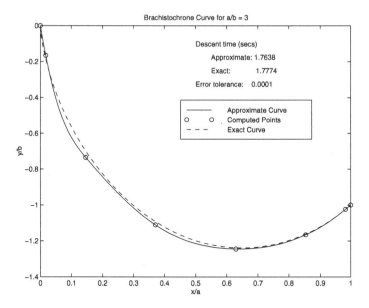

Figure 13.6. Brachistochrone Curve for $\frac{a}{b} = 3$

for the interpolation points constant and allowing the y values to vary expresses the integral as a function of the y values at a finite number of points. Then a multi-dimensional search function such as **fmins** can optimize the choice of Y values. Before carrying out this process for the brachistochrone problem it is convenient to change variables so that $x = X/a$ and

$$Y(X) = b[x + y(x)] \qquad 0 \le x \le 1$$

$$y(0) = y(1) = 0$$

Then the descent integral becomes

$$t = \frac{a}{\sqrt{2gb}} \int_0^1 \sqrt{\frac{1 + (b/a)^2[1 + y'(x)]^2}{x + y}} \, dx$$

For any selected set of interpolation points, the spline routines **spc** and **spltrp** can evaluate $y(x)$ and $y'(x)$ needed in the integrand, and function **gcbpwf** can be used to perform Gaussian integration. An optimization search employing **fmins** produces the curve heights yielding an approximation to the brachistochrone as shown in Figure 13.6.

13.6.1   Program Output and Code

Script File brachist

```
 1: % Example: brachist
 2: % ~~~~~~~~~~~~~~~~~
 3: % This program determines the shape of a
 4: % smooth curve down which a particle can slide
 5: % in minimum possible time. The analysis
 6: % employs a piecewise cubic spline to define
 7: % the curve by interpolation using a fixed set
 8: % of base point positions. The curve shape
 9: % becomes a function of the heights chosen at
10: % the base points. These heights are determined
11: % by computing the descent time as a function
12: % of the heights and minimizing the descent
13: % time by use of an optimization program. The
14: % Nelder and Mead unconstrained search
15: % procedure is used to compute the minimum.
16: %
17: % User m functions called:
18: %      chbpts, brfaltim, fltim, gcbpwf,
19: %      bracifun, spc, spltrp, genprint
20: %-------------------------------------------------
21:
22: fprintf(...
23: '\nBRACHISTOCHRONE DETERMINATION BY NONLINEAR');
24: fprintf('\n         OPTIMIZATION SEARCH \n\n');
25: fprintf(...
26:   'Press [Enter] to begin the optimization\n');
27: pause; fprintf(['\nPlease wait. The ',...
28:   'calculation takes a while\n']);
29: clear all; clf;
30: global cbp cwf cofs n xc yc a b b_over_a ...
31:        grav nparts nquad nfcls
32:
33: % Initialize
34: a=30; b=10; grav=32.2; nparts=1; nquad=60;
35: tol=1e-4; n=6;
36: b_over_a = b/a;
37:
38: [cbp,cwf]=gcbpwf(0,1,nquad,nparts);
39: xc=chbpts(0,1,n); xc=xc(:);
40: y0=5*sin(pi*xc); xc=[0;xc;1];
```

```
41:
42: % Calculate results from the exact solution
43: [texact,xexact,yexact]=brfaltim(a,b,grav,100);
44:
45: % Perform minimization search for
46: % approximate solution
47: nfcls=0; flp=flops;
48: yfmin=fmins('fltim',y0,[0,tol,tol]);
49: flp=flops-flp;
50:
51: % Evaluate final position and approximate
52: % descent time
53: Xfmin=xc; Yfmin=Xfmin+[0;yfmin(:);0];
54: tfmin=a/sqrt(2*grav*b)*fltim(yfmin(:));
55:
56: % Summary of calculations
57: fprintf('\n\nBrachistochrone Summary');
58: fprintf('\n----------------------');
59: fprintf('\n\nNumber of function calls:   ');
60: fprintf('%g',nfcls);
61: fprintf(  '\nNumber of FLOPS (millions): ');
62: fprintf('%g',flp/1e6);
63: fprintf('\n');
64:
65: % Plot results comparing the approximate
66: % and exact solutions
67: xplot=linspace(0,1,100);
68: yplot=spline(Xfmin,Yfmin,xplot);
69: plot(xplot,-yplot,'-',Xfmin,-Yfmin,'o', ...
70:      xexact/a,-yexact/b,'--');
71: xlabel('x/a'); ylabel('y/b'); % grid
72: title(['Brachistochrone Curve for ', ...
73:       'a/b = ',num2str(a/b)]);
74: text(.5,-.1,'Descent time (secs)')
75: text(.55,-.175,['Approximate: ', ...
76:      num2str(tfmin)])
77: text(.55,-.25,['Exact:            ', ...
78:      num2str(texact)]);
79: text(.5,-.325, ...
80:    sprintf('Error tolerance:    %g',tol));
81: legend('Approximate Curve', ...
82:        'Computed Points','Exact Curve',3);
83: figure(gcf);
84: %pauz; fprintf('\nPress [Enter] to continue\n');
85: %genprint('brachist');
```

Function brfaltim

```
 1: function [tfall,xbrac,ybrac]=brfaltim ...
 2:                             (a,b,grav,npts)
 3: %
 4: %
 5: % [tfall,xbrac,ybrac]=brfaltim(a,b,grav,npts)
 6: % ~~~~~~~~~~~~~~~~~~~~~~~~~~~~~~~~~~~~~~~~~~~~~
 7: %
 8: % This function determines the descent time
 9: % and a set of points on the brachistochrone
10: % curve passing through (0,0) and (a,b).
11: % The curve is a cycloid expressible in
12: % parametric form as
13: %
14: %    x=k*(th-sin(th)),
15: %    y=k*(1-cos(th))    for 0<=th<=thf
16: %
17: % where thf is found by solving the equation
18: %
19: %    b/a=(1-cos(thf))/(thf-sin(thf)).
20: %
21: % Once thf is known then k is found from
22: %
23: %    k=a/(th-sin(th)).
24: %
25: % The exact value of the descent time is given
26: % by
27: %
28: %    tfall=sqrt(k/g)*thf
29: %
30: % a,b   - final values of (x,y) on the curve
31: % grav  - the gravity constant
32: % npts  - the number of points computed on
33: %         the curve
34: %
35: % tfall - the time required for a smooth
36: %         particle to to slide along the curve
37: %         from (0,0) to (a,b)
38: % xbrac - x points on the curve with x
39: %         increasing to the right
40: % ybrac - y points on the curve with y
41: %         increasing downward
42: %
```

```
43: % User m functions called: bracifun
44: %-------------------------------------------------
45:
46: % The ratio b/a is passed to bracifun as a
47: % global variable.
48: global b_over_a
49:
50: b_over_a=b/a; th=fzero('bracifun',.1);
51: k=a/(th-sin(th)); tfall=sqrt(k/grav)*th;
52: if nargin==4
53:    thvec=(0:npts-1)'*(th/(npts-1));
54:    xbrac=k*(thvec-sin(thvec));
55:    ybrac=k*(1-cos(thvec));
56: end
```

Function bracifun

```
1: function val=bracifun(th)
2: %
3: % val=bracifun(th)
4: % ~~~~~~~~~~~~~~~~
5: %
6: % A zero of this function determines
7: % the end points of the brachistochrone curve.
8: %
9: % th        - the angle parameterizing the
10: %              brachistochrone
11: %
12: % val       - the function value which becomes
13: %              zero when the correct value of th
14: %              is found using function fzero
15: %
16: % b_over_a - the value of b/a for the curve
17: %
18: % User m functions called:  none
19: %-------------------------------------------------
20:
21: global b_over_a
22: val=cos(th)-1+b_over_a*(th-sin(th));
```

Function fltim

```
 1: function t=fltim(y)
 2: %
 3: % t=fltim(y)
 4: % ~~~~~~~~~~
 5: %
 6: % This function evaluates the time descent
 7: % integral for a spline curve having heights
 8: % stored in y.
 9: %
10: % y - vector defining the curve heights at
11: %       interior points corresponding to base
12: %       positions in xc
13: %
14: % t - the numerically integrated time descent
15: %       integral evaluated by use of base points
16: %       cbp and weight factors cwf passed as
17: %       global variables
18: %
19: % User m functions called: spc, spltrp
20: %-------------------------------------------------
21:
22: global xc cofs nparts bp wf nfcls cbp cwf ...
23:          b_over_a
24:
25: nfcls=nfcls+1; x=cbp;
26:
27: % Generate coefficients used in spline
28: % interpolation
29: yc=[0;y(:);0]; cofs=spc(xc,yc,[3,3],[0,0]);
30:
31: y=spltrp(x,cofs,0); yp=spltrp(x,cofs,1);
32:
33: % Evaluate the integrand
34: f=(1+(b_over_a*(1+yp)).^2)./(x+y); f=sqrt(f);
35:
36: % Evaluate the integral
37: t=cwf(:)'*f(:);
```

Function gcbpwf

```
 1: function [cbp,cwf]=gcbpwf ...
 2:                   (xlow,xhigh,nquad,mparts)
 3: %
 4: % [cbp,cwf]=gcbpwf(xlow,xhigh,nquad,mparts)
 5: % ~~~~~~~~~~~~~~~~~~~~~~~~~~~~~~~~~~~~~~~~~~~
 6: %
 7: % This function computes base points, cbp, and
 8: % weight factors, cwf, in a composite
 9: % quadrature formula which integrates an
10: % arbitrary function from xlow to xhigh by
11: % dividing the interval of integration into
12: % mparts equal parts and integrating over each
13: % part using a Gauss formula requiring nquad
14: % function values.  Suppose a function f(x) is
15: % being integrated by use of the composite
16: % formula. Denote the base points and weight
17: % factors of the standard Gauss formula as bp
18: % and wf. Then the integration can be
19: % accomplished with the following algorithm:
20: %
21: %    x=b
22: %    integral( f(x)*dx ) =
23: %    x=a
24: %
25: %          j=n k=m
26: %      d1*sum sum( wf(j)*fun(a1+d*k+d1*bp(j)) )
27: %          j=1 k=1
28: %
29: % where
30: %   bp are base points
31: %   wf are weight factors
32: %   n = nquad,  m = mparts
33: %   d = (b-a)/m, d1 = d/2, a1 = a-d1
34: %
35: % The base points bp and weight factors wf are
36: % first generated by eigenvalue methods.
37: %
38: % The same calculation can be performed with a
39: % single summation using the composite base
40: % points and weight factors as follows:
41: %
42: %    x=b
```

```
43: %      integral( f(x)*dx ) =
44: %      x=a
45: %
46: %                        j=nquad*mparts
47: %                        sum( cwf(j)*fun( cbp(j) )
48: %                        j=1
49: %
50: % User m functions called:  none
51: %-----------------------------------------------
52:
53: % Compute base points and weight factors for a
54: % single interval
55: u=(1:nquad-1)./sqrt((2*(1:nquad-1)).^2-1);
56: [vc,bp]=eig(diag(u,-1)+diag(u,1));
57: [bp,k]=sort(diag(bp)); wf=2*vc(1,k)'.^2;
58:
59: % Evaluate the composite base points and weight
60: % factors in terms of bp and wf
61: d=(xhigh-xlow)/mparts;  d1=d/2;
62: dbp=d1*bp(:); dwf=d1*wf(:);  dr=d*(1:mparts);
63: cbp=dbp(:,ones(1,mparts))+ ...
64:     dr(ones(nquad,1),:)+(xlow-d1);
65: cwf=dwf(:,ones(1,mparts)); cwf=cwf(:);cbp=cbp(:);
```

Function chbpts

```
 1: function x=chbpts(xmin,xmax,n)
 2: %
 3: % x=chbpts(xmin,xmax,n)
 4: % ~~~~~~~~~~~~~~~~~~~~~
 5: % Determine n points with Chebyshev spacing
 6: % between xmin and xmax.
 7: %
 8: % User m functions called:  none
 9: %-----------------------------------------------
10:
11: x=(xmin+xmax)/2+((xmin-xmax)/2)* ...
12:    cos(pi/n*((0:n-1)'+.5));
```

Appendix A

List of MATLAB Routines
with Descriptions

Table A.1. List of MATLAB Routines with Descriptions

Routine	Chapter	Description
adams2	4	Function to determine coefficients in the Adams-type formulas used to solve $y'(x) = f(x, y)$.
adamsex	4	Driver program illustrating use of function **adams2** for determining coefficients in the explicit or implicit Adams-type formulas used for differential equation solution.
allprop	5	Function to compute area, centroidal coordinates, and second moments of area (inertial moments) for a spline curve.
animate	9	Function to draw an animated plot of data values stored in an array.
arcprop	5	Function to compute area properties of a circular arc.
areaprop	5	Function used to determine geometrical properties of any area bounded by straight lines and circular arcs.
arearun	5	Driver program to compute the area, centroidal coordinates, moments of inertia, and product of inertia for any area bounded by straight lines and circular arcs.
		continued on next page

Routine	Chapter	Description
as1	5	Lower x integration limit used with function **triplint**.
as2	5	Upper x integration limit used with function **triplint**.
aspiral	2	Computes geometrical properties of a spiral curve.
assemble	10	Assembles global stiffness matrix and mass matrix for a plane truss structure.
beamresp	9	Function to evaluate the time dependent displacement and bending moment in a constant cross-section simply-supported beam which is initially at rest when a harmonically varying moment is suddenly applied at the right end.
besf	5	Companion function for script runcases.
bisect	1	Determines a root of a function by interval halving.
bmvardep	11	Function to compute the shear, moment, slope, and deflection in a variable depth beam.
brachist	13	Determines the shape of a smooth curve along which a particle can slide in minimum possible time.
bracifun	13	Determines the end points of the brachistochrone curve.
brfaltim	13	Determines the descent time and a set of points on the brachistochrone curve passing through $(0,0)$ and (a,b).
bs1	5	Lower y integration limit used with function **triplint**.
bs2	5	Upper y integration limit used with function **triplint**.
cabldefl	13	Computes the static equilibrium position for a cable of rigid weightless links having concentrated loads applied at the joints and the outside of the last link.

continued on next page

Routine	Chapter	Description
cablemk	7	Function to form the mass and stiffness matrices for a cable.
cablenl	8	Driver program for numerical integration of the matrix differential equations for the nonlinear dynamics of a cable fixed at both ends.
cablinea	7	Program which uses modal superposition to compute the dynamic response of a cable suspended at one end and free at the other.
cablsolv	13	Computes the equilibrium position for a cable composed of rigid weightless links with loads applied at the frictionless joints.
canimate	7	Function to draw an animated plot of cable deflection data values.
cartstrs	12	Function which uses values of the complex stress functions to evaluate cartesian stress components relative to the (x, y) axes.
cauchint	12	Performs numerical evaluation of a Cauchy integral.
cauchtst	12	Solves a mixed boundary value problem for the interior of a circle by numerical evaluation of a Cauchy integral.
cbfreq	9	Driver program for computing approximate natural frequencies of a uniform depth cantilever beam using finite difference and finite element methods.
cbfrqfdm	9	Function to compute approximate cantilever beam frequencies by the finite difference method.
cbfrqfem	9	Function to determine natural frequencies of a uniform depth cantilever beam by the finite element method.

continued on next page

Routine	Chapter	Description
cbfrqnwm	9	Function to determine cantilever beam frequencies by Newton's method.
cbpwf6	5	Function to compute base points and weight factors in a composite quadrature formula which integrates an arbitrary function by dividing the interval of integration into equal parts and integrating over each part using a Gauss formula requiring six function values.
chbpts	2, 13	Function to compute points with Chebyshev spacing.
chopsine	6	Sample code extract.
colbuc	10	Determines the Euler buckling load for a slender column of variable cross section.
crc2crc	12	Determines the circle or straight line into which a circle or straight line maps under a linear fractional transformation.
crclovsq	5	Script file defining a half annulus above a square which contains a square hole. Used as input for script file **arearun**.
crnrtest	5	Example showing effect of corner points on spline curve approximation of a general geometry.
crosmat	5	Computes the vector cross product.
crossdat	10	Data specification for a cross-shaped truss.
crvprp3d	2	Computes the primary differential properties of a three-dimensional curve parameterized in the form $R(t)$.
crvprpsp	2	Computes spline interpolated values for coordinates, base vectors and curvature obtained by passing a spline curve through data values.
		continued on next page

continued on next page

Routine	Chapter	Description
cubrange	2, 5, 8, 10, 12-13	Determines limits for a square- or cube-shaped region for plotting data values to an undistorted scale.
deislner	7	Driver program using implicit second or fourth order integrators to compute the linearized dynamical response of a cable suspended at one end and free at the other end.
derivtrp	4	Function which computes coefficients to interpolate derivatives by finite differences.
dife	5	Function to differentiate evenly spaced data.
drawtrus	10	Draws a truss defined by nodal coordinates.
dvdcof	4	Function which uses divided differences to compute coefficients for polynomial interpolation by the Newton form of the interpolating polynomial.
dvdtrp	4	Performs polynomial interpolation using the Newton form of the interpolating polynomial.
ecentric	12	Determines the bilinear transformation which maps the region $1 \leq |z| \leq r$ onto an eccentric annulus.
eigc	10	Computes eigenvalues and eigenvectors for the problem $Kx = \lambda Mx$.
eigsym	10	Computes the eigenvalue of the constrained eigenvalue problem.
eilt	10	Computes the moment of inertia along a linearly tapered circular cross section and uses that value to produce the product EI.
elipcyl	12	Computes the flow field around an elliptic cylinder.
elipdisk	12	Computes a rational function mapping $|\zeta| \leq 1$ onto an elliptical disk.

continued on next page

Routine	Chapter	Description
elipdplt	12	Plots contour lines showing how a circular disk maps onto an elliptic disk.
eliphole	12	Determines curvilinear coordinate stresses around an elliptic hole in a plate uniformly stressed at infinity.
elipinvr	12	Inverts the transformation of the prescribed function.
elipprop	5	Driver program to compute the area, centroidal coordinates, and inertial moments of a rotated ellipse.
elipsoid	2	Plots an ellipsoid having semi-diameters a, b, and c.
elmstf	10	Forms the stiffness matrix for a truss element and computes the member volume.
elpmaxst	12	Example which plots the stress concentration around an elliptic hole as a function of the semi-diameter ratio.
endfl	13	Computes how much the position of the outer end of the last link deviates from the desired position when a force applied acts at the cable end.
equamo	8	Function forming nonlinear equations of motion for a cable fixed at both ends and loaded only by gravity forces.
eventime	8	Function to compute cable position coordinates for a series of evenly spaced time values.
examplmo	9	Function which evaluates the response caused when a downward load at the middle and an upward load at the free end is applied.
expc	5	Companion function for script runcases.
extload	11	Function which computes the shear, moment, slope, and deflection in a uniform depth Euler beam loaded by a series of concentrated and ramp loads.

continued on next page

Routine	Chapter	Description
fbot	9	Function used to solve Laplace's equation for a rectangle.
fftaprox	6	Sample code extract for Fourier series approximation.
finidif	4	Program using truncated Taylor series to compute finite difference formulas approximating derivatives of arbitrary order which are interpolated at an arbitrary set of base points not necessarily evenly spaced.
flopex	3	Driver program used to determine the number of floating point operations required to perform several familiar matrix calculations.
floptest	3	Determine the FLOP counts for various matrix operations.
fltim	13	Evaluates the time descent integral for a spline curve having specified heights.
fnc	1	Function employed in root finding example.
forcresp	3	Sample code extract illustrating the solution of matrix differential equations.
fouaprox	6	Performs Fourier series approximation.
fouseris	6	Driver program to illustrate the convergence rate of Fourier series approximations derived by applying the FFT to a general function which may be specified either by piecewise linear interpolation in a data table or by analytical definition in a function given by the user.
fousum	6	Function to sum the Fourier series of a real valued function.
frud	9	Function employing modal superposition to solve $mx'' + kx = f_1 \cos(\omega t) + f_2 \sin(\omega t)$

continued on next page

Routine	Chapter	Description
frus	2	Computes the points on the surface of a conical frustum which has its axis along the z axis.
frusdist	13	Determines the point on the surface of a conical frustum which is closest to another point outside the surface.
frustum	13	Creates points defining a frustum of a cone having its axis in the z direction.
fsphere	5	Integrand used in function **triplint**.
gaussint	5	Generates Gaussian base points and weight factors of arbitrary order.
gcbpwf	12, 13	Computes base points and weight factors in a composite quadrature formula.
genprint	2-13	System dependent function which saves a plot to a file.
gquad	5	Function used to evaluate the integral of a function over specified integration limits. The numerical integration is performed using a composite Gauss integration rule.
gquad6	5	Function used to evaluate the integral of a function over arbitrary integration limits. The numerical integration is performed using a composite 6-point Gauss integration rule.
gquad10	5	Function used to evaluate the integral of a function over specified integration limits. The numerical integration is performed using a composite 10-point Gauss integration rule.
grule	5	Function to compute Gauss base points and weight factors.
gtop	9	Function used in solution of Laplace's equation for a rectangle.

continued on next page

Routine	Chapter	Description
heat	9	Function to evaluate transient heat conduction in a slab which has the left end insulated and has the right end subjected to a temperature variation.
hmpf	5	Companion function for script runcases.
hsmck	6	Function to solve $m\,y_h''(t)+c\,y_h'(t)+k\,y_h(t)=0$ subject to initial conditions of $y_h(0)=y_0$ and $y_h'(0)=v_0$.
imptp	6	Function defining a piecewise linear function resembling the ground motion of the earthquake which occurred in 1940 in the Imperial Valley of California.
initdefl	9	Defines the linearly interpolated initial deflection configuration.
jnft	6	Function computing integer order Bessel functions of the first kind computed by use of the Fast Fourier Transform (FFT).
lapcrcl	12	Solves Laplace's equation inside a circle of unit radius.
laplarec	9	Program which uses Fourier series to solve Laplace's equation in a rectangle.
laprec	9	Function which sums the series solving Laplace's equation in a rectangle.
lineprop	5	Function to compute area property contributions associated with a polyline.
linfrac	12	Determines the linear fractional transformation to map any three points in the z-plane into any three points in the w-plane.
lintrp	2, 4, 6, 8-12	Function which performs piecewise linear interpolation through data.

continued on next page

Routine	Chapter	Description
lntrp	B	Alternate version of function **lintrp** which works with MATLAB version 4.x.
logf	5	Companion function for script runcases.
makcrcsq	5	Function used to create data for a geometry involving half of an annulus placed above a square containing a square hole.
makratsq	3	Driver program to create a rational function map of a unit disk onto a square.
mapsqr	12	Maps the interior of a circle onto the interior of a square using a rational function of approximate form.
matlbdat	4	Example illustrating the use of splines to draw the word MATLAB.
mbvp	3	Function used to solve a mixed boundary value problem for a function which is harmonic inside the unit disk, symmetric about the x axis, and has boundary conditions involving function values on one part of the boundary and zero gradient elsewhere.
mbvprun	3	Driver program to analyze a mixed boundary value problem for a function harmonic inside a circle.
mbvtest	12	Determines a function which is harmonic for $|z| < 1$ with specified end conditions.
mckde2i	7	Function using a second order implicit integrator to solve the matrix differential equation $mx'' + cx' + kx = f(t)$ where m, c, and k are constant matrices and $f(t)$ is an externally defined function.
mckde4i	7	Function using a fourth order implicit integrator to solve the matrix differential equation $mx'' + cx' + kx = f(t)$ where m, c, and k are constant matrices and $f(t)$ is an externally defined function.

continued on next page

Routine	Chapter	Description
membran	3	Function to compute the transverse deflection of a uniformly tensioned membrane subjected to uniform pressure.
missdis	13	Determines the error measure indicating how much a target is missed for a particular initial inclination angle of the projectile.
mom	8	Function to compute the driving moment needed to produce an exact solution in a forced pendulum response example.
motion	2	Function to animate the motion of a vibrating string.
ndbemrsp	9	Function to evaluate the nondimensional displacement and moment in a constant cross-section simply-supported beam.
numlist	B	Lists a file with preceding line numbers.
oneovrei	11	Computes $1/EI$ by piecewise linear interpolation.
oneovx	1	Function defining the integrand passed to **simpson** in root calculation example.
output	4	Function to print results for **finidif**.
pauz	2-3, 5-7, 9, 12-13	Pause statement which lets you use the print menu during the pause.
pinvert	8	Function defining the equation of motion for the pendulum example.
platecrc	12	Computes coefficients in the series expansions which defined the Kolosov-Muskhelishvili stress functions for a plate having a circular hole of unit radius.
plft	9	Function used to solve Laplace's equation for a rectangle.
ploteasy	2	Plot function with a simple argument list.

continued on next page

Routine	Chapter	Description
plotjrun	6	Driver program to compute integer order Bessel functions.
plotmotn	8	Function to plot the cable time history.
plotsave	9	Function to plot errors in frequencies computed by two approximate methods.
plterror	7	Plot error measures showing how different integrators and time steps compare with the exact solution using a modal response.
pltxmidl	8	Function to plot horizontal position of midpoint in nonlinear cable example.
polflip	12	Evaluates function **polyval** with the order of elements reversed.
polgnrun	5	Example program which illustrates the computation of geometrical properties of a volume generated by revolving a polygon about the z-axis.
polgnvol	5	Computes the area and volume properties for a polygonal area in the xz-plane which is rotated about the z-axis.
polhdplt	5	Makes a surface plot of an arbitrary polyhedron.
polhdrun	5	Driver program for routine **polhedrn**.
polhedrn	5	Determines the volume, centroidal coordinates and inertial moment for an arbitrary polyhedron.
polyplot	2	Program illustrating how the location of interpolation points affects the accuracy and smoothness of polynomial approximations.
polyterp	4	Function to interpolate through n data points using a polynomial of degree $n - 1$.
		continued on next page

Routine	Chapter	Description
polyxy	5	Computes the area, centroidal coordinates, and inertial moments of an arbitrary polygon.
prnstres	10	Computes principal stresses and principal stress directions for a three-dimensional stress state.
projcteq	8, 13	Defines the equation of motion for a projectile loaded by gravity and atmospheric drag.
prun	8	Driver program for dynamics of an inverted pendulum analyzed using **ode45**.
pvibs	9	Computes the forced harmonic response of a pile buried in an oscillating elastic medium.
pyramid	5	Determines geometrical properties of a pyramid.
qrht	9	Function used to solve Laplace's equation for a rectangle.
quadit3d	5	Computes the specified iterated integral.
ratcof	3, 12	Function determining coefficients to approximate a rational function.
raterp	3	Function performing rational function interpolation using coefficients from function **ratcof**.
read	1, 2, 5, 6, 8, 9	Function to interactively read up to 20 variables from single line.
readv	4	Function to input a vector of specified length.
rec2polr	12	Transforms cartesian stress components to polar coordinate stresses.
recstrs	9	Function employing point matching to obtain an approximate solution for torsional stresses in a Saint Venant beam of rectangular cross section.
rector	9	Driver program for point matching solution of torsional stresses in a Saint Venant beam of rectangular cross section.

continued on next page

Routine	Chapter	Description
rgdbodmo	2	Transforms coordinates (x, y, z) to new coordinates (X, Y, Z) by rotating and translating the reference frame.
rkdestab	8	Program to plot the boundary of the region of the complex plane governing the maximum step size which may be used for stability of a Runge-Kutta integrator of arbitrary order.
rnbemimp	9	Program analyzing an impact dynamics problem for an elastic Euler beam of constant cross section which is simply supported at each end.
rootest	1	Instructional program illustrating nested function calls. The base of natural logarithms (e) is approximated by finding a value of x which makes the chosen function equal zero.
rotapoly	5	Plots a partial surface of revolution produced by rotating a polygon.
rotatran	2	Creates a rotation matrix.
runcases	5	Driver program for comparing results from several numerical integration methods.
rundfl	9	Example program which analyzes wave motion of a string having arbitrary initial deflection.
runimpv	6	Driver program for the earthquake dynamics example.
runplate	12	Computes stresses around a hole in a plate by use of the complex variable method of Kolosov-Muskhelishvili.
runpvibs	9	Driver program for function **pvibs**.
runtors	12	Example showing torsional stress computation for a beam of square cross section using conformal mapping and a complex stress function.

continued on next page

Routine	Chapter	Description
runtraj	13	Example program which integrates the differential equations governing two-dimensional motion of a projectile subjected to gravity loading and atmospheric drag.
setup	9	Function used in example involving the Laplace equation for a rectangle.
shftprop	5	Function to compute area properties for a set of axes rotated relative to the original axes.
shkbftss	6	Function used in earthquake example to determine the steady state solution of a scalar differential equation.
shkstrng	2	Function computing the motion of a string having one end fixed and the other end shaken harmonically.
simpson	1, 5	Function to integrate by Simpson's rule.
simpsum	5	Special Simpson's rule function concerning a matrix.
sincof	9	Calculates sine coefficients.
sine	6	Function specifying all or part of a sine wave.
sinetrp	4	Driver program illustrating cubic spline approximation of $\sin(x)$, its first two derivatives, and its integral.
sinfft	9	Function determining coefficients in the Fourier sine series of a general real valued function.
slabheat	9	Program illustrating the temperature variation in a one-dimensional slab with the left end insulated and the right end given a temperature variation $\sin(\omega t)$.
smotion	9	Function to animate the motion of a vibrating string.
		continued on next page

Routine	Chapter	Description
sngf	11	Evaluates singularity functions.
spc	4, 5, 12, 13	Function to compute a matrix containing coefficients needed to perform a piecewise cubic interpolation among data values. The output from this function is used by function **spltrp** to evaluate the cubic spline function, its first two derivatives, or the function integral from $x(1)$ to an arbitrary upper limit.
spcurv2d	4, 5	Function to tabulate points on a spline curve connecting data points on a general curve.
splaprop	5	Function to compute geometrical properties of an area bounded by a spline curve.
splined	2	Evaluates the first or second derivative of the piecewise cubic interpolation curve defined by the intrinsic spline function in MATLAB.
splinerr	2	Calculates the binormal and curvature error for the spiral example.
spltrp	4, 5, 12, 13	Function for cubic spline interpolation using a data array obtained by first calling function **spc**. This function also integrates and differentiates.
sqmp	3	Function to evaluate the Schwarz-Christoffel transformation mapping $\mathrm{abs}(z) \leq 1$ inside a square of side length two.
sqrtsurf	12	Illustrates the discontinuity in the function $w = \sqrt{z^2 - 1}$.
sqtf	5	Companion function for script runcases.
squarat	12	Maps either the interior of a circle onto the interior of a square, or maps the exterior of a circle onto the exterior of a square using a rational function.
		continued on next page

Routine	Chapter	Description
squarmap	2	Function to evaluate the conformal mapping produced by the Schwarz-Christoffel transformation mapping abs(z) $\leq$ 1 inside a square.
squarrun	2	Driver program to plot the mapping of a circular disk onto the interior of a square by the Schwarz-Christoffel transformation.
srfex	2	Example which plots three annuli intersecting a spike.
strfun	12	Evaluates complex stress functions using series coefficients from function **platecrc**.
stringmo	2	Driver program to illustrate motion of a string having one end subjected to harmonic oscillation.
strvib	9	Sums the Fourier series for the string motion.
stwav	9	Computes the dynamic response of a vibrating string released from rest with initial deflection defined by piecewise linear interpolation.
sumser	9	Function used in a beam dynamics example to sum a double Fourier series.
surfesy	13	Easy interface for function **surf**.
swcsq10	12	Square map approximations pertaining to truncated Schwarz-Christoffel transformations.
swcsqmap	12	Evaluates power series approximations for mapping either the inside of a circle onto the inside of a square, or mapping the outside of a circle onto the outside of a square.
topde	8	Defines the equation of motion for a symmetrical top.
toprun	8	Example which analyzes the response of a spinning conical top.

continued on next page

Routine	Chapter	Description
topview	12	Plots a surface from the top to show the coordinate lines of the surface.
torstres	12	Computes torsional stresses in a beam such that $|\zeta| \leq 1$ is mapped onto the beam cross section by a specified function.
traject	8, 13	Integrates the dynamical equations for a projectile subjected to gravity loading and atmospheric drag.
trapsum	10	Evaluates a function using trapezoidal rule.
trifacsm	10	Determines an upper triangular matrix u such that $u^T u = a$ where a is symmetric and positive definite.
triplint	5	Example of triple integration on inertial moment of a sphere.
trisub	3	Function to solve $LX = B$, where L is lower triangular.
trusvibs	10	Analyzes natural vibration modes for a general plane pin-connected truss using the direct stiffness method.
udfrevib	7	Function to compute undamped natural frequencies, modal vectors, and time response by modal superposition.
ulbc	9	Function used in example on solution of Laplace's equation for a rectangle.
unsymerr	8	Function to compute an error measure in nonlinear cable dynamics example.
vdb	11	Calculates the shear, moment, slope, and deflection of a variable depth indeterminate beam subjected to complex loading and general end conditions.
versn	3	Returns the version number of MATLAB.

continued on next page

Routine	Chapter	Description
vibfit	13	Demonstrates the use of the Nelder and Mead multi-dimensional function minimization method to determine an equation for $y(t)$ which depends nonlinearly on several parameters.
vibfun	13	Evaluates the least square error for a set of vibration data.

MATLAB Utility Functions

Function genprint

```
1: function genprint(fname,append)
2: %
3: % genprint(fname,append)
4: % ~~~~~~~~~~~~~~~~~~~~~~~
5: % This function saves a plot to a file.  If
6: % the file exists, it is erased first unless
7: % the append option is specified.
8: %
9: % fname  - name of file to save plot to
10: %          without a filename extension
11: % append - optional, if included plot is
12: %          appended to file fname
13: %
14: % SYSTEM DEPENDENT ROUTINE
15: %
16: % User m functions called:  none
17: %-----------------------------------------------
18:
19: %...Define these appropriately
20: ext=['.eps']; % filename extension to use
21: opt=['eps'];  % option for print command
22:
23: %...Append extension to filename
24: file_name=[fname,ext];
25:
26: %...Determine computer type
27: system_type=computer;
28:
29: %...Use correct command for different systems
30: if strcmp(system_type(1:2),'PC')
```

```
31:    erase_cmd=['delete ', file_name];
32: elseif strcmp(system_type(1:3),'SGI')
33:    erase_cmd=['!rm ', file_name];
34: else
35:    disp(' ');
36:    disp('Unknown system type in genprint');
37:    break;
38: end
39:
40: % Save to encapsulated postscript file
41: if nargin == 1
42:    if exist(file_name)==2
43:      eval(erase_cmd);
44:    end
45:    eval(['print -d',opt,' ',fname]);
46: else
47:    eval(['print -d',opt,' -append ',fname]);
48: end
```

Function gquad6

```
1: function area=gquad6(fun,xlow,xhigh,mparts)
2: %
3: % area = gquad6(fun,xlow,xhigh,mparts)
4: % ~~~~~~~~~~~~~~~~~~~~~~~~~~~~~~~~~~~~~~~~
5: % This function determines the area under an
6: % externally defined function fun(x) between
7: % limits xlow and xhigh. The numerical
8: % integration is performed using a composite
9: % Gauss integration rule.  The whole interval
10: % is divided into mparts subintervals and the
11: % integration over each subinterval is done
12: % with a six point Gauss formula which
13: % involves base points bp and weight factors
14: % wf.  The normalized interval of integration
15: % for the bp and wf constants is -1 to +1.
16: % The algorithm is structured in terms of a
17: % parameter mquad = 6 which can be changed
18: % along with bp and wf to accommodate a
19: % different order formula.  The composite
20: % algorithm is described by the following
21: % summation relation:
22: %
```

```
23: % x=b
24: % integral( f(x)*dx ) =
25: % x=a
26: %
27: %           j=n k=m
28: %        d1*sum sum( wf(j)*fun(a1+d*k+d1*bp(j)) )
29: %           j=1 k=1
30: %
31: % where d = (b-a)/m, d1 = d/2, a1 = a-d1,
32: %       m = mparts, and n = nquad.
33: %
34: % User m functions called:  argument fun
35: %-----------------------------------------------
36:
37: % The weight factors are
38: wf=[ 1.71324492379170d-01; ...
39:      3.60761573048139d-01;...
40:      4.67913934572691d-01];
41: wf=[wf;wf([3 2 1])];
42:
43: % The base points are
44: bp=[-9.32469514203152d-01; ...
45:     -6.61209386466265d-01;...
46:     -2.38619186083197d-01];
47: bp=[bp;-bp([3 2 1])];
48:
49: d=(xhigh-xlow)/mparts;
50: d2=d/2; nquad=length(bp);
51: x=(d2*bp)*ones(1,mparts)+ ...
52:   (d*ones(nquad,1))*(1:mparts);
53: x=x(:)+(xlow-d2);
54: fv=feval(fun,x); wv=wf*ones(1,mparts);
55:
56: area=d2*(wv(:)'*fv(:));
```

Function lintrp

```
1: function y=lintrp(xd,yd,x)
2: %
3: % y=lintrp(xd,yd,x)
4: % ~~~~~~~~~~~~~~~~~
5: % This function performs piecewise linear
6: % interpolation through data values stored in
```

```
 7: % xd, yd, where xd values are arranged in
 8: % nondecreasing order. The function can handle
 9: % discontinuous functions specified when two
10: % successive values in xd are equal. Then the
11: % repeated xd values are shifted by a tiny
12: % amount to remove the discontinuities.
13: % Interpolation for any points outside the range
14: % of xd is also performed by continuing the line
15: % segments through the outermost data pairs.
16: %
17: % xd,yd - data vectors defining the
18: %           interpolation
19: % x       - matrix of values where interpolated
20: %           values are required
21: %
22: % y       - matrix of interpolated values
23: %
24: % NOTE:   This routine is dependent on MATLAB
25: %         Version 5.x function interp1q.  A
26: %         Version 4.x solution can be created
27: %         by renaming routine lntrp.m to
28: %         lintrp.
29: %
30: %-------------------------------------------------
31:
32: xd=xd(:); yd=yd(:); [nx,mx]=size(x); x=x(:);
33: xsml=min(x); xbig=max(x);
34: if xsml<xd(1)
35:   ydif=(yd(2)-yd(1))*(xsml-xd(1))/(xd(2)-xd(1));
36:   xd=[xsml;xd]; yd=[yd(1)+ydif;yd];
37: end
38: n=length(xd); n1=n-1;
39: if xbig>xd(n)
40:    ydif=(yd(n)-yd(n1))*(xbig-xd(n))/ ...
41:        (xd(n)-xd(n1));
42:    xd=[xd;xbig]; yd=[yd;yd(n)+ydif];
43: end
44: k=find(diff(xd)==0);
45: if length(k)~=0
46:    n=length(xd);
47:    xd(k+1)=xd(k+1)+(xd(n)-xd(1))*1e3*eps;
48: end
49: y=reshape(interp1q(xd,yd,x),nx,mx);
```

Function Intrp

```
 1: function y=lintrp(xd,yd,x)
 2: %
 3: % y=lintrp(xd,yd,x)
 4: % ~~~~~~~~~~~~~~~~~
 5: % This function performs piecewise linear
 6: % interpolation through data defined by vectors
 7: % xd,yd. The components of xd are presumed to
 8: % be in nondecreasing order. Any point where
 9: % xd(i)==xd(i+1) generates a jump
10: % discontinuity. For points outside the data
11: % range, the interpolation is based on the
12: % lines through the outermost pairs of points
13: % at each end.
14: %
15: % x     - vector of values for which piecewise
16: %           linear interpolation is required
17: % xd,yd - vectors of data values through which
18: %           interpolation is performed
19: % y     - interpolated function values for
20: %           argument x
21: %
22: % User m functions required: none
23: %-------------------------------------------------
24:
25: x=x(:); xd=xd(:); yd=yd(:);
26: y=zeros(length(x),1);
27: xmin=min(x); xmax=max(x); nd=length(xd);
28:
29: if xmax > xd(nd)
30:   yd=[yd; yd(nd)+(yd(nd)-yd(nd-1))/ ...
31:       (xd(nd)-xd(nd-1))*2*(xmax-xd(nd)))];
32:   xd=[xd;2*xmax-xd(nd)]; nd=nd+1;
33: end
34:
35: if xmin < xd(1)
36:   yd=[yd(1)+(yd(2)-yd(1))/(xd(2)-xd(1))* ...
37:       (xmin-xd(1));yd];
38:   xd=[xmin;xd]; nd=nd+1;
39: end
40:
41: for i=1:nd-1
42:   xlft=xd(i); ylft=yd(i);
```

```
43:    xrht=xd(i+1); yrht=yd(i+1); dx=xrht-xlft;
44:    if dx~=0, s=(yrht-ylft)/dx;
45:      y=y+(x>=xlft).*(x<xrht).* ...
46:        (ylft+s*(x-xlft));
47:    end
48:  end
49:
50:  k=find(x==xd(nd));
51:  if length(k)>0, y(k)=yd(nd); end
```

Function numlist

```
1:  function numlist(fname)
2:  %
3:  % numlist(fname)
4:  % ~~~~~~~~~~~~~~
5:  % This function lists a file with preceding
6:  % line numbers.
7:  %
8:  % fname  - name of file
9:  %
10: % User m functions called:  none
11: %--------------------------------------------------
12:
13: fid=fopen(fname,'rt'); n=0;
14: while feof(fid)==0
15:   line=fgetl(fid); n=n+1;
16:   N=num2str(n);
17:   if n<10
18:     N=[N,'   : '];
19:   elseif N<100
20:     N=[N,' : '];
21:   else
22:     N=[N,': '];
23:   end
24:   disp([N,line])
25: end
26: fclose(fid);
```

Function pauz

```
1: function pauz(strng)
```

```
 2: %
 3: % pauz(strng)
 4: % ~~~~~~~~~~~
 5: % On some systems MATLAB will not let you
 6: % invoke the print menu option when using a
 7: % pause statement. This routine gets around
 8: % that problem by invoking another MATLAB
 9: % function (input) which does not have that
10: % characteristic.
11: %
12: % User m functions called:  none
13: %------------------------------------------------
14:
15: if nargin==0, strng=' '; end
16: dumy=input(strng,'s');
```

Function read

```
 1: function [a1,a2,a3,a4,a5,a6,a7,a8,a9,a10, ...
 2:           a11,a12,a13,a14,a15,a16,a17,a18, ...
 3:           a19,a20]=read(labl)
 4: %
 5: % [a1,a2,a3,a4,a5,a6,a7,a8,a9,a10,a11,a12, ...
 6: %  a13,a14,a15,a16,a17,a18,a19,a20]=read(labl)
 7: %~~~~~~~~~~~~~~~~~~~~~~~~~~~~~~~~~~~~~~~~~~~~~~~~
 8: %
 9: % This function reads up to 20 variables on one
10: % line. The items should be separated by commas
11: % or blanks. Using more than 20 output
12: % variables will result in an error.
13: %
14: % labl                - Label preceding the
15: %                       data entry.  It is set
16: %                       to '? ' if no value of
17: %                       labl is given.
18: % a1,a2,...,a_nargout - The output variables
19: %                       which are created
20: %                       (cannot exceed 20)
21: %
22: % A typical function call is:
23: % [A,B,C,D]=read('Enter values of A,B,C,D: ')
24: %
25: % User m functions required: none
```

```
26: %------------------------------------------------
27:
28: if nargin==0, labl='? '; end
29: n=nargout;
30: str=input(labl,'s'); str=['[',str,']'];
31: v=eval(str);
32: L=length(v);
33: if L>=n, v=v(1:n);
34:    else, v=[v,zeros(1,n-L)]; end
35: for j=1:nargout
36:    eval(['a',int2str(j),'=v(j);']);
37: end
```

Function readv

```
1: function [v,l]=readv(n)
2: %
3: % v=readv(n)
4: % ~~~~~~~~~~~
5: % This function inputs a vector of length n.
6: % If fewer than n values are given, then the
7: % vector is padded with zeros.
8: %
9: %    n  - number of values to be input
10: %    v  - a vector of length n. If fewer than n
11: %         values were input, the final values
12: %         of v are set to zero
13: %    l  - the number of values actually read
14: %
15: % User m functions called:  none
16: %------------------------------------------------
17:
18: str=input('? > ','s');
19: str=['[',str,']'];
20: v=eval(str); l=length(v);
21: if l>n, v=v(1:n); end
22: if l<n, v=[v,zeros(1,n-l)]; end
```

Function simpson

```
1: function ansr=simpson(funct,a,b,neven)
2: %
```

```
 3: % ansr=simpson(funct,a,b,neven)
 4: % ~~~~~~~~~~~~~~~~~~~~~~~~~~~~~~
 5: %
 6: % This function integrates "funct" from
 7: % "a" to "b" by Simpson's rule using
 8: % "neven+1" function values.  Parameter
 9: % "neven" should be an even integer.
10: %
11: % Example use:  ansr=simpson('sin',0,pi/2,4)
12: %
13: % funct    -  character string name of
14: %             function integrated
15: % a,b      -  integration limits
16: % neven    -  an even integer defining the
17: %             number of integration intervals
18: % ansr     -  Simpson rule estimate of the
19: %             integral
20: %
21: % User m functions called: argument funct
22: %-----------------------------------------------
23:
24: ne=max(2,2*round(.1+neven/2)); d=(b-a)/ne;
25: x=a+d*(0:ne); y=feval(funct,x);
26: ansr=(d/3)*(y(1)+y(ne+1)+4*sum(y(2:2:ne))+...
27:      2*sum(y(3:2:ne-1)));
```

Function spc

```
 1: function splmat=spc(x,y,i,v,icrnr)
 2: %
 3: % splmat=spc(x,y,i,v,icrnr)
 4: % ~~~~~~~~~~~~~~~~~~~~~~~~~~
 5: % This function computes matrix splmat
 6: % containing coefficients needed to perform a
 7: % piecewise cubic interpolation among data
 8: % values contained in vectors x and y. The
 9: % output from this function is used by function
10: % spltrp to evaluate the cubic spline function,
11: % its first two derivatives, or the function
12: % integral from x(1) to an arbitrary upper
13: % limit.
14: %
15: %  x    - vector of abscissa values arranged
```

```
16: %              in increasing order. The number of
17: %              data values is denoted by n.
18: %
19: % y    - vector of ordinate values
20: %
21: % i    - a two component vector [i1,i2].
22: %              Parameters i1 and i2 refer to left
23: %              end and right end conditions,
24: %              respectively. These equal 1, 2,
25: %              or 3 as explained below.
26: %
27: % v    - a two component vector [v1,v2]
28: %              containing end values of y'(x)
29: %              or y''(x)
30: %
31: %              if i1=1: y'(x(1)) is set to v(1)
32: %                  =2: y''(x(1)) is set to v(1)
33: %                  =3: y''' is continuous at x(2)
34: %              if i2=1: y'(x(n)) is set to v(2)
35: %                  =2: y''(x(n)) is set to v(2)
36: %                  =3: y''' is continuous at x(n-1)
37: %
38: %              Note: When i1 or i2 equal 3 the
39: %                    corresponding values of vector
40: %                    v should be zero
41: %
42: % icrnr - A vector of indices identifying
43: %              interior points where slope
44: %              discontinuities are to be generated
45: %              by requiring y''(x) to equal zero.
46: %              Vector icrnr should have components
47: %              lying between 2 and n-1. The vector
48: %              can be omitted from the argument list
49: %              if no slope discontinuities occur.
50: %
51: % User m functions called:  none
52: %-------------------------------------------------
53:
54: x=x(:); y=y(:);
55: n=length(x); a=zeros(n,1); b=a; c=a;
56: d=a; t=a; if nargin < 5, icrnr=[]; end
57: ncrnr=length(icrnr);
58: i1=i(1); i2=i(2);
59: v1=v(1); v2=v(2);
60:
```

```
61: % Form the tridiagonal system to solve for
62: % second derivative values at the data points
63:
64: n1=n-1; j=2:n1;
65: hj=x(j)-x(j-1); hj1=x(j+1)-x(j);
66: hjp=hj+hj1; a(j)=hj./hjp; wuns=ones(n-2,1);
67: b(j)=2*wuns; c(j)=wuns-a(j);
68: d(j)=6.*((y(j+1)-y(j))./hj1- ...
69:        (y(j)-y(j-1))./hj)./hjp;
70:
71: % Form  equations for the end conditions
72:
73: % slope specified at left end
74: if i1==1
75:    h2=x(2)-x(1); b(1)=2; c(1)=1;
76:    d(1)=6*((y(2)-y(1))/h2-v1)/h2;
77:
78: % second derivative specified at left end
79: elseif i1==2
80:    b(1)=1; c(1)=0; d(1)=v1;
81:
82: % not a knot condition
83: else
84:    b(1)=1;
85:    a(1)=hj(1)/hj(2);
86:    c(1)=-1-a(1);
87:    d(1)=0;
88: end
89:
90: % slope specified at right end
91: if i2==1
92:    hn=x(n)-x(n-1); a(n)=1; b(n)=2;
93:    d(n)=6.*(v2-(y(n)-y(n-1))/hn)/hn;
94:
95: % second derivative specified at right end
96: elseif i2==2
97:    a(n)=0; b(n)=1; d(n)=v2;
98:
99: % not a knot condition
100: else
101:    a(n)=1;
102:    b(n)=-(x(n-1)-x(n-2))/(x(n)-x(n-2));
103:    c(n)=-(x(n)-x(n-1))/(x(n)-x(n-2));
104: end
105:
```

```
106:  % Adjust for slope discontinuity
107:  % specified by lcrnr
108:
109:  if ncrnr > 0
110:    zro=zeros(ncrnr,1);
111:    a(icrnr)=zro; c(icrnr)=zro;
112:    d(icrnr)=zro; b(icrnr)=ones(ncrnr,1);
113:  end
114:
115:  % Solve the tridiagonal system
116:  % for t(1),...,t(n)
117:
118:  bb=diag(a(2:n),-1)+diag(b)+diag(c(1:n-1),1);
119:  if a(1)~=0, bb(1,3)=a(1); d(1)=0;    end
120:  if c(n)~=0, bb(n,n-2)=c(n); d(n)=0; end
121:  t=bb\d(:);
122:
123:  % Save polynomial coefficients describing the
124:  % cubics for each interval.
125:
126:  j=1:n-1; k=2:n;
127:  dx=x(k)-x(j); dy=y(k)-y(j); b=t(j)/2;
128:  c=(t(k)-t(j))./(6*dx); a=dy./dx-(c.*dx+b).*dx;
129:  int=(((c.*dx/4+b/3).*dx+a/2).*dx+y(j)).*dx;
130:  int=[0;cumsum(int)]; n1=n-1;
131:  ypn=dy(n1)/dx(n1)+dx(n1)*(2*t(n)+t(n1))/6;
132:
133:  % The columns of splmat contain the
134:  % following vectors
135:  % [x,y,x_coef,x^2_coef,x^3_coef,integral_coef]
136:
137:  splmat=[ ...
138:    [ x(1),  y(1),  a(1),   0,   0,   0      ];
139:    [ x(j),  y(j),  a,      b,   c,   int(j) ];
140:    [ x(n),  y(n),  ypn,    0,   0,   int(n) ] ];
```

Function spcurv2d

```
1:  function [xout,yout,sout]= ...
2:           spcurv2d(xd,yd,nseg,ncrnr)
3:  %
4:  % [xout,yout]=spcurv2d(xd,yd,nseg,ncrnr)
5:  % ~~~~~~~~~~~~~~~~~~~~~~~~~~~~~~~~~~~~~~~
```

```
 6: %
 7: % This function tabulates points xout, yout on
 8: % a spline curve connecting data points xd, yd
 9: % on the cubic spline curve
10: %
11: %  xd,yd      - input data points
12: %  nseg       - number of tabulation intervals
13: %                 used per spline segment
14: %  ncrnr      - point indices where corners are
15: %                 required
16: %  xout,yout  - output data points on the
17: %                 spline curve. The number of
18: %                 points returned equals
19: %
20: %                 nout=(nd-1)*nseg+1.
21: %
22: % User m functions called:  spc, spltrp
23: %-------------------------------------------------
24:
25: nd=length(xd); sd=(1:nd)';
26: if nargin==2; nseg=10; end
27: if nargin<=3, ncrnr=[]; end;
28: nout=(nd-1)*nseg+1; sout=linspace(1,nd,nout);
29: if norm([xd(1)-xd(nd),yd(1)-yd(nd)]) < 100*eps
30:   yp=(yd(2)-yd(nd-1))/2;
31:   iend=[1;1]; vy=[yp;yp];
32:   xp=(xd(2)-xd(nd-1))/2; vx=[xp;xp];
33: else
34:   iend=[3;3]; vy=[0;0]; vx=vy;
35: end
36: matx=spc(sd,xd,iend,vy,ncrnr);
37: maty=spc(sd,yd,iend,vx,ncrnr);
38: xout=spltrp(sout,matx,0);
39: yout=spltrp(sout,maty,0);
```

Function spltrp

```
 1: function f=spltrp(x,mat,ideriv)
 2: %
 3: % f=spltrp(x,mat,ideriv)
 4: % ~~~~~~~~~~~~~~~~~~~~~~~
 5: % This function performs cubic spline
 6: % interpolation using data array mat obtained
```

```
7:  % by first calling function spc.
8:  %
9:  % x       - the vector of interpolation values
10: %           at which the spline is to be
11: %           evaluated.
12: % mat     - the matrix output from function spc.
13: %           This array contains the data points
14: %           and the polynomial coefficients
15: %           needed to define the piecewise
16: %           cubic curve connecting the data
17: %           points.
18: % ideriv  - a parameter specifying whether
19: %           function values, derivative values,
20: %           or integral values are computed.
21: %           Taking ideriv equal to 0, 1, 2, or
22: %           3, respectively, returns values of
23: %           y(x), y'(x), y''(x), or the integral
24: %           of y(x) from the first data point
25: %           defined in array mat to each point
26: %           in vector x specifying the chosen
27: %           interpolation points. For any points
28: %           outside the original data range,
29: %           the interpolation is performed by
30: %           extending the tangents at the first
31: %           and last data points.
32: %
33: % User m functions called:  none
34: %-------------------------------------------------
35:
36: % identify interpolation intervals
37: % for each point
38: [nrow,ncol]=size(mat);
39: xx=x(:)'; xd=mat(2:nrow);
40: xd=xd(:); np=length(x); nd=length(xd);
41: ik=sum(xd(:,ones(1,np)) < xx(ones(nd,1),:));
42: ik=1+ik(:);
43: xk=mat(ik,1); yk=mat(ik,2); dx=x(:)-xk;
44: ak=mat(ik,3); bk=mat(ik,4);
45: ck=mat(ik,5); intk=mat(ik,6);
46:
47: % obtain function values at x
48: if ideriv==0
49:   f=((ck.*dx+bk).*dx+ak).*dx+yk; return
50:
51: % obtain first derivatives at x
```

```
52: elseif ideriv==1
53:    f=(3*ck.*dx+2*bk).*dx+ak; return
54:
55: % obtain second derivatives at x
56: elseif ideriv==2
57:    f=6*ck.*dx+2*bk; return
58:
59: % obtain integral from xd(1) to x
60: elseif ideriv==3
61:    f=intk+(((ck.*dx/4+bk/3).*dx+ ...
62:      ak/2).*dx+yk).*dx;
63: end
```

Bibliography

[1] M. Abramowitz and I.A. Stegun. *Handbook of Mathematical Functions with Formulas, Graphs, and Mathematical Tables.* National Bureau of Standards, Applied Math. Series #55. Dover Publications, 1965.

[2] J. H. Ahlberg, E. N. Nilson, and J. L. Walsh. *The Theory of Splines and Their Applications.* Mathematics in Science and Engineering, Volume 38. Academic Press, 1967.

[3] J. Albrecht, L. Collatz, W. Velte, and W. Wunderlich, editors. *Numerical Treatment of Eigenvalue Problems*, volume 4. Birkhauser Verlag, 1987.

[4] E. Anderson, Z. Bai, C. Bischof, J. Demmel, J. Dongarra, J. Du Croz, A. Greenbaum, S. Hammarling, A. McKenney, S. Ostrouchov, and D. Sorensen. *LAPACK User's Guide.* SIAM, Philadelphia, 1992.

[5] F. Arbabi and F. Li. Macroelements for variable-section beams. *Computers & Structures*, 37(4):553–559, 1990.

[6] B. A. Barsky. *Computer Graphics and Geometric Modeling Using Beta-splines.* Computer Science Workbench. Springer-Verlag, 1988.

[7] K. J. Bathe. *Finite Element Procedures in Engineering Analysis.* Prentice-Hall, 1982.

[8] E. Becker, G. Carey, and J. Oden. *Finite Elements, An Introduction.* Prentice-Hall, 1981.

[9] F. Beer and R. Johnston, Jr. *Mechanics of Materials.* McGraw-Hill, second edition, 1992.

[10] K. S. Betts. Math packages multiply. *CIME Mechanical Engineering*, pages 32–38, August 1990.

[11] K.E. Brenan, S.L. Campbell, and L.R. Petzold. *Numerical Solu-tion of Initial-Value Problems in Differential-Algebraic Equations.* Elsevier Science Publishers, 1989.

[12] R. Brent. *Algorithms for Minimization Without Derivatives.* Prentice-Hall, 1973.

[13] P. Brown, G. Byrne, and A. Hindmarsh. VODE: A variable co-efficient ODE solver. *SIAM J. Sci. Stat. Comp.*, 10:1038–1051, 1989.

[14] G. Carey and J. Oden. *Finite Elements, Computational Aspects.* Prentice-Hall, 1984.

[15] B. Carnahan, H.A. Luther, and J. O. Wilkes. *Applied Numerical Methods.* John Wiley & Sons, 1964.

[16] F. E. Cellier and C. M. Rimvall. Matrix environments for contin-uous system modeling and simulation. *Simulation*, 52(4):141–149, 1989.

[17] B. Char, K. Geddes, G. Gonnet, and S. Watt. *MAPLE User's Guide*, chapter First Leaves: A Tutorial Introduction to MAPLE. Watcom Publications Ltd., Waterloo, Ontario, 1985.

[18] R. V. Churchhill, J. W. Brown, and R. F. Verhey. *Complex Vari-ables and Applications.* McGraw-Hill, 1974.

[19] Column Research Committee of Japan. *Handbook of Structural Stability.* Corona Publishing Company, Tokyo, 1971.

[20] S. D. Conte and C. de Boor. *Elementary Numerical Analysis: An Algorithmic Approach.* McGraw-Hill, third edition, 1980.

[21] J. W. Cooley and J. W. Tukey. An algorithm for the machine calculation of complex fourier series. *Math. Comp.*, 19:297–301, 1965.

[22] R. R. Craig Jr. *Structural Dynamics.* John Wiley & Sons, 1988.

[23] J. K. Cullum and R. A. Willoughby, editors. *Large Scale Eigen-value Problems*, chapter High Performance Computers and Algo-rithms From Linear Algebra, pages 15–36. Elsevier Science Pub-lishers, 1986. by J. J. Dongarra and D. C. Sorensen.

[24] J. K. Cullum and R. A. Willoughby, editors. *Large Scale Eigen-value Problems*, chapter Eigenvalue Problems and Algorithms in Structural Engineering, pages 81–93. Elsevier Science Publishers, 1986. by R. G. Grimes, J. G. Lewis, and H. D. Simon.

[25] P. J. Davis and P. Rabinowitz. *Methods of Numerical Integration.* Computer Science and Applied Mathematics. Academic Press, Inc., second edition, 1984.

[26] C. de Boor. *A Practical Guide to Splines,* volume 27 of *Applied Mathematical Sciences.* Springer-Verlag, 1978.

[27] J. Dennis and R. Schnabel. *Numerical Methods for Unconstrained Optimization and Nonlinear Equations.* Prentice-Hall, 1983.

[28] J. Dongarra, E. Anderson, Z. Bai, A. Greenbaum, A. McKenney, J. Du Croz, S. Hammerling, J. Demmel, C. Bischof, and D. Sorensen. LAPACK: A portable linear algebra library for high performance computers. In *Supercomputing 1990.* IEEE Computer Society Press, 1990.

[29] J. Dongarra, J. Du Croz, I. Duff, and S. Hammarling. A set of level 3 basic linear algebra subprograms. Technical report, Argonne National Laboratory, Argonne, Illinois, August 1988.

[30] J. Dongarra, P. Mayes, and G. R. di Brozolo. The IBM RISC System/6000 and linear algebra operations. Technical Report CS-90-122, University of Tennessee Computer Science Department, Knoxville, Tennessee, December 1990.

[31] J. J. Dongarra, J. Du Croz, I. Duff, and S. Hammarling. A set of level 3 basic linear algebra subprograms. *ACM Transactions on Mathematical Software,* December 1989.

[32] J. J. Dongarra, J. Du Croz, S. Hammarling, and R. Hanson. An extended set of fortran basic linear algebra subprograms. *ACM Transactions on Mathematical Software,* 14(1):1–32, 1988.

[33] J.J. Dongarra, J.R. Bunch, C.B. Moler, and G.W. Stewart. *LINPACK User's Guide.* SIAM, Philadelphia, 1979.

[34] T. Driscoll. Algorithm 756: A MATLAB toolbox for Schwarz-Christoffel mapping. *ACM Transactions on Mathematical Software,* 22(2), June 1996.

[35] A. C. Eberhardt and G. H. Williard. Calculating precise cross-sectional properties for complex geometries. *Computers in Mechanical Engineering,* Sept./Oct. 1987.

[36] W. Flugge. *Handbook of Engineering Mechanics.* McGraw-Hill, 1962.

[37] G. Forsythe and C. B. Moler. *Computer Solution of Linear Algebraic Systems.* Prentice-Hall, 1967.

[38] G. Forsythe and W. Wasow. *Finite Difference Methods for Partial Differential Equations*. John Wiley & Sons, 1960.

[39] G. E. Forsythe, M. A. Malcolm, and C. B. Moler. *Computer Methods for Mathematical Computations*. Prentice-Hall, 1977.

[40] R. L. Fox. *Optimization Methods for Engineering Design*. Addison-Wesley Publishing Company, 1971.

[41] B. S. Garbow, J. M. Boyle, J. Dongarra, and C. B. Moler. *Matrix Eigensystem Routines — EISPACK Guide Extension*, volume 51 of *Lecture Notes in Computer Science*. Springer-Verlag, 1977.

[42] C. W. Gear. *Numerical Initial Value Problems in Ordinary Differential Equations*. Prentice-Hall, 1971.

[43] J. M. Gere and S. P. Timoshenko. *Mechanics of Materials*. Wadsworth, Inc., second edition, 1984.

[44] J. Gleick. *Chaos: Making a New Science*. Viking, 1987.

[45] G.H. Golub and J. M. Ortega. *Scientific Computing and Differential Equations: An Introduction to Numerical Methods*. Academic Press, Inc., 1992.

[46] G.H. Golub and C.F. Van Loan. *Matrix Computations*. Johns Hopkins University Press, second edition, 1989.

[47] D. Greenwood. *Principles of Dynamics*. Prentice-Hall, 1988.

[48] R. Grimes and H. Simon. New software for large dense symmetric generalized eigenvalue problems using secondary storage. *Journal of Computational Physics*, 77:270–276, July 1988.

[49] R. Grimes and H. Simon. Solution of large, dense symmetric generalized eigenvalue problems using secondary storage. *ACM Transactions on Mathematical Software*, 14(3):241–256, September 1988.

[50] P. Henrici. *Discrete Variable Methods in Ordinary Differential Equations*. John Wiley & Sons, 1962.

[51] P. Henrici. *Applied Complex Analysis*, volume 3. John Wiley & Sons, 1986.

[52] E. Horowitz and S. Sohni. *Fundamentals of Computer Algorithms*. Computer Science Press, 1978.

[53] T. J. Hughes. *The Finite Element Method — Linear Static and Dynamic Finite Element Analysis*. Prentice-Hall, 1987.

[54] J. L. Humar. *Dynamics of Structures*. Prentice-Hall, 1990.

[55] L.V. Kantorovich and V.I. Krylov. *Approximate Methods of Higher Analysis*. Interscience Publishers, 1958.

[56] W. Kerner. Large-scale complex eigenvalue problems. *Journal of Computational Physics*, 85(1):1–85, 1989.

[57] H. Kober. *Dictionary of Conformal Transformations*. Dover Publications, 1957.

[58] E. Kreyszig. *Advanced Engineering Mathematics*. John Wiley & Sons, Inc., 1972.

[59] C. Lanczos. *Applied Analysis*. Prentice-Hall, 1956.

[60] L. Lapidus and J. Seinfeld. *Numerical Solution of Ordinary Differential Equations*. Academic Press, 1971.

[61] C. Lawson and R. Hanson. *Solving Least Squares Problems*. Prentice-Hall, 1974.

[62] C. Lawson, R. Hanson, D. Kincaid, and F. Krogh. Basic linear algebra subprograms for fortran usage. *ACM Transactions on Mathematical Software*, 5:308–325, 1979.

[63] Y. T. Lee and A. A. G. Requicha. Algorithms for computing the volume and other integral properties of solids, i. known methods and open issues. *Communications of the ACM*, 25(9), 1982.

[64] I. Levit. A new numerical procedure for symmetric eigenvalue problems. *Computers & Structures*, 18(6):977–988, 1984.

[65] J. A. Liggett. Exact formulae for areas, volumes and moments of polygons and polyhedra. *Communications in Applied Numerical Methods*, 4, 1988.

[66] J. Marin. Computing columns, footings and gates through moments of area. *Computers & Structures*, 18(2), 1984.

[67] L. Meirovitch. *Analytical Methods in Vibrations*. Macmillan, 1967.

[68] L. Meirovitch. *Computational Methods in Structural Dynamics*. Sijthoff & Noordhoff, 1980.

[69] C. Moler and G. Stewart. An algorithm for generalized matrix eigenvalue problems. *SIAM Journal of Numerical Analysis*, 10(2):241–256, April 1973.

[70] C.B. Moler and C.F. Van Loan. Nineteen dubious ways to compute the exponential of a matrix. *SIAM Review*, 20:801–836, 1979.

[71] N.I. Muskhelishvili. *Some Basic Problems of the Mathematical Theory of Elasticity.* P. Noordhoff, Groninger, Holland, 4th edition, 1972.

[72] N.I. Muskhelishvili. *Singular Integral Equations.* P. Noordhoff, Groninger, Holland, 2nd edition, 1973.

[73] Z. Nehari. *Conformal Mapping.* McGraw-Hill, 1952.

[74] D. T. Nguyen and J. S. Arora. An algorithm for solution of large eigenvalue problems. *Computers & Structures*, 24(4):645–650, 1986.

[75] J. M. Ortega. *Matrix Theory: A Second Course.* Plenum Press, 1987.

[76] J. M. Ortega and W. C. Rheinboldt. *Iterative Solution of Nonlinear Equations in Several Variables.* Academic Press, 1970.

[77] B. Parlett. *The Symmetric Eigenvalue Problem.* Prentice-Hall, 1980.

[78] M. Paz. *Structural Dynamics: Theory & Computation.* Van Nostrand Reinhold Company, 1985.

[79] R. Piessens, E. de Doncker-Kapenga, C.W. Uberhuber, and D.K. Kahaner. *QUADPACK: A Subroutine Package for Automatic Integration*, volume 1 of *Computational Mathematics.* Springer-Verlag, 1983.

[80] P. Prenter. *Splines and Variational Methods.* John Wiley & Sons, 1975.

[81] W. H. Press, B. P. Flannery, S. A. Teukolsky, and W. T. Vetterling. *Numerical Recipes: The Art of Scientific Computing.* Cambridge University Press, 1986.

[82] J. R. Rice. *The Approximation of Functions, Volumes 1 and 2.* Addison-Wesley, 1964.

[83] R. J. Roark and W. C. Young. *Formulas for Stress and Strain.* McGraw-Hill, 1975.

[84] Scientific Computing Associates, Inc., New Haven, CT. *CLAM User's Guide*, 1989.

[85] N. S. Sehmi. *Large Order Structural Analysis Techniques.* John Wiley & Sons, New York, 1989.

[86] L. Shampine and M. Gordon. *Computer Solutions of Ordinary Differential Equations: The Initial Value Problem.* W. H. Freeman, 1976.

[87] B. T. Smith, J. M. Boyle, J. Dongarra, B. S. Garbow, Y. Ikebe, V. C. Klema, and C. Moler. *Matrix Eigensystem Routines — EISPACK Guide*, volume 6 of *Lecture Notes in Computer Science.* Springer-Verlag, 1976.

[88] I. S. Sokolnikoff. *Mathematical Theory of Elasticity.* McGraw-Hill, 1946.

[89] M. R. Spiegel. *Theory and Problems of Vector Analysis.* Schaum's Outline Series. McGraw-Hill, 1959.

[90] M. R. Spiegel. *Theory and Problems of Complex Variables.* Schaum's Outline Series. McGraw-Hill, 1967.

[91] R. Stepleman, editor. *Scientific Computing*, chapter ODEPACK, A Systemized Collection of ODE Solvers. North Holland, 1983. by A. Hindmarsh.

[92] G. W. Stewart. *Introduction to Matrix Computations.* Academic Press, 1973.

[93] G. Strang. *Introduction to Applied Mathematics.* Cambridge Press, 1986.

[94] G. Strang. *Linear Algebra and Its Applications.* Harcourt Brace Jovanovich, 1988.

[95] The MathWorks Inc. *MATLAB User's Guide.* The MathWorks, Inc., South Natick, MA, 1991.

[96] The MathWorks Inc. *The Spline Toolbox for Use With MATLAB.* The MathWorks, Inc., South Natick, MA, 1992.

[97] The MathWorks Inc. *The Student Edition of MATLAB For MS-DOS Personal Computers.* The MATLAB Curriculum Series. Prentice-Hall, Englewood Cliffs, NJ, 1992.

[98] S. Timoshenko. *Engineering Mechanics.* McGraw-Hill Book Company, fourth edition, 1956.

[99] S. Timoshenko and D. H. Young. *Advanced Dynamics.* McGraw-Hill Book Company, 1948.

[100] L. H. Turcotte and H. B. Wilson. *Computer Applications in Mechanics of Materials Using MATLAB.* Prentice-Hall, 1998.

[101] C. Van Loan. A survey of matrix computations. Technical Report CTC90TR26, Cornell Theory Center, Ithaca, New York, October 1990.

[102] G. A. Watson, editor. *Lecture Notes in Mathematics*, volume 506. Springer-Verlag, 1975. *An Overview of Software Development for Special Functions* by W. J. Cody.

[103] R. Weinstock. *Calculus of Variations: With Applications to Physics and Engineering*. Dover Publications, 1974.

[104] D. W. White and J. F. Abel. Bibliography on finite elements and supercomputing. *Communications in Applied Numerical Methods*, 4:279–294, 1988.

[105] J. H. Wilkinson. *Rounding Errors in Algebraic Processes*. Prentice-Hall, 1963.

[106] J. H. Wilkinson. *The Algebraic Eigenvalue Problem*. Oxford University Press, 1965.

[107] J. H. Wilkinson and C. Reinsch. *Handbook for Automatic Computation, Volume II: Linear Algebra*. Springer-Verlag, 1971.

[108] H. B. Wilson. *A Method of Conformal Mapping and the Determination of Stresses in Solid-Propellant Rocket Grains*. PhD thesis, Dept. of Theoretical and Applied Mechanics, University of Illinois, Urbana, IL, February 1963.

[109] H. B. Wilson and G. S. Chang. Line integral computation of geometrical properties of plane faces and polyhedra. In *1991 ASME International Computers in Engineering Conference and Exposition*, Santa Clara, CA, August 1991.

[110] H. B. Wilson and K. Deb. Inertial properties of tapered cylinders and partial volumes of revolution. *Computer Aided Design*, 21(7), September 1989.

[111] H. B. Wilson and K. Deb. Evaluation of high order single step integrators for structural response calculation. *Journal of Sound and Vibration*, 141(1):55–70, 1991.

[112] H. B. Wilson and D. S. Farrior. Computation of geometrical and inertial properties for general areas and volumes of revolution. *Computer Aided Design*, 8(8), 1976.

[113] H. B. Wilson and D. S. Farrior. Stress analysis of variable cross section indeterminate beams using repeated integration. *International Journal of Numerical Methods in Engineering*, 14, 1979.

[114] H. B. Wilson and S. Gupta. Beam frequencies from finite element and finite difference analysis compared using MATLAB. *Sound and Vibration*, 26(8), 1992.

[115] H. B. Wilson and J. L. Hill. Volume properties and surface load effects on three dimensional bodies. Technical Report BER Report No. 266-241, Department of Engineering Mechanics, University of Alabama, Tuscaloosa, Alabama, 1980. U.S. Army Engineer Waterways Experiment Station, Vicksburg, MS, 1980.

[116] S. Wolfram. *A System for Doing Mathematics by Computer.* Addison-Wesley, 1988.

[117] C. R. Wylie. *Advanced Engineering Mathematics.* McGraw-Hill, 1966.

[118] D. Young and R. Gregory. *A Survey of Numerical Mathematics, Volume 1 and 2.* Chelsea Publishing Co., 1990.

Index